THE BIOLOGY
OF *EUGLENA*

Volume I
*General Biology
and Ultrastructure*

CONTRIBUTORS TO THIS VOLUME

EUGENE C. BOVEE

DENNIS E. BUETOW

J. R. COOK

THEODORE L. JAHN

LELAND P. JOHNSON

JAMES B. LACKEY

GORDON F. LEEDALE

BLAINE H. LEVEDAHL

BARRY W. WILSON

THE BIOLOGY
OF *EUGLENA*

Edited by DENNIS E. BUETOW

DEPARTMENT OF PHYSIOLOGY AND BIOPHYSICS
UNIVERSITY OF ILLINOIS
URBANA, ILLINOIS

Volume I

General Biology
and Ultrastructure

1968

ACADEMIC PRESS   New York and London

ACADEMIC PRESS, INC.
111 Fifth Avenue, New York, New York 10003

United Kingdom Edition published by
ACADEMIC PRESS, INC. (LONDON) LTD.
Berkeley Square House, London W.1

LIBRARY OF CONGRESS CATALOG CARD NUMBER: 66-14645

PRINTED BY THE ST. CATHERINE PRESS, LTD., BRUGES, BELGIUM

LIST OF CONTRIBUTORS

Numbers in parentheses indicate the pages on which the authors' contributions begin.

EUGENE C. BOVEE (45), Department of Zoology, University of California, Los Angeles, California

DENNIS E. BUETOW (109), Department of Physiology and Biophysics, University of Illinois, Urbana, Illinois

J. R. COOK (243), Department of Zoology, University of Maine, Orono, Maine

THEODORE L. JAHN (45), Department of Zoology, University of California, Los Angeles, California

LELAND P. JOHNSON (1), Department of Biology, Drake University, Des Moines, Iowa

JAMES B. LACKEY (27), University of Florida, Gainesville, Florida*

GORDON F. LEEDALE (185), Department of Botany, Leeds University, Leeds, England

BLAINE H. LEVEDAHL (315), Department of Chemistry, University of Alabama, Birmingham, Alabama

BARRY W. WILSON (315), Department of Poultry Husbandry, University of California, Davis, California

* *Present Address:* Melrose, Florida

v

PREFACE

A comprehensive work on the biology of *Euglena* should cover the subject in all its aspects from taxonomy and ecology to biochemistry. It should also include a consideration of the role and use of *Euglena* in modern cell biology experimentation. This two-volume treatise was planned to be such a work. The literature on *Euglena* is so widely scattered that it was thought that it would be helpful, at the very least, to have available an inclusive set of references in one treatise.

A brief survey of the modern literature on *Euglena* will quickly indicate the wide variety of biological experimentation being done on these organisms. This is the result of course of the unique taxonomic position held by this genus. The obvious animal-like characteristics as well as the obvious plantlike characteristics of members of the genus *Euglena* make it the object of research in many laboratories.

Especial attention is given currently to the biology of the *Euglena* chloroplast. This attention is reflected by the amount of space that is given to the chloroplast in this treatise, particularly in Volume II. Inevitably, in so active a field, disagreements arise. It is hoped that all views are covered. Chapter 10 by Schiff and Epstein, which appears in Volume II, is an updated reprint of the article published in the 24th Symposium of the Society for Developmental Biology, "Reproduction: Molecular, Subcellular, and Cellular" (M. Locke, ed., Academic Press, 1966). This very good article on the *Euglena* chloroplast appeared during the early planning stages of this work. It was felt that it should be included here for complete coverage.

Each topic in "The Biology of *Euglena*" is reviewed in its historical context and development, but emphasis is placed on the current literature. The efforts of the authors as well as the efforts of that unsung group, the typists, are certainly appreciated. Especial thanks are due my wife, Mary Kathleen, who carefully checked all the manuscripts for typographical errors.

DENNIS E. BUETOW

Urbana, Illinois
May, 1968

vii

CONTENTS

Chapter 4. **Morphology and Ultrastructure of** *Euglena*

DENNIS E. BUETOW

Chapter 5. **The Nucleus in** *Euglena*

GORDON F. LEEDALE

Chapter 6. **The Cultivation and Growth of** *Euglena*

J. R. COOK

Chapter 7. Synthetic and Division Rates of *Euglena*: A Comparison with Metazoan Cells

BARRY W. WILSON AND BLAINE H. LEVEDAHL

CONTENTS OF VOLUME II
Biochemistry

CHAPTER 1

THE TAXONOMY, PHYLOGENY, AND EVOLUTION OF THE GENUS *EUGLENA*

Leland P. Johnson

I. Introduction

Euglena has long been an enigma to many biologists. The "plantlike" or "animal-like" characteristics present a universal taxonomic problem to neophytes in biology as well as to taxonomists little versed with the protists. The genus possesses a large number of stable taxonomic forms, and inhabits a great diversity of ecological niches including: fresh, brackish; acid, alkaline; aerobic, anaerobic; and tropical waters. Some species form blooms on snow and some are found in the soil. Michajlow (1965) has described parasitic forms in the gut of Copepoda. In addition, Wenrich (1924) has described *Euglenamorpha hegneri* from the gut of tadpoles. These attributes, plus the formation of colorless strains and the absence of verified sexuality in the genus,* offer special problems to the taxonomists and evolutionists.

* *Editor's Note:* for a discussion of early claims of meiosis in *Euglena* see Chapter 5, Section V.

1

II. Description of *Euglena*

The genus *Euglena* Ehrenberg, 1830 is characterized by great diversity in shape and number of chloroplasts; a grass green color; the presence or absence of pyrenoids; the presence of paramylon, an iodophobic polysaccharide; two flagella, one (extending beyond the anterior opening of the gullet) with a flagellar swelling (a photosensitive structure at level of the eyespot), and a second (lacking a swelling and contained within the reservoir) which may adhere to the flagellar swelling so that the flagella appear as one bifurcated structure (both possess blepharoplasts); a rhizoplast, demonstrated in some species; a reservoir (into which the contractile vacuole empties) opening ventrally via a gullet opposite the location of the eyespot; a stigma or eyespot composed of hematochrome granules situated at the junction of the reservoir and the gullet; a relatively large spherical nucleus with an endosome located near the center of the body; numerous scattered mitochondria; the presence or absence of hematochrome bodies free in cytoplasm; a body form which may be nearly spherical, elongated, spindle-shaped, ridged, spirally twisted, and which always possesses some metabolic movement; a pellicle or periplast with striae; muciferous bodies often located below the pellicle parallel to the striae; asexual reproduction by longitudinal division of a trophozoite or in cysts; and phototropic or heterotrophic nutrition but never holozoic nutrition. Colorless forms have been described for many species. Figure 1 demonstrates a *Euglena* characterizing the typical morphology used in the taxonomy and classification of species.

III. Problems in Identification

Size and shape of a given species vary greatly, for example, lengths of *E. ehrenbergii* were listed as 107–300 μ by Chu (1947), and 190–400 μ by Johnson (1944). Pringsheim (1956) suggested that organisms of *E. ehrenbergii* described by various authors comprise several taxonomic units, since such extremes did not occur in his clone cultures.

Identification is further compounded since body shapes in a species may be circular or flattened, and rectangular in outline with the body twisted as in *E. acus* and *E. spirogyra* (Fig. 3B,I). Most species of *Euglena* possess evident spiral striae on the periplast. This is accompanied by a spiral body torsion. The more rigid forms best demonstrate this, but the torsion also occurs in metabolic forms such as *E. granulata*. It can be observed to undergo metaboly in which spiral ridges are evident.

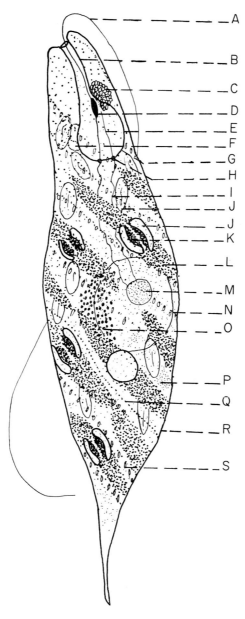

Fig. 1. *Euglena*, typical structures: A, locomotor flagellum; B, gullet; C, stigma; D, photoreceptor; E, internal flagellum (usually adheres to photoreceptor); F, contractile vacuole; G, reservoir; H, blepharoplast (kinetosome); I, rhizoplast (reported in a few species); J, paramylon (free, and sheath around pyrenoid); K, pyrenoid; L, chromatophore (chloroplast); M, endosome; N, nucleus; O, hematochrome granules (in red forms); P, mitochondria; Q, stria; R, pellicle; S, muciferous body (may or may not be present).

Argyrophilic structures on the periplast which react with vital stains (methylene blue, cresyl blue, and neutral red) occur in many species of *Euglena*, but are most evident in *E. spirogyra* and *E. granulata*. The number and size vary with the stage in the life cycle and external media. They may be absent, located on each stria, or on some order of alternating striae. Leedale *et al.* (1965) demonstrated such variations in pellicular granules in *E. spirogyra*.

Mucous bodies, located below the pellicle, also vary significantly in number and size in a given species. They respond to a variety of vital stains: neutral red, methylene blue, brilliant cresyl blue, ruthenium red, and react to iodine. In *E. velata* the bodies are rodlike and in *E. granulata* they are more fusiform.

The bifurcated condition of the *Euglena* flagellum observed in most living, as well as fixed and stained, preparations is associated with the shorter flagellum adhering to the locomotor flagellum. The apparent fusion occurs at the photoreceptor. Hollande (1942) demonstrated the double flagellar condition in *E. mutabilis*. The flagellar swelling, the photoreceptor, is associated with the presence of a stigma. Colorless forms, *Astasia longa*, lacking an eyespot are without flagellar swellings, and *E. quartana* and species of *Khawkinea* with a stigma possess the swelling. The flagellar length may also vary significantly. A flagellum may be lost from *E. rubra* and a new one regrown within 15 minutes. Variations in length occur not infrequently in this manner. Rhizoplasts extending from the blepharoplast to the nucleus are present in *E. fracta* Johnson (1956).

Chromatophores exhibit great diversity in shape and number within the genus. The intracellular chromatophore varies markedly in shape from the released chromatophore free in the outside medium. Fixation and staining also modify chromatophore shape and size. Pringsheim (1956) deemed the presence of pyrenoids a diagnostic characteristic in species determination. The pyrenoids may be absent in *E. acus*, naked in *E. deses*, or sheathed with paramylon in *E. gracilis*. An inner pyrenoid has also been described in *Euglena* by Iyengar (1962).

The polysaccharide reserve substance, paramylon, present in the euglenoids, does not stain with iodine (see also Vol. II, Chapter 7). The paramylon grains vary in number, size, shape, and location within the cell. They may be rodlike, as in *E. acus* and *E. ehrenbergii*, enlarged annular forms, as in *E. spirogyra*, or small bodies dispersed in the cytoplasm. In others, e.g., *E. gracilis*, annular paramylon bodies may be associated with the pyrenoid of the chromatophore. Many possess both paramylon bodies associated with the chromatophore and bodies free in the cytoplasm, e.g., *E. gracilis* and *E. sanguinea*.

A number of species of *Euglena* are available in pure culture for morphological and physiological study. The protozoologists (1958) and botanists, Starr (1964), list strains of *Euglena* available in the culture collection at Indiana University, Bloomington, Indiana. Jahn (1946) and Huber- Pestalozzi (1955) identified nutritional types of *Euglena*. These are seldom used taxonomically but it is common knowledge that numerous organisms develop physiological strains that maintain morphological identity with the original clone.

Gojdics (1953) described and illustrated 155 species of *Euglena*, Huber-Pestalozzi (1955) described and duplicated original illustrations of 101 species of *Euglena*, and Pringsheim (1956) reported over 200 species but accepted only 56 in a critical list of species. Numerous species have been described since 1956 and the total number of *Euglena* species described approximates 250. Great variation occurs in the description of new species by authors, the number of organisms on which a species is based, and the care used in preparation of figures to illustrate the *Euglena*. Johnson (1944), Gojdics (1953), and Pringsheim (1956) cite the need for studying clones under varied environments and obtaining detailed information concerning stages in *Euglena* life histories. Chu (1947) developed a soil–water culture method which proved successful for most *Euglena* and taxonomists presently use this medium.

IV. Asexuality and Speciation

Asexually reproducing organisms do not form interbreeding populations and cannot be tested for genetic similarity by the orthodox methods applied to bisexual species. To date, no valid evidence has been presented that sexual reproduction occurs in the genus *Euglena*. Leedale (1958a,b, 1959, 1962) described the occurrence of mitosis, amitosis, and meiosis in Euglenineae. Although meiosis was observed, no evidence of sexual reproduction was reported. Chromosomal counts of Leedale (1958a) substantiate the suggestion by Pringsheim (1956) that polyploidy may be expected. Leedale (1958a) lists chromosomal numbers as follows: *E. spirogyra*, 86; *E. viridis*, 42; and *E. gracilis*, approximately 45. Bělǎr (1926) lists chromosomal numbers after Dangeard 1902 as follows: *E. pisciformis*, 12–15; *E. viridis*, 30 or more; *E. geniculata*, 25–30; *E. splendens* 35-40; and *E. proxima*, over 50.

Although the chromosomal number has been commonly used in species determination in both plants and animals, knowledge concerning the composition of the bases and their sequences in DNA may prove to be more valuable. The percentages of guanine and cytosine varied greatly between species of protozoa as reported by Gibson (1966). Schildkraut *et al.* (1962)

found similar variations and suggested the G–C content may be an aid in the study of the phylogenetic origin and relation of algae and protozoa.

The problems of speciation are complex since each organism is reproductively isolated, both from its parent and its sister clone. Yet each daughter organism theoretically and normally possesses the same hereditary composition. A clone population is a true species in that all organisms are genetically alike, as would be the case in vegetative reproduction within plants or parthenogenesis in animals. If a mutation occurs in a clone, is it a new species? Arbitrary distinctions ultimately are the criteria for species determination in asexually reproducing organisms such as *Euglena*. Although many species of *Euglena* have been grown as clones under uniform conditions, few studies have been performed in which environmental factors have been modified. It is not known how many of the described species of *Euglena* are ecological species due to diverse environments. Corliss (1962) discussed some of these problems of protozoological nomenclatural practices in light of the new International Code of Zoological Nomenclature.

Problems of speciation in *Euglena* and the bacteria have many similarities. Physiological strains exist in morphological identical populations, yet *Euglena* exhibits much greater structural diversity than bacteria. This is associated with a larger size and greater organelle differentiation. Buchanan (1954) discussed the complications involved in speciation of asexually reproducing bacteria and fungi. He recommended the establishment of a type center where active research could be maintained, and stated that morphological characteristics will continue to be the primary criterion for species differentiation. He suggested that such a center could act as a repository for permanent slides of type specimens. A similar center could be responsible for the collection and maintenance of *Euglena* species under standard conditions, study of clones under modified conditions, and maintenance of nutritional and physiological studies and ultramicroscopic studies as well as a complete literature of the group.

Until adequate information is available concerning the structure and function of asexually reproducing forms under both standard and modified conditions, species determination in the genus *Euglena* is subject to the whims of the observer and whether he is a "splitter" or a "lumper." It may well be advantageous for the student of asexual forms, such as *Euglena*, to convert from genus and species connotation to a computerized system, or use the computerized image as a more objective description. The primary description and identification should include structural and functional characteristics observed under standard conditions. Secondary criteria for identification should include changes in form and function in relation to stated environmental modifications.

V. Classification and Affinities

Commonly used textbooks in protozoology, Hall (1953) and Kudo (1966), include a system of classification patterned after Doflein and Reichenow (1929). The Committee on Taxonomy and Taxonomic Problems of the Society of Protozoologists (1964) recommended categories of classification for the protozoa. The portions directly related to the euglenoids follow:

"Phylum PROTOZOA Goldfuss, 1838 *emend.* von Siebold, 1845
  Subphylum I. SARCOMASTIGOPHORA Honigberg & Balamuth, 1963
    Flagella, pseudopodia, or both types of locomotory organelles; single type of nucleus except in developmental stages of certain Foraminiferida; typically no spore formation; sexuality, when present, essentially syngamy.
  Superclass I. MASTIGOPHORA Diesing, 1866
    One or more flagella typically present in trophozoites; solitary or colonial; asexual reproduction basically by symetrogenic binary fission; sexual reproduction unknown in many groups; nutrition phototrophic, heterotrophic, or both.
  Class 1. PHYTOMASTIGOPHOREA Calkins, 1909
    Typically with chromatophores; if chromatophores lost secondarily, relationship to pigmented forms clearly evident; commonly only one or two emergent flagella; amoeboid forms frequent in some groups; sexual reproduction known with certainty in few orders; mostly free-living.
  Order 8. EUGLENIDA Bütschli, 1884
    Typically one or two flagella emerging from anterior reservoir; green chromatophores of various shapes, absent in some species; typically metabolic changes of body form but no amoeboid movement; food reserve paramylum.
  Suborder (1) EUGLENINA Bütschli, 1884
    Flagellar sheath not swollen at base; phototrophic or osmotrophic.
  Suborder (2) PERANEMATINA Hollande, 1942
    Two flagella, one trailing; flagella thickened at base; colorless; phagotrophic or osmotrophic.
  Suborder (3) PETALOMONADINA Hollande, 1942
    One or two flagella swollen at base; colorless; body compressed and rigid; phagotrophic."

G. F. Leedale (1966), in a discussion on the classification of Protozoa, recommended at the Second International Conference on Protozoology, London, England, that the phytoflagellates remain in the Protozoa. It was generally agreed that the definitive classification used by the botanists as published by Christensen (1962) should be substituted for the present system used by protozoologists. It is:

Eucaryota
  Contophora
    Chlorophyta, Division
      Euglenophyceae, Class
        Euglenales, Order
        Peranematales, Order

Most botanists use the categories of Christensen at the division level and below. Chadefaud (1962) is an exception and placed the euglenoids with the brown algae rather than with the green algae. Scagel *et al.* (1965) placed the euglenoids in a separate division and indicated they were not closely related to any other group.

Klebs (1883) placed the euglenoids with the algae on the basis of green color, and the formation of a palmella stage. Smith (1950) stated that the euglenoids constitute a well-defined series with an evolution toward algal organization in which there are no higher types than palmelloid colonies. He further compared the similarity of pigments of the euglenoids with those of other forms.

Chadefaud (1962) divided the Phycophytea into the red, brown, and green algae. The euglenoids were included with the brown, the Pyrrophycees. Chadefaud (1937, 1938) had earlier pointed out the similarity of euglenoids and the dinoflagellates on the basis of cytoplasmic structures, primarily the similarity of gullets. The presence of chlorophyll a and b, and the absence of chlorophyll c in *Euglena* may deny an affinity with the brown algae, but at least one xanthine (Krinsky, 1964) has been reported from *E. gracilis*. Studies on pigment synthesis and spectral analysis of euglenoids and other algal groups as exemplified by the works of Goodwin and Jamikorn (1954), Krinsky and Goldsmith (1960), and Green (1963) may reveal information concerning affinities of the euglenoids. Scagel *et al.* (1965) and many others have placed the euglenoids in a separate division, stating the group is one of the most primitive. He recognized only one class, the Euglenophyceae.

Fritsch (1935) reported the presence of a stigma not only in *Euglena* but also in motile cells of the green and brown algae. Are the *Euglena* derived from motile zoospores or gametes that have been maintained vegetatively? Is asexual reproduction in *Euglena* a derived condition or may it be of independent origin? Stebbins (1960) reviewed problems of the origin of sexuality from asexuality and vice versa in lower forms. Answers to these problems in the euglenoids will depend upon determination of affinities. Data concerning pigmentation and metabolism may aid in solving the phylogeny of the euglenoids, but lack of information concerning possible evolution of these and other characteristics may hide answers to the origin of asexuality in the group.

A. ORDER EUGLENIDA

The euglenoid order has been divided into varied family categories. Hollande (1952) included the families: Euglenidae, Peranemidae, Anisonemidae, and Petalomonidae. The colorless *Astasia* were included with the

Euglenidae. The family Euglenidae was divided into three subfamilies, Eutreptiinae having two functional flagella, Eugleninae having one functional flagellum, and Euglenamorphinae having three or more flagella. Kudo (1966) listed the families Euglenidae, Astasiidae, and Anisonemidae; the last two included the colorless forms. Christensen (1962) listed three families, the Eutreptiaceae, Euglenaceae as the type family, and Menoidiaceae. Chadefaud (1962) approximates Kudo but substitutes the Peranemenes in lieu of the Anisonemidae. The system proposed by Hollande (1952) most nearly demonstrated natural relationships, especially in the colorless forms. The use of subfamilies Eutreptiinae and Eugleninae is preferable to giving each family status. It is questionable if two external flagella versus one external flagellum is a sufficient characteristic to differentiate families when two flagella, one external and one internal, occur in the genus *Euglena*. The two groups are most similar in other respects.

B. Family Euglenidae

Major affinities within the family Euglenidae of the zoologists or Euglenaceae of the botanists have been variously categorized phylogenetically and taxonomically. Klebs (1883) divided the Euglenaceae into the Euglenae and Astasiae. The genera *Euglena*, *Phacus*, *Eutreptia*, *Ascoglena*, *Trachelomonas*, and *Colacium* were placed with the Euglenae. The colorless genera under the Astasiae were *Astasia*, *Rhabdomonas*, *Menoidium*, *Peranema*, and *Anisonema*. The Eugleninae of Hollande (1952) included the following genera: *Euglena*, *Menoidium*, *Ascoglena*, *Trachelomonas*, *Strombomonas*, *Phacus*, *Lepocinclis*, *Colacium* and close relatives. Pringsheim (1956) listed 11 closely related genera among the green Euglenaceae, *Euglena*, *Colacium*, *Eutreptia*, *Eutreptiella*, *Phacus*, *Lepocinclis*, *Cryptoglena*, *Trachelomonas*, *Strombomonas*, *Ascoglena*, and *Klebsiella*, and called them a natural group.

Genera bearing chromatophores in the family in addition to *Euglena* are illustrated in Fig. 2. All occur in a variety of habitats and are commonly found except for *Ascoglena* and *Klebsiella* which are rare, *Ascoglena* having been recorded only by Stein (1878). *Lepocinclis* (Fig. 2A) and *Phacus* (Fig. 2B) are rigid and exhibit no or slight metaboly. The remaining genera illustrated show metabolic movement. Metaboly can be observed in *Trachelomonas* (Fig. 2G) and *Strombomonas* (Fig. 2H) when the lorica is removed. This may occur accidentally due to pressure of a cover glass. *Eutreptia* (Fig. 2C) and *Colacium* (Fig. 2D) lack a restraining envelope.

Eutreptia (Fig. 2C) possesses strong metaboly, two flagella, and paramylon centers in the region of the nucleus from which chromatophore ribbons

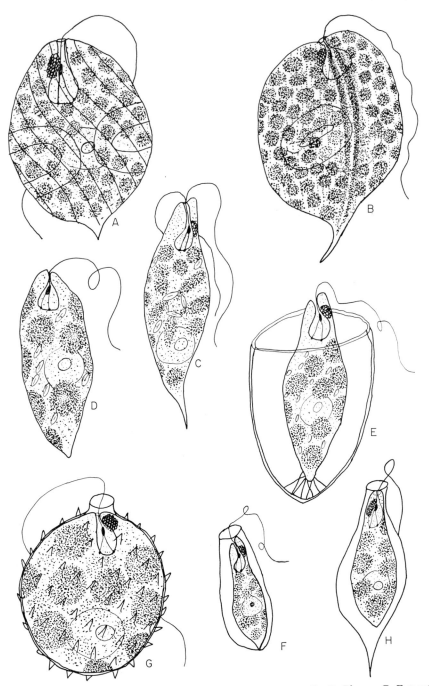

Fig. 2. Typical genera in the family Euglenidae: A, *Lepocinclis;* B, *Phacus;* C, *Eutreptia;* D, *Colacium;* E, *Ascoglena;* F, *Klebsiella;* G, *Trachelmonas;* H, *Strombomonas.*

radiate. The ribbons may become separated and appear as elongated spindles or discs. It has been mistaken for *Euglena viridis* Perty (1852). Mainx (1927a,b) considered it primitive. Chu (1947) suggested that *Euglena viridis* is primitive and is the form from which other species of *Euglena* have been derived. Pringsheim (1956) held that chromatophores without pyrenoids and the paramylon arrangement of *E. viridis* forms were not primitive in *Eutreptia* or *Euglena viridis*. The morphology of the chromatophore reported for *E. viridis* is ambiguous. This is due in part to the lack of critical descriptions of the chromatophore by many observers. Whether primitive or derived forms, *Eutreptia viridis* and *Euglena viridis* are most similar. One mutation could account for the major difference between the genera. A reduction in length of one flagellum or growth in length of one flagellum could cause the transition from *Euglena* to *Eutreptia* or vice versa.

Lepocinclis (Fig. 2A) and *Phacus* (Fig. 2B) possess many common characteristics: small disc chromatophores without pyrenoids, flagellar arrangement typical of *Euglena*, rigid body, and spiral striations. The body in cross section through *Phacus pyrum* approaches a scalloped circle. Slight body movement occurs in each genus, Pochmann (1953). Spiral torsion is not uncommon in species of both genera. Rigid and metabolic *Euglena* also exhibit the spiral torsion. No sharp line of demarcation can be drawn between the genera *Euglena* and *Phacus*. The organism *E. tripteris* possesses characteristics indistinguishable from the genus *Phacus*. The position of the gullet opening varies in species of *Phacus* (Fig. 2B); it is similar to the medial opening of *Lepocinclis* (Fig. 2A) in some, and opens ventrally (as typical of *Euglena*) in others.

The genera *Colacium* (Fig. 2D), *Strombomonas* (Fig. 2H), and *Trachelomonas* (Fig. 2G) possess dislike chromatophores with inner pyrenoids in most species. Pyrenoids are sheathed with paramylon in *Colacium*. *Trachelomonas hispida* have chromatophores bearing naked innerpyrenoids. Inner pyrenoids have also been reported in *E. pringsheimii* by Iyengar (1962). *Euglena cyclopicola* and *Colacium vesiculosum* are considered to be the same by Pringsheim (1956). The *E. cyclopicola* described by Johnson (1944) lacks a stalk and possesses only a thin mucous layer. *Colacium arbuscula* possesses elongated stalks. No sharp lines of demarcation occur between the motile, sessile, and stalked forms of *Euglena* and *Colacium*. The naked protoplasts of *Trachelomonas* and *Strombomonas* are indistinguishable from *Euglena*. It is not unlikely that free *Trachelomonas* protoplasts have been described as a species of *Euglena*.

The genera *Ascoglena* (Fig. 2E) and *Klebsiella* (Fig. 2F) have a sessile stage and possess a protoplast similar to *E. cyclopicola*, *Colacium*, and naked *Trachelomonas*. Pyrenoids may or may not be present on dislike chromatophores. *Ascoglena* and *Klebsiella* possess urnlike loricas rather than a muciferous covering. The similarity of the protoplast structure within these genera and *Trachelomonas* may be evidence for a common ancestry.

It seems plausible to suggest ancestors within the genus *Euglena* for the related genera. A line of evolution may lead to the genus *Phacus* from a form similar to *E. tripteris*, another to *Lepocinclis* via a more globular rigid form related to the *E. acus* forms. *Colacium*, through a series of variations, could give rise to *Ascoglena* and *Klebsiella*, the lorica-bearing forms. Wolken (1961), following Hutner and Provasoli (1951; Hutner, 1955), suggested that the genus *Euglena* was derived from *Chlamydomonas* via the genus *Eutreptia*.

The parasitic genera *Euglenamorpha* and *Hegneria* of Hollande's (1952) subfamily *Euglenamorphinae* resemble *Eutreptia* in possessing more than one functional flagellum, but lack the central orientation of paramylum and chromatophore extensions. The three or more flagella of these genera could be derived from either the single- or double-flagellated condition. The parasitic environment, the gut of a tadpole, could well house other heterotrophic forms, and it might be expected that species of *Euglena* may develop symbiotic relationships. Hall (1931) described *E. leucops* as a parasite in *Stenostomum*.

The family *Euglenidae* or *Euglenacea* constitutes a natural unit. Its origin is shrouded in incomplete information, yet possible evolutionary relationships appear. Sufficient data concerning artificial loss of color have accrued through study of euglenoid genera to explain the derivation of forms such as *Astasia* and *Khawkinea*.

C. Affinities within the Genus *Euglena*

Affinities within the genus *Euglena* are yet to be defined with assurance. Lemmerman (1913) separated the *Euglena* into three major groups on the basis of chromatophore morphology as shown in the tabulation.

Group	Chromatophore structure	Example
I	Ribbonlike	*E. elongata; E. terricola*
II	Star-shaped	*E. viridis; E. sanguinae*
III	Dishlike	*E. acus; E. deses*

Chu (1947) and Gojdics (1953) also used chromatophore structure in defining major groups. Chu identified four kinds of chromatophores according to the tabulation:

Type	Chromatophore characteristics	Example
1	Stellate; with pyrenoid or pyrenoids and numerous associated paramylon grains located centrally near nucleus	*E. viridis*
2	Stellate; with pyrenoids surrounded by double paramylon caps; distributed in parietal region of body	*E. sanguinea*
3	Discoid; with pyrenoids surrounded by paramylon caps	*E. gracilis*
4	Lens-shaped; without pyrenoids; distributed in parietal region of the body	*E. acus*

Chu (1947) considered *Euglena* of type 1 to be primitive and those of type 4 to be highest in the evolutionary scale.

Gojdics (1953) divided the genus *Euglena* into eight main groups on the basis of chromatophore structure as shown in the tabulation. Gojdics

Group	Chromatophore characteristics	Example
A	Band-shaped	*E. elongata; E. viridis*
B	Urn-shaped or curved plate–shaped	*E. nana; E. mutabilis*
C	Reticulate	*E. reticulata*
D	Numerous, small and discoid	*E. cyclopicola; E. acus*
E	Over 6μ long; oval or round; with pyrenoids, but lacking paramylon sheaths	*E. deses*
F	Bear paramylon-sheathed pyrenoids	*E. granulata; E. gracilis*
G	Spindle-shaped	*E. splendens*
H	Unlike those of Groups A–G	*E. guntheri*

reviewed and evaluated previously published relationships within the genus and sister genera. She did not suggest an evolutionary series.

Pringsheim (1956) presented a critical analysis of affinities within the genus *Euglena*. Using the work of Chu, he refined chromatophore shapes and correlated them with body shape and metaboly as shown in the tabulation.

Group	Type species	Tentative name of group (Taxa or subgenus)
I	*E. acus*	Rigidae
II	*E. proxima*	Lentiferae
III	*E. gracilis*	Catilliferae
IV	*E. viridis*	Radiatae
V	*E. deses*	Serpentes
VI	*Astasia*	Limpidae

Jahn (1946) described two major nutritional groups within the genus *Euglena:* phototrophic and heterotrophic; holozoic has not been observed. All green *Euglena* were phototrophic. Photoautotrophic forms were capable of utilizing ammonium and nitrate compounds, the photomesotrophic forms required amino acids, and photometatrophic utilized peptones or proteins as a nitrogen source. The heterotrophic forms were similarly subdivided. Under varied conditions *E. gracilis* has been categorized under all six headings. Birdsey and Lynch (1962) reported *E. gracilis* incapable of growing on a nitrate source even though it can reduce nitrate to nitrite.

Until a breakthrough occurs involving knowledge on all *Euglena* species comparable to what is know concerning the morphology and physiology of *E. gracilis*, major affinities within the genus must be based on gross morphological characteristics. The major categories of Pringsheim (1956) are most reliable and are presently recommended. His critical analysis of *Euglena* species for the most part is excellent but assumes information on forms which he has not studied.

1. *Rigidae*

Group I of Pringsheim (1956) is characterized by an inflexible body. Although bending and bulging occurs in organisms within the group, true metaboly does not exist. A cylindrical colorless posterior tip is present in all species of the group. Most species of the genera *Lepocinclis* and *Phacus* exhibit a similar process. It is typically shorter and may be absent in species of *Lepocinclis*. The chromatophores are small and disc-shaped; Pringsheim (1956) called them lens-shaped. Pyrenoids have not been demonstrated. The paramylon bodies are elongated rods or large annular structures. The organisms have been observed only in the trophozoite stage. Muciferous bodies have not been reported in the smooth forms, but an adhesive substance associated with the posterior spine allows attachment to vegetation and other solid materials. Pellicular papillae are obvious in *E. spirogyra*. Variations in the papillae and muciferous bodies were described by Leedale *et al.* (1965).

Representative organisms in this group are illustrated in Fig. 3. Pringsheim (1956) reported detailed descriptions and evaluations of the literature on the following species: *E. acus* Ehrenberg, 1830, *E. tripteris* (Dujardin) Klebs, 1883, *E. spirogyra* Ehrenberg, 1838, and *E. oxyuris* Schmarda, 1846.

A possible evolutionary series exists between *E. acus* (Fig. 3A) and *E. tripteris* (Fig. 3D). Both three- and four-sided forms occurred in *E. acus* var. *angularis* Johnson, 1944 (Fig. 3B). Size ranges were within those of *E. acus*, 52–175 μ by 8–18 μ. *Euglena trisulcata* Johnson, 1944 (Fig. 3C)

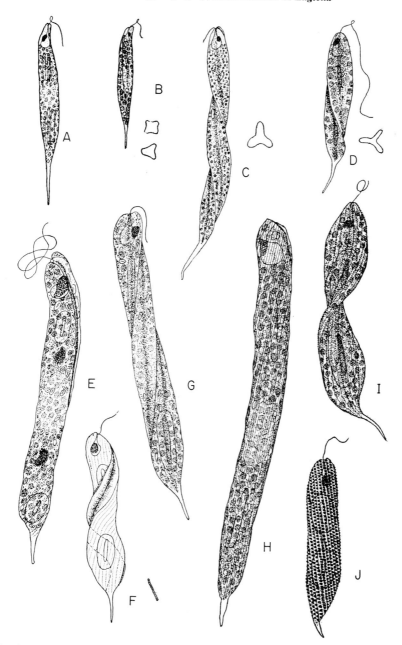

Fig. 3. *Euglena* of the Rigidae group: A, *E. acus;* B, *E. acus* var. *angularis;* C, *E. trisulcata;* D, *E. tripteris;* E, *E. oxyuris;* F, *E. oxyuris;* G, *E. oxyuris;* H, *E. antefossa;* I, *E. spirogyra* var. *suprema;* J, *E. spirogyra* var. *marchica.*

possessed the three-ridge condition of *E. tripteris*. The ridges were less pronounced and the body was narrower, intermediate in width between *E. acus* and *E. tripteris*. *Euglena trisulcata* was longer than *E. tripteris* and the chromatophores were slightly smaller than in either *E. acus* or *E. tripteris*.

The relationships of the remaining species illustrated in Fig. 3, one to the other and to the *E. acus* and *E. tripteris* complex is not as easily postulated. *Euglena antefossa* Johnson, 1944 (Fig. 3H) possessed rodlike paramylon, similar to the first group and unlike the larger annular paramylon bodies of *E. oxyuris* Schmarda (1846) (Fig. 3E,F,G) and *E. spirogyra* Ehrenberg, 1838 (Fig. 3I,J). Forms of *E. oxyuris* were ridged, grooved, and sometimes flattened. Transverse sections through the body of *E. oxyuris* (Fig. 3G) approached outlines typical of intermediates between *E. acus* var. *angularis* and *E. trisulcata*. The form of *E. oxyuris* (Fig. 3E) was circular in outline at the posterior end. It resembled *E. antefossa* in this respect, but possessed annular rather than rodlike paramylon bodies. Grooved forms of *E. oxyuris* (Fig. 3F,G) are intermediate between *E. tripteris* forms and *E. antefossa*. Evolution of this sequence is not improbable providing changes in paramylon bodies occurred. Equally probable would be a giant *E. acus* giving rise to *E. antefossa* and ultimately to variations observed in *E. oxyuris*.

The derivation of *E. spirogyra* (Fig. 3I,J) could have been from either forms like *E. oxyuris* or *E. antefossa* by the development of protuberances on the pellicle, providing changes in shape of body or paramylon bodies occurred. Affinities between *E. oxyuris* and *E. spirogyra* may be closer in light of the similar paramylon bodies and flattened bodies that occur in varieties of each species. Studies involving X-irradiation, and subsequent clone-culturing of these species with modification of environments may shed light on evolutionary affinities.

2. *Lentiferae*

Group II, the Lentiferae, according to Pringsheim (1956) possessed chromatophores similar to the Rigidae, but the chromatophores were usually larger and thinner at the margins. Pyrenoids were not present, and free paramylon bodies were small granules or elongated rods. The body was metabolic, a characteristic which definitely separated the Lentiferae from the Rigidae. The Lentiferae lacked a posterior endpiece. Pringsheim placed three species in this group: *E. proxima* Dangeard, 1901 (Fig. 4C), a similar species *E. variabilis* Klebs, 1883 and *E. ehrenbergii* Klebs, 1883 (Fig. 4A,B). The latter species possessed a body shape, paramylon bodies, and chromatophores closely allied to the larger species of the Rigidae.

Are chromatophores bearing pyrenoids primitive or derived? If such chromatophores are derived, can the pyrenoids be lost secondarily? The size

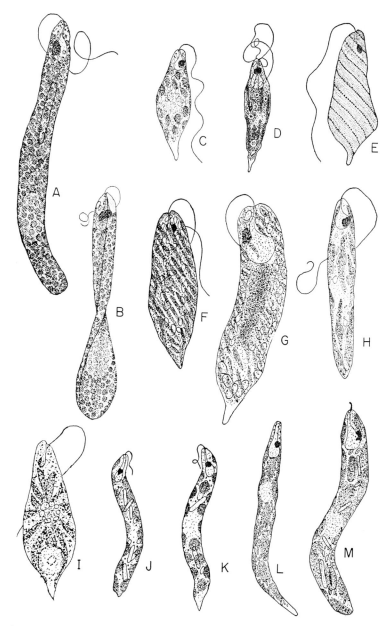

Fig. 4. Euglena of the Lentiferae, Catilliferae, Radiatae, and Serpentes groups:
A, *E. ehrenbergii;* B, *E. ehrenbergii;* C, *E. proxima;* D, *E. granulata;* E, *E. granulata;*
F, *E. velata;* G, *E. rubra;* H, *E. gracilis;* I, *E. viridis;* J, *E. intermedia;* K, *E. intermedia;*
L, *E. deses;* M, *E. deses.*

and shape of the chloroplast of *E. proxima* is more like members of the Catilliferae than of the Rigidae or *E. ehrenbergii*. The loss of a rigid periplast from an organism like *E. antefossa* (Fig. 3H) may result in a body form similar to *E. ehrenbergii* (Fig. 4A). Internal organelles of the two species are nearly identical, including rod-shaped paramylon bodies. The blunt or rounded posterior end possessed by *E. ehrenbergii* is significantly different from the Rigidae but may be associated with a change in the rigidity of the periplast.

Euglena acusformis Schiller, 1925, a marine form, appears to have similarities to both *E. acus* (Fig. 3A) and *E. proxima* (Fig. 4C). If such a freshwater form were to be postulated, *E. proxima* may be closely related to forms in the Rigidae rather than the Catilliferae. *Euglena proxima* may be derived from a form like *E. acus*, and *E. ehrenbergii* may be derived from one of the larger ametabolic species such as *E. antefossa*.

3. *Catilliferae*

The many species in the Catilliferae were characterized by a fusiform body, metaboly, and chromatophores bearing a centrally located double pyrenoid, (Fig. 1K) surrounded on each side with a saucer-shaped paramylon cap from which the group received its name. Paramylon bodies were also free in cytoplasm. The chromatophores were platelike, or lobed, and from a side view appeared spindle-shaped. In some, the lobes of chromatophores appeared to be separated from the pyrenoid-bearing portion. The chromatophores of some species approached the stellate chromatophore condition of Group IV, the Radiatae. Muciferous bodies were often present; encystment and palmella stages were prevalent.

The chromatophores of the Catilliferae, according to Pringsheim (1956), were as follows: outlines smooth, rounded, or polygonal in *E. gracilis* (Fig. 4H); lobed in *E. velata* (Fig. 4F); star-shaped in *E. olivacea;* and fringed in *E. pisciformis*, *E. granulata* (Fig. 4D,E), and *E. sanguinea*. He further emphasized the retraction of the fringed condition with the slightest injury. He found similarities and a size sequence in a series *E. pisciformis* to *E. caudata* and reported that *E. gracilis* (Fig. 4H) was not closely related to other species. An *E. sanguinea* group was reported to possess a complicated system of chromatophores, in which the chromatophores were bands radiating from a pyrenoid center, some of which were bifurcate and parallel to the periplast markings. Fine colorless filaments interconnected lobes of the chromatophores.

A series of organisms of increasing size from *E. pisciformis* Klebs, 1883 to *E. caudata* Hübner (1886) possessed the typical chromatophore, spindle-shaped in side view and platelike in face view, in the living trophozoite. *Euglena granulata* (Klebs) Schmitz (1884) (Fig. 4D,E) can easily be distinguished from the other members of the series by the rows of granules on the periplast and the presence of globular muciferous bodies arranged in rows parallel to the granules. The granules may be the same size or in alternating rows of large and small granules (Fig. 4E). Chromatophores of *E. caudata* often are lobed in fixed and stained preparations. Problems exist in differentiating several of the larger forms including *E. caudata, E. velata, E. sociabilis,* and *E. polymorpha.* The similarity of the chromatophores in these forms and those of the *E. sanguinea* group give evidence of natural relationships.

Pringsheim (1956) suggested that *E. nana* Johnson, 1944 is a variety of *E. pisciformis.* This seems unlikely on two bases: the relatively rigid body and chromatophores without pyrenoids or paramylon bodies. Except for the large chromatophore, *E. nana* has greater affinities with the Rigidae, or except for the rigid body has greater similarity to *E. proxima,* one of the Lentiferae. Forms such as *E. proxima* and *E. anabaena* of the Catilliferae exhibit great similarity except for chromatophore–pyrenoid relationships. The Lentiferae and Catilliferae may have affinities at this level.

Euglena rubra Hardy (1911) (Fig. 4G) is suggested as a synonym for *E. sanguinea* by Pringsheim (1956). It resembles *E. sanguinea* in general body form, muciferous bodies, and general alignment of chromatophores. The chromatophores in *E. rubra* do not bear pyrenoids as evidence by Heidenhain's iron hematoxylin preparations. Heidenhain's staining, preceded by fixation in warm Schaudinn's fluid, allows demonstration of pyrenoids in *E. sanguinea.* It is suggested that *E. rubra* is a form constituting a true species, not of synonym standing. It could have been derived from a form like *E. sanguinea* through loss of pyrenoids and separation of chromatophore bands into units. The red coloration is independent of the age of the culture. Disappearance of blooms is usually associated with parasitism and not an indication of old age of the population as suggested by Pringsheim (1956). He indicated that a great ambiguity existed among green and red forms having similarities with *E. sanguinea.* Pringsheim added to the confusion by describing new species, *E. magnifica* and *E. laciniata,* closely allied to the *E. granulata* and *E. sanguinea* complex. The chromatophores described in the new species were similar to those of *E. granulata* and *E. caudata.* The thickness and markings of the periplast were not significantly different from the periplast of *E. granulata* or *E. sanguinea.*

4. *Radiatae*

The Radiatae possessed chromatophores with ribbons or bands radiating from one, two, or three centers. The centers were associated with pyrenoids and bodies of paramylon. The ribbons may be separated from the main body of the chromatophore, appearing as individual chromatophores without pyrenoids. The organisms possessed average metaboly, pellicles with delicate striae, and usually muciferous bodies.

Euglena viridis Ehrenberg, 1830 (Fig. 41) possesses one center from which chromatophore bands radiate. Muciferous bodies are spherical in form and parallel periplast striae. There are usually four rows of these bodies. *Euglena pseudoviridis* Chadefaud, 1937 lacks muciferous bodies. This may be a stage in the life cycle of *E. viridis*.

A species having two centers of paramylon is characterized by *E. geniculata* Dujardin, 1841. The body form may be as in *E. viridis* or more elongated and slender. Taxonomists may have described forms of *E. geniculata* as a species in the group Serpentes because of chromatophore separations. *Euglena tristella* Chu, 1947 possesses three paramylon centers. This species is similar in general appearance and in chromatophore structure to the *E. sanguinea* group of the Catilliferae. The Radiatae may be intermediate between the Catilliferae and Serpentes.

5. *Serpentes*

As reported by Pringsheim (1956), the *Euglena* of the Serpentes possessed a relatively long body. Locomotion occurred by creeping and highly metabolic action. It was serpentlike. A short flagellum occurred which often may be lost. The chromatophores were disc-shaped or plate-shaped. The chromatophores were larger than those reported in the Ridigae and Lentiferae, but smaller than those of *E. gracilis*, *E. pisciformis*, and *E. granulata* of the Catilliferae. A pyrenoid located centrally can be demonstrated with Heidenhain's staining. Paramylon caps are absent. The group included *E. deses* Ehrenberg (1833), *E. mutabilis* Schmitz (1884), and *E. intermedia* (Klebs) Schmitz (1884). Figure 4J,K,L,M illustrates typical forms.

Many variations occur within the species *E. deses*. Lefévre (1931) described the body shape as flat in an alkaline or neutral medium, but cylindrical in an acid medium. At times, mucus may be expelled with the formation of a sheath around the body. *Euglena mutabilis* is similar but possesses fewer chromatophores, four to eight. *Euglena intermedia* resembles *E. deses* except for smaller chromatophores. *Euglena deses* (Fig. 4L,M) and *E. intermedia* (Fig. 4J,K) illustrate this similarity. Pringsheim suggested that the two species are in reality one, *E. deses*.

The Serpentes could be derived from the Lentiferae through the development of pyrenoids and larger chromatophores. The body movement of the Serpentes is not unlike that of *E. ehrenbergii*. An *E. deses* could be derived from *E. geniculata* of the Radiatae if the paramylon centers were lost. Paramylon bodies differ, being more rodlike in the Serpentes than in Catilliferae or Radiatae.

6. *Limpidae*

The colorless forms are placed in Group VI. This group is not a natural subgenus of *Euglena*. Colorless forms exist which have direct affinities with each of the five subgenera. Other genera of Euglenidae produce colorless forms and Pringsheim (1956) suggested that Globatae should be applied to colorless *Lepocinclis*.

Spontaneous and artificial apoplastidy has often been observed in which *E. gracilis* develops into a colorless form similar to *A. longa*. Wild strains lack a stigma. Artificial clones may retain a part of the stigma following loss of chlorophyll. Jahn (1946) included colorless euglenids bearing a stigma in the genus *Khawkinea*. Blum *et al.* (1965) report significant physiological differences between an apochlorotic race of *E. gracilis* var. *bacillaris*, strain SM-L1, and *A. longa* (Jahn). The colorless forms of *Euglena* presently are identified as *Astasia*, *Khawkinea*, or *Euglena* sp. *hyalina*. Phylogenetically, it would be ideal to identify the colorless forms under the *Euglena* species name from which the colorless forms are derived. The work of Blum *et al.* (1965) would lead one to question if this is possible in naturally occurring colorless forms. So that natural affinities may be pursued, it is recommended that colorless species related to *Euglena* be designated *E.* (species) var. *Khawkinea* when affinities to *Euglena* are known. Organisms of unknown affinity should be placed in the genera *Astasia* and *Khawkinea* under a heading *Incertae sedis* until ancestral relationships are determined.

D. Recommended Classification

Accepting the class Phytomastigophora of the Committee on Taxonomy (1964) as artificial, and that a higher rank should be accorded the euglenoids, it is nevertheless recommended that categories and endings of the Committee on Taxonomy be followed for the sake of communication until more natural affinities are understood. The recommended classification of the euglenoids at order and lower levels follow:

Order.   Euglenida Bütschli 1884
  Suborder.   Euglenina Bütschli, 1884
    Family.   Euglenidae (Klebs) *emend.* Hollande, 1952
      Subfamily.   Eutreptiinae Hollande, 1952
        Genus.   *Eutreptia* Perty, 1852
      Subfamily.   Eugleninae Hollande, 1952
        Genus.   *Euglena* Ehrenberg, 1830
          Subgenera.   Rigidae Pringsheim, 1956
                    Lentiferae Pringsheim, 1956
                    Catilliferae Pringsheim, 1956
                    Radiatae Pringsheim, 1956
                    Serpentes Pringsheim, 1956

Incertae sedis (Recommended to include only colorless forms
for which affinities to a green form are not known)
Subfamily.   Euglenamorphinae Hollande, 1952
Genus.   *Euglenamorpha* Wenrich, 1924
Family.   Menoidiidae [Modified from botanical Menoidiceae (Christensen,
(1962)]
Genus.   *Menoidium* Perty, 1852
Suborder.   Peranematina Hollande, 1942
Suborder.   Petalomonadina Hollande, 1942

E. POSSIBLE AFFINITIES

Authors suggesting affinities within the family Euglenidae and genus
Euglena use chromatophore structure to indicate relationships. When
possible, other factors are included. No primitive chromatophore has
been accepted. The subgenera of Pringsheim (1956), Rigidae, Lentiferae,
Catilliferae, Radiatae, Serpentes, and Limpidae, constitute the most solid
base for seeking natural affinities. Little is known concerning the origin or
loss of pyrenoids on chromatophores, the development of lobes and bands,
or the separation of chromatophore fragments into separate units.

The Rigidae, having dislike chromatophores which lack pyrenoids, are
exemplified by *E. acus*, *E. tripteris*, *E. oxyuris*, *E. antefossa*, and *E. spirogyra*.
Closely related genera *Lepocinclis* and *Phacus* have similar chromatophores
and rigid bodies. The Lentiferae include *E. ehrenbergii* which is similar
to the Rigidae in respect to chloroplasts and paramylon bodies, but possesses
a sluggish and active metaboly. Although body shapes are more varied,
the subgenera Rigidae and Lentifera and the genera *Lepocinclis* and *Phacus*
may constitute a group with natural affinities. The species *E. proxima* of the
Lentiferae does not appear to be closely related to the Rigidae because of
chromatophore size, body shape, and shape of paramylon bodies, but
Schiller (1925) indicated *E. acusformis* to be an intermediate form between
E. acus and *E. proxima*.

Organisms of the subgenus Catilliferae bear pyrenoids, usually double,
on the chromatophores. It is a large and varied group. Iyengar (1962)
described *E. pringsheimii* which bears inner pyrenoids. This was the first
verified record of true inner pyrenoids in *Euglena*. *Colacium*, *Trachelomonas*,
and *Strombomonas* have inner pyrenoids and *Ascoglena* is reported with
or without pyrenoids. These genera and the Catilliferae may be related,
possibly through the genus *Colacium*.

Colacium, because of the sessile form and the mucus covering is postulated
as closely allied to algae. Assuming it to be primitive, a simple phylogenetic
series may be postulated. The chromatophore of *Colacium* bears an inner
pyrenoid. *Ascoglena*, *Trachelomonas*, and related forms are easily derived
by addition of a test or lorica. *Euglena proxima* lacks pyrenoids. The loss

of pyrenoids from *Colacium* may result in a form like *E. proxima*. The development of less metabolic forms accompanied with small chromatophores is readily postulated to include the Lentiferae and Rigidae subgenera of *Euglena* and the genera *Lepocinclis* and *Phacus*.

The pyrenoids may have been retained as in *E. pringsheimii* and modified to produce the sheathed pyrenoids of the Catilliferae or modified without developing paramylon caps as in the Serpentes group of *Euglena*. Chromatophore structure was explored extensively by organisms of the Catilliferae. The chromatophore may be smooth, lobed or fringed, and banded or in ribbons which may separate from the main body of the chromatophore. Species with chromatophores bearing band-shaped processes may be related to the Radiatae including *E. viridis*, *E. geniculata*, and *E. tristella*, with possible affinities to *Eutreptia* via *Euglena viridis*. These show similarities to the *E. sanguinea* forms of Catilliferae. Species similar to *E. rubra* should be retained in the Catilliferae even though pyrenoids are not present.

The Serpentes possess large disc-shaped to shield-shaped chromatophores each bearing a pyrenoid free of paramylon sheaths. It appears to be a natural group. Their origin may have been from the Radiatae through separation of chromatophore bands bearing pyrenoids.

Colorless forms, the Limpidae of Pringsheim (1956), should be placed with the species from which they were derived. The genera *Astasia* and *Khawkinae* under *Incertae sedis* would be retained for forms in which affinities to the green species have been lost.

Broad evolutionary patterns of the Euglenidae have been postulated based on a few relatively stable characteristics. These must be verified or denied with additional information on the locomotor apparatus, chromosomal counts, electron microscopic studies, biochemical analyses, irradiation and mutation, and nutrition studies. Data in the chapters that follow may aid in identifying questions concerning phylogeny and evolution of the genus *Euglena*. The euglenoids and the genus *Euglena* offer unique organisms for developing an understanding of the evolutionary processes in asexually reproducing organisms.

VI. Note Added in Proof

Leedale discusses taxonomy in relation to phylogeny of *Euglena* and includes excellent references [Leedale, G. F. (1967). *Ann. Rev. Microbiol.* **21**, 31].

References

Bělăr, K. (1926). "Der Formwechsel der Protistenkerne." Fischer, Jena.

Birdsey, E. C., and Lynch, V. H. (1962). *Science* **137**, 763.

Blum, J. S., Sommer, J. R., and Kahn, V. (1965). *J. Protozool.* **12**, 202.

Buchanan, R. E. (1954). *Ann. N. Y. Acad. Sci.* **60**, 6.

Bütschli, O. (1883–1887). Protozoa. *In* "Kassen und Ordnungen des Thier-Reichsl", (H. Bronn, ed.), Abt. III. Leipzig, p. 1098.

Chadefaud, M. (1937). *Botaniste* **28**, 85.

Chadefaud, M. (1938). *Rev. Algologique* **11**, 189.

Chadefaud, M. (1962). "Les Végétaux Non Vascularies (Cryptogamie)," Vol. I. Masson, Paris.

Christensen, T. (1962). *In* "Botanik," Vol. II, No. 2, pp. 1–178. Masson, Paris.

Chu, S. P. (1947). *Sinensia* **17**, 75.

Committee on Taxonomy and Taxonomic Problems of the Society of Protozoologists (1964). "A Revised Classification of the Phylum Protozoa." *J. Protozool.* **11**, 7.

Corliss, J. O. (1962). *J. Protozool.* **9**, 307.

Doflein, R., and Reichenow, E. (1929). "Lehrbuch der Protozoenkunde." Fischer, Jena.

Ehrenberg, C. G. (1830). *Physik. Abhandl. Kgl. Akad. Wiss. Berlin, 1830*, p. 1.

Ehrenberg, C. G. (1833). *Physik. Abhandl. Kgl. Akad. Wiss. Berlin 1833*, p. 145.

Ehrenberg, C. G. (1838). "Die Infusiontierchen als vollkommene Organismen." Voss, Leipzig.

Fritsch, F. E. (1935). "The Structure and Reproduction of the Algae," pp. 721–756. Macmillan, New York.

Gibson, I. (1966). *J. Protozool.* **13**, 650.

Gojdics, M. (1953). "The Genus *Euglena*." Univ. of Wisconsin Press, Madison, Wisconsin.

Goodwin, T. W., and Jamikorn, M. (1954). *J. Protozool.* **1**, 216.

Green, J. (1963). *Comp. Biochem. Physiol.* **9**, 313.

Hall, R. P. (1953). "Protozoology." Prentice-Hall, Englewood Cliffs, New Jersey.

Hall, S. R. (1931). *Biol. Bull.* **60**, 327.

Hardy, A. D. (1911). *Vict. Nat.* **27**, 215.

Hollande, A. (1942). *Arch. Zool. Exptl. Gen.* **83**, 1.

Hollande, A. (1952). *In* "Traité De Zoologie," (P. P. Grassé) Fasc. I, pp. 238–284. Masson, Paris.

Hübner, E. F. W. (1886). *Progr. Realgymn. Stralsund.* **20**.

Huber-Pestalozzi, G. (1955). *In* "Die Binnengewasser" (A. Theinemann, ed), Vol. 16: Das Phytoplankton der Süsswassers. Part 4. Schweizerbart'sche Verlagbuchhandlung, Stuttgart.

Hutner, S. H. (1955). *In* "Biochemistry and Physiology of Protozoa" (S. H. Hutner and A. Lwoff, eds.), Vol. II, pp. 1–40. Academic Press, New York.

Hutner, S. H., and Provasoli, L. (1951). *In* "Biochemistry and Physiology of Protozoa" (A. Lwoff, ed.), Vol. I, pp. 27–28. Academic Press, New York.

Iyengar, M. O. P. (1962). *Arch. Mikrobiol.* **42**, 322.

Jahn, T. L. (1946). *Quart. Rev. Biol.* **21**, 246.

Johnson, L. P. (1944). *Trans. Am. Microscop. Soc.* **63**, 96.

Johnson, L. P. (1956). *Trans. Am. Microscop. Soc.* **75**, 271.

Klebs, G. (1883). *Untersuch. Botan. Inst. Tübingen* **1**, 233.

Krinsky, N. I. (1964). *Plant Physiol.* **39**, 441.

Krinsky, N. I., and Goldsmith, T. H. (1960). *Arch. Biochem. Biophys.* **91**, 271.

Kudo, R. R. (1966). "Protozoology," 5th ed. Thomas, Springfield, Illinois.

Leedale, G. F. (1958a). *Nature* **181**, 502.

Leedale, G. F. (1958b). *Arch. Mikrobiol.* **32**, 2.

Leedale, G. F. (1959). *Cytologia (Tokyo)* **24**, 213.

Leedale, G. F. (1962). *Arch. Mikrobiol.* **42**, 237.

Leedale, G. F. (1966). Discussion of Classification of Protozoa at Second International Conference on Protozoology, London. *J. Protozool.* **13**, 189.

Leedale, G. F., Meeuse, B. J. D., and Pringsheim, E. G. (1965). *Arch. Mikrobiol.* **50**, 133.

Lefèvre, M. (1931). *In* "Cryptogamiques," pp. 343–354. Mangin, Paris.

Lemmerman, E. (1913). *In* "Die Süsswasserflora Deutschlands, Österreichs und der Schweiz" (A. Pascher, ed.), Vol. II, pp. 115–174. Fischer, Jena.

Mainx, F. (1927a). *Arch. Protistenk.* **60**, 305.

Mainx, F. (1927b). *Arch. Protistenk.* **60**, 355.

Michajlow, W. (1965). *Acta Parasitol. Polon.* **13**, 313.

Perty, M. (1852). "Zur Kenntnis kleinster Lebensformen." Bern.

Pochmann, A. (1953). *Planta* **42**, 478.

Pringsheim, E. G. (1956). *Nova Acta Leopoldina* **18**, 1.

Protozoologists (1958). "A Catalogue of Laboratory Strains of Free-Living and Parasitic Protozoa." (The Committee on Cultures, L. Provasoli, Chairman) *J. Protzool.* **5**, 1.

Scagel, R. F., Bandoni, R. J., Rouse, G. E., Schofield, W. B., Stein, J. R., and Taylor, T. M. C. (1965). "An Evolutionary Survey of the Plant Kingdom." Wadsworth, Belmont, California.

Schmarda, L. K. (1846). "Kleine Beiträge zur Naturgeschichte der Infusorien," Wien.

Schildkraut, C. L. Mandel, M., Levisohn, S., Smith-Sonneborn, J. E., and Marmur, J, (1962). *Nature* **196**, 795.

Schiller, J. (1925). *Arch. Protistenk* **53**, 59.

Schmitz, F. (1884). *Jahrb. Wiss. Bot.* **15**. 1.

Smith, G. M. (1950). "The Fresh Water Algae of the United States," 2nd ed. McGraw-Hill. New York.

Starr, R. C. (1964). *Am. J. Botany* **51**, 1013.

Stebbins, G. L. (1960). *In* "The Evolution of Life" (S. Tax, ed.). Univ. of Chicago Press, Chicago, Illinois.

Stein, F. (1878). *In* "Der Organismus der Infusionsthiere," Article 3. Leipzig.

Wenrich, D. H. (1924). *Biol. Bull.* **47**, 149.

Wolken, J. J. (1961). "*Euglena*." Inst. Microbiol., Rutgers Univ. Press, New Brunswick, New Jersey.

CHAPTER 2

ECOLOGY OF *EUGLENA*

James B. Lackey

I. Introduction

An ecological consideration of a structurally and physiologically related plant or animal group may reveal wide differences in environmental responses. The genus *Euglena* is such a related group, and has received a number of monographic treatments. The most recent of these are by Gojdics (1953), Pringsheim (1956), and Huber-Pestalozzi (1955). The monographs of Gojdics and Huber-Pestalozzi are based more on reviews than on observation of living specimens, although the former has a great many firsthand observations. Gojdics recognizes 154 or more species, while Huber-Pestalozzi lists 101. Pringsheim, basing much of his work on direct observation and much of it on soil–water cultures, recognizes 56 species. It is interesting that so many different species develop within what appears to be a single type of environment. He notes several different environments viz., salt versus fresh water and heavily fertilized or polluted versus slightly polluted or unfertilized water.

The occurrence of easily recognized species such as *E. pisciformis*, *E. gracilis*, or *E. mutabilis* can actually be shown to embrace a very wide range of environments. I have found *E. mutabilis* in alkaline waters, acid mine waters, soft-water ponds, and iron seeps. It occurs to a limited extent in brackish water and there is some question as to whether or not it is widespread in salt water. Both Pringsheim and Gojdics recognize this when they list the sources from which the various species were reported. There is often doubt, however, as to whether or not two or more workers have recorded the same species. It has been shown all too often that recognition of a species may require very careful study, and the limnologist or ecologist who may be concerned with a large list of genera and species may not have sufficient time to identify all species carefully. The well-defined species are easily noted, but less familiar or less frequently occurring species may be subject to some confusion. The environment from which the latter species are reported, therefore, may not always be their actual environment.

At least the genus is a fairly well-defined one. If a worker specifies *Euglena* it is reasonably certain that he is discussing a *Euglena*. Among the salt or brackish-water species there might be some confusion with regard to *Eutreptia* and *Eutreptiella*, but both are distinctive, not only in their two flagella, but in several other noticeable ways. In fresh water, some species of *Lepocinclis* might be confused with *Euglena*, but these are also soon recognized.

II. The Paucity of Widespread Ecological Observations

A. THE INCOMPLETE RECORD

One difficulty in determining the ecological status of the various species of *Euglena* is that so many species have been recorded only a few times.

The three monographs mentioned above have done much to eliminate taxonomic confusion, and for those species that occur widely and in large numbers, environmental factors of occurrence are often cited.

It seems inevitable that the most dense populations of a microorganism occur in environments most nearly approaching an optimum for the species. I had been familiar with *Euglena mutabilis* for perhaps 12 years before encountering dense populations, and regarded it as a well-defined species. The occasional occurrences were in a variety of environments, however and it was not until dense populations were found in streams, containing sulfuric acid from coal-mining operations (Lackey, 1938) that some appreciation of its optimum environment was gained. These were field observations and it is by no means certain that the optimum environment is that of these small, shallow, acid streams, but they certainly represent the nearest approach to it we have observed.

This also illustrates that in field observations the uncontrolled multiple factor situation prevents the assignment of true values to any particular factor such as pH per se, ferrous or ferric ions, the form of sulfur present, predation, antibiotics, etc. It is true that we can sometimes determine effects of a specific factor such as a massive kill of *Euglena* by chytrids, but this is the exception, not the rule.

Because of confusion as to the validity of many species, and because of the multiple factor situations in field studies, ecological relationships are almost unknown for many species. *Euglena sciotensis* (Lackey, 1939) is an example. This species has been accepted as valid by Gojdics, Huber-Pestalozzi, and Pringsheim but seems to be not widely known. It is distinctive in appearance so the inference is that it occurs infrequently. It is common in Ohio Valley streams, however, especially the Scioto (Ohio) River, and has been found once in the estuarine waters of the Sacramento River in California. Occurrence elsewhere has been very limited, although the soft-water lakes and streams of Florida have been repeatedly examined for euglenoids. It might be inferred that this is a hard-water species, but one for which optimum conditions have not been described. Certainly it did not attain dense populations in some Ohio valley polluted situations where other species such as *E. pisciformis* bloomed. This example illustrates the need for careful recording of bloom situations—such records may indicate important ecological factors.

It appears that reasons why not more is known of the ecology of many *Euglena* species are threefold: (a) confusion as to the validity of certain species; (b) infrequent records of the occurrence of many species; and (c) lack of detailed ecological observations including records of occurrence.

B. The Need for Laboratory Studies

One avenue of further information lies in laboratory cultures of *Euglena* species. Ecology is properly a field study, but it is strongly buttressed by laboratory studies of taxonomy (cytology) and culture, the latter embracing more than just nutrition. The soil–water method of Pringsheim (1946) is an excellent method for maintaining growth cultures from which unialgal or axenic cultures can be developed. In nature, normal growth of any species probably follows a unimodal curve on which a high population represents the nearest attainment of a whole group optimal ecological factors.

Less favorable conditions are indicated by fewer numbers on either side of the curve, and the whole curve may be wiped out by some catastrophic factor. With axenic cultures we can investigate, one at a time, a whole series of conditions—light, H_2S content, the presence of a certain amino acid, cobalamin content and so on, as well as a combination of factors.

Very few species of *Euglena* have been grown axenically. Pringsheim (1956) has grown *E. anabaena, E. deses, E. gracilis, E. pisciformis, E. spirogyra, E. stellata,* and *E. viridis* under these conditions. The Indiana University Culture Collection (Starr, 1964) lists 10 species, but not all of these and their many strains are axenic. It is tedious and often unrewarding work to secure either type of culture. At the Phelps Laboratory, University of Florida, Gainesville, Florida, attempts to establish cultures of *E. clara* over a 6-month period were unsuccessful despite a constant abundance of the organism in an experimental waste treatment plant at the laboratory.

Euglena gracilis is very easily grown and is widely used in experimental work. *Biological Abstracts* currently records about 20 papers on this species to one on other species. The result is a precise knowledge of its nutritional requirements (treated in Chapter 6) with records of its reactions to many other factors. In fact, its autecology is on a sound basis, Nevertheless, application of these findings to its field occurrence must be cautiously interpreted, simply because no amount of laboratory work can consider the combinations of variables existing in the field. There is an urgent need for continued and expanded studies of *Euglena* species in axenic culture. The ecological findings that have thus been established for *E. gracilis* are not necessarily true for other species.

III. Habitats

The habitats of *Euglena* species are as varied as might be expected for such a large number of species. They can be classified on the basis of salinity as salt, brackish, and fresh water. With regard to physical substrate they are found in plankton, as neuston, at sediment–water interfaces, on mud banks, in the living (*Aufwuchs*), on water plant stems, on snow, and as epizoites. As to light conditions, some occur in direct sunlight where they are exposed to ultraviolet, some at somewhat lower depths but with bright illumination, and some, notably those that crawl, on the bottom in greatly reduced light. These living conditions and others will be discussed one at a time.

A. Species of Saltwater *Euglena*

Nine species from salt water were recorded by Gojdics (1953). Pringsheim does not accept *E. acusiformis, E. baltica, E. gojdicsae, E. interrupta, E. limosa, E. reticulata, E. salina, E. vangoori,* or *E. vermiformis* as valid species, leaving only *E. obtusa* and *E. variabilis* as saltwater species. Neither of these was originally (nor has been since?) reported from salt water, and therefore if *E. obtusa* is not synonomous with *E. vangoori,* and if *E. variabilis* is not synonomous with *E. gojdicsae,* we are left with no saltwater species unless *E. limosa* and *E. obtusa* are synonyms. Nevertheless, three different species of *Euglena* can be found in saltwater habitats with regularity and often in considerable numbers.

There is certainly an abundance of *E. mutabilis* (*E. vermiformis*) in salt water, if we accept the taxonomic description in recent monographs. In San Francisco Bay it is found in large numbers in two locations where sewage enters the Bay. It seems too much, however, to expect this same *Euglena* to exist in huge numbers in coal mine streams having a pH of 2.5, and in salt-laden alkaline ocean water. It seems more reasonable that two species, closely related morphologically, are involved. The saltwater form *(E. vermiformis)* is generally somewhat longer, and usually has about eight chromatophores. The considerable smaller *E. vermiformis* described by Carter (1938) has not been found in these saltwater situations. On the other hand, *E. mutabilis* from acid streams usually has two to four chromatophores, and occurs in a wide range of sizes. The occurrence of *E. limosa,* whatever its taxonomic position, has been well documented by Bracher (1929; *E. deses,* 1937) and by Fraser (1932) on mud banks exposed by tidal action the estuaries of English rivers.

Of the other two saltwater species one is certainly related to *E. fenestrata* Elenkin as reported by Gojdics. (I do not agree with Pringsheim that this is a synonym for *E. obtusa*). The second is presumably new, but has been noted in manuscript. These three different species, whatever their taxonomic position, are common on silty bottoms of inshore marine waters where there is decomposition of organic matter such as sewage or algae. A fourth, related to *E. sanguinea* by its abundant hematochrome and its shape, occurs

less frequently, but is especially common in San Francisco Bay. None of
these four occurs everywhere. I have found none of them in the Eel Pond
at Woods Hole, Massachusetts, and only two records of *E. vermiformis*
around Oahu Island in Hawaii. It would be interesting to observe the species
of *Euglena* reported by Vorhies (1937) from Great Salt Lake, Utah.

Various others tolerate salt water for various periods of time. Included
are *E. viridis*, *E. pisciformis*, and *E. deses* which are actively in motion
in weak to strong concentrations of sea water. Apparently none of them is
viable for long or multiplies in water of even low salinity.

B. Species of Freshwater *Euglena*

Many of the better known freshwater species seem to be cosmopolitan.
Thus *E. oxyuris* has been reported from England, Western Europe, Australia,
and Asia. It is a common endemic in a pool on the campus of the University
of Hawaii (Honolulu, Oahu Island, Hawaii), the only other freshwater
species I have found on this island being *E. clara*, *E. gracilis*, *E. mutabilis*,
and *E. pisciformis*. *Euglena pisciformis* and *E. viridis* are also reported from
many parts of the world, the latter probably ranking in the literature next
to *E. gracilis*. *Euglena gracilis* is now widely cultured axenically, however,
and its experimental use accounts for its frequent mention. Unfortunately,
a record is often only for the genus *Euglena* and not for a species, otherwise
our knowledge of the distribution of many species would be greater.

C. Colonization of New Habitats

Speculation as to how *E. oxyuris* might have reached the island of Oahu
is interesting. The abundant specimens observed were all near the maximum
reported size, 400 μ, and were all free of chytrid parasites. Schlichting (1960)
found two or more species of *Euglena* carried by ducks, but whether or not
the colonizing *Euglena* were present at the place from which the ducks came
is not known. There are few land birds that might have carried the organism
while migrating although it might have arrived on the feet of golden plovers
which stop at Oahu between the Arctic and South America and which
occur inland. Unfortunately, I know of no records of *E. oxyuris* in either
of these two places, although if cosmopolitan, it should occur there. This
species is not known to form cysts, a further argument against transport by
birds. One could make out a better case for the presence of marine species
in Hawaii: drifting debris, seaweeds, migrating animals. Only one record of
a marine species was made, however. For both of these groups it should
be specified that search was made only from September 1st through November
24th, however, more than 100 freshwater and seawater collections were made.

IV. Nature of the Aqueous Habitat

The freshwater species are typical still-water forms—inhabiting pools, ponds, and lakes—although many instances of river occurrence are known. Thus I found 24 species in the Scioto River of Ohio in a 2-year study of that stream. Some freshwater species become temporarily neustonic in habit. The Cedar Swamp at Woods Hole, Massachusetts, often developed a bloom of *E. polymorpha* (along with *Gonyostomum semen*) on its surface; and *E. sanguinea*, coloring the surface of ponds in northern Alabama brick red to brilliant red, actually looked as if it had been dusted on the surface when viewed from a low angle. These blooms were in the top inch of water, and could be aggregated by sweeping the surface in one direction with a stick. This neustonic habit was temporary, however; in late afternoon and evening most of the euglenas had migrated deeper, often to the bottom. Lund (1942) found 18 species of *Euglena* in the marginal debris of five ponds he investigated in England. All of these, and all of the Scioto River species (except *E. sciotensis*) were the most readily recognized forms. Difficulty in recognition of many species probably keeps them from being termed "common."

ARE THERE PLANKTONIC SPECIES OF *EUGLENA*?

It is doubtful that truly planktonic species of *Euglena* exist. Pringsheim (1956) says *E. tiscae*, *E. variabilis*, *E. spirogyra*, and *E. tripteris*, "are all found swimming." In my experience, *E. acus*, *E. spirogyra*, and *E. tripteris* are as often found in the interface material as in the water, and the last-mentioned two are often seen creeping by the slow torsion of their cells, sometimes with no sign of the flagellum. I have more often found *E. pisciformis*, *E. gracilis*, *E. clara*, and *E. velata* swimming vigorously and tending to form surface blooms. Indeed, the first two will follow the small circle of light from a substage lamp so quickly when a slide is moved across the stage that it is necessary to kill them in order to obtain an accurate count.

Most of the species with which I am familiar, both saltwater and freshwater, tend to be interface dwellers, e.g., *E. deses*, *E. ehrenbergii*, *E. mutabilis*, *E. oxyuris*, and others. This may involve a causal relationship between weak flagellar action, or a very short flagellum, and the size or weight of the body. Certainly for general population assessment, it is best to stir the bottom debris and sediment slightly where the water is perhaps 6 inches deep, and collect the stirred-up material. Collection and centrifugation of a water sample from the polishing pond at the University of Florida sewage treatment plant usually yields predominantly *E. viridis*, *E. clara*, and *E. pisciformis*. When a fragment of bottom sludge floats up, however, its interface surface is usually a tangle of filamentous blue-green algae and diatoms with numerous *E. tripteris, E. spirogyra, E. deses, E. fusca,* and often other species attached.

V. Factors Inherent in Habitats

A. CURRENT

Little is known about current action. *Euglena sciotensis* appears to be a potamophile form, but has been reported only a few times by European workers. Most others seem independent of current or still water, as long as they can maintain position, however, one is more likely to collect *Euglena* from a pool than from a stream.

B. LIGHT

The amount of light available seems important. *Euglena spirogyra* and varieties of *E. ehrenbergii* occur more frequently in shaded woodland pools, while *E. sanguinea* and related red varieties occur at the surface of sunlit ponds where they would seem to be susceptible to injury by ultraviolet. In some of these ponds the summer afternoon temperatures may approach 40°C, while the woodland pools may be cooler by 5–10°C. Too little, however, is known of euglenoid–temperature relationships in natural conditions. Jahn (1946) discussed the periodicity of euglenas as related to light intensity, and there is little question that a diurnal variation in hematochrome distribution and top-to-bottom migration occurs in the red euglenas. Laboratory investigations of responses to light have been rather inconclusive.

C. SOFT WATER AND HARD WATER

Soft-water situations generally seem to have the same species as hard water. This supports the argument for independence of pH, but it also indicates that there is tolerance of wide CO_2 and CO_3 levels. Interestingly, Fritsch (1931) quoted Tannreuther as saying that *E. gracilis* grows well in an alkaline medium and Skadowsky as stating that the species has an optimum pH of just below 4.0. However if one examines situations in which large quantities of *Pistia, Eichornia, Azolla,* or duckweeds are growing, very few euglenas of any kind are customarily found. This is true even when there is enough light admitted for photosynthesis. The answer is probably found in the ability of these floating plants to remove a large moiety of available phosphorus.

D. DISSOLVED OXYGEN

Relationships of *Euglena* to dissolved oxygen seem to have been little investigated. Photosynthetic organisms should be independent, unless they occur in dense populations so that the oxygen is depleted at night, or unless there is some balancing-out of effects by nonphotosynthetic organisms also present. Dense populations of *E. gracilis* in a tall cylinder maintain a rather uniform suspension which will persist thus for days with no apparent night-time mortality. It is also seen that dense cultures of *E. gracilis* that have lost their chlorophyll, persist for many days, either in an open container, or beneath a thick film of paraffin oil, despite no source of oxygen other than diffusion. Paraffin oil does not completely inhibit oxygen diffusion into the water beneath, however.

Euglena relatives are usually facultative anaerobes. *Peranema* and *Entosiphon* glide freely through the oxygen-free area under a cover glass, while bacteria form a dense ring around the margin where the oxygen enters by diffusion. These and other euglenoid genera are often abundant in interface material, even if it is black and nasally sulfuretted. The greater number of *Euglena* species are found in situations presumed to contain dissolved oxygen, however and presumably while they tolerate low oxygen tensions, oxygenated situations are nearer optimal. The saltwater species of *Euglena* are not found free in the water, but crawling on the surface of the interface material. If exposed by low tide, they enter the interface where the material is black and probably anaerobic only a millimeter or so below the surface. Here they may be found along with colorless euglenoids, certain zooflagellates, certain blue-green algae, and sulfur bacteria belonging to the Beggiatoales. Transferred to clear oxygenated seawater on a slide, they live several days in a hanging drop. Evidently, adaptability to a wide range of oxygen tensions is a characteristic of *Euglena*.

VI. *Euglena* and Pollution

A. ANIMAL EGESTA AND EXCRETIONS

It seems well demonstrated that a requirement for *Euglena* growth is dissolved organic matter, including cobalamin. Discussion of nutritional requirements properly appears in Chapter 6 which deals with exact needs. *Euglena* species have often been mentioned as indicator organisms for degrees of organic pollution, and within limits, rightfully so. However, the matter tends to have a quantitative basis rather than a qualitative one, i.e., the number of each of several species is determined rather than the number of different species. This phenomenon is simply a nutritive response just as large populations of ciliates develop where bacteria serving as food are abundant.

It is well known that ponds and situations polluted by cattle droppings and urine tend to develop large populations of euglenoids—*Euglena*, *Trachelomonas*, *Cryptoglena*, *Phacus*, etc., as well as Volvocales and other organisms. The same thing happens to a domestic sewage situation, but not immediately—raw sewage, diluted, does not develop numerous euglenoids until after considerable bacterial action has materially changed the sewage, i.e., downstream. (Raw sewage put into an oxidation pond develops enormous numbers of bacteria, Chlorophyceae, certain Volvocales and a few, sometimes many, *E. pisciformis*.) In the downstream area various species of *Euglena* develop, often in great numbers, and eventually, moving downstream, a wide spectrum of *Euglena* species can be found. Along with them are various species of Volvocales but, rather curiously, few *Trachelomonas*. *Strombomonas* species, which appear to be potamophiles, may be abundant farther downstream.

This different behavior on the part of *Euglena* species toward domestic sewage as compared to animal pollution is hardly explicable but there are some possibilities. One is the different nature of the animal fecal matter, which contains a larger amount of cellulose and cellulose derivatives than human sewage and perhaps more reduced nitrogenous matter. This is in line with the statement by Leedale *et al.* (1965) that ammonium magnesium phosphate improved development of *E. spirogyra* in soil–water cultures. Human sewage contains a greater diversity of other substances also. Possibly it is more nearly degraded to orthophosphate and nitrate as is indicated by the biota of oxidation ponds. In any case, there is a stronger reaction in euglenoids generally, and some species of *Euglena*, to animal sewage containing reduced nitrogen than to human sewage. This is very easily checked by examining material from the margin of a pond polluted by cattle, and the region of a stream below a sewer outfall. It is generally recognized by workers. Professor Maxwell S. Doty (University of Hawaii) stated, "We go into that region for euglenoids because there are pig farms there."

B. SPECIFIC TYPES OF POLLUTION

1. *Acids*

Some *Euglena* species show diverse reactions to particular kinds of pollution. The reaction of *E. mutabilis* to acid coal-mine wastes (Lackey, 1941) has been mentioned above. Flying over these small streams in West Virginia was like looking down on a bright green branching tree. Values for pH as low as 0.9 still showed living euglenas. This same species is often abundant in iron seeps, or in the gelatinous ferric hydroxide along the edges of small streams. The question remains as to whether the organism responds to the sulfate ion, or to the iron in these two environments.

2. *Citrus Wastes*

Euglena gracilis is often common in lagoons used to increase biological reaction time for citrus wastes (Lackey and Morgan, 1960). At Plymouth, Florida, it appeared early in the lagoons and the last 5 of 12 in series were green with this species and a much smaller population of *Chlorogonium*. The high numbers were ineffective in materially reducing the waste problem, although the citrate level was somewhat reduced. In vegetable cannery waste lagoons, a mixed group of Euglenophyceae and Volvocales sometimes appears with *E. pisciformis* as a dominant. This is also true of paper and pulp mill lagoons. Generally, the higher the biochemical oxygen demand of a waste, the less the chance of finding *Euglena* in the environment.

3. *Silt and Debris*

Euglena may be smothered by silt or fibrous material such as asbestos wastes. Some species form a mucilaginous outer layer to which the silt adheres, and within this cocoon they are viable for several days at least. This is evidently one function of their muciferous bodies. The tidal flats of the Plym estuary at Plymouth, England are periodically flushed with fine clay from upstream, yet are a habitat for several species of *Euglena* and many colorless euglenoids. This is interesting in view of the statements of Pringsheim (1956) relative to the use of clay in culture media.

4. *Radioactivity*

Euglena is unaffected by many radionuclides in its environment. Observations in a seepage area from radioactive waste pits at Oak Ridge showed several species of this genus (as well as other algae and protozoa) that could have been exposed to massive doses of radiation for several years. Yet not one abnormal form was found. A small pond near one of the facilities was so "hot" that the tolerated human dosage was obtained in 1 hour. This pond, all summer long, supported a good population of what I at first called *E. fusca*, but was in reality *E. limnophila*. However the individuals were much longer—a rather uniform 200 μ. It is interesting to speculate whether or not this was a tetraploid, and if so, whether or not it resulted from radiation. A detailed survey of the Clinch River which received wastes from the Oak Ridge installations showed (Lackey, 1957) a normal and abundant microbiota, including various species of *Euglena*.

Dense populations of *E. gracilis* exposed to as much as 10^6 r from ^{60}CO showed no more than normal die-off in our laboratory, and their reproductive rates seemed unimpared. A colorless euglenoid, *Entosiphon sulcatum*, was subjected to ^{60}CO radiation (Lackey and Bennett, 1962) for 60 generations. Four clones were used and 24-hour division rates were plotted. During this time viability was normal and the division rate was unaffected. Exposed to tagged radionuclides, *E. gracilis* concentrated some of these several hundred times the amount in the environment. It must be remembered that many microorganisms concentrate a given stable isotope many times over its presence in the environment, and that we are dealing here more with parts per billion than with parts per million.

Altogether there are many types of pollution to which various species of *Euglena* react, and in various ways. Many of the pollutants are catastrophic, such as the hot water that almost sterilizes a stretch of the White River above Indianapolis, Indiana, or the sudden minor floods that load the Licking River with excessive silt above Covington, Kentucky. Regardless of the destruction of these populations, if *Euglena* species existed there before the catastrophe, they quickly return. It should be remembered that some species of *Euglena* are among the first of the protists to appear after organic pollution of a water.

VII. Specific Environments

There are certain specific environments that favor *Euglena* populations. One of these is the commonly used hay infusion. When such an infusion is cooled, brought into equilibrium, and inoculated with a small amount of detritus from a pond, it often develops thriving cultures of microbiota, including *E. acus, gracilis*, and *E. pisciformis*. *Euglena gracilis* readily becomes colorless under such conditions and so does *E. acus* to a lesser degree. The colorless form of *E. acus* should not be confused with often dense populations of *Cyclidiopsis* which may appear, however. This loss of chlorophyll is a prime example of ability to live either as an autotroph or as a heterotroph. Indeed, the normal existence of many *Euglena* species is a combination of these modes, which we recognize in using organic compounds in *Euglena* culture media.

The reactions are sharply defined, however. For years *Peranema* and *Entosiphon* were grown on tap water and autoclaved wheat grains. If *E. deses, E. pisciformis*, or *E. gracilis* were also introduced, they lived for a while but did not multiply. When, however, a layer of paraffin oil was introduced, *E. gracilis* multiplied, but lost its chlorophyll. This was not controlled experimental work, but has some interesting connotations.

A. Tree Hole *Euglena*

Another specific location for certain euglenoids is tree holes (Lackey, 1940). Normally this water is brown, probably as a result of tannin compounds, and restrictive, and contains not too many species; but *E. fusca* is common. The only dense population of *Trachelomonas reticulata* I have ever seen was in a tree hole. In swamps in which the dominant trees are Tupelo gum *(Nyssa aquatica)*, the dark areas within the expanded bulbous base of the tree that are open to the surrounding water often contains good populations of *E. velata* and several other species, whereas there are few in the outside waters. No explanation is offered for this, but it is significant in that tree hole situations are often breeding places for certain mosquito species. *Anopheles barberi* breeding was unknown until larvae were found (Shields and Lackey, 1938) in the bases of Tupelo gums. In such situations the euglenoids are an important food for the mosquito larvae. There is some question, however, as to the degree to which they are digested. Viable euglenoids pass through the digestive tracts of some mosquito larvae which, according to Senior-White (1928), use them as food. They also serve as food for frog larvae (tadpoles) and pass through the digestive tract in a viable state; at least some live within the utricles of Utricularia (Hegner, 1926).

B. Snow

Another specific habitat is reported by Kiener (1944) who found patches of snow on the high plains of Nebraska colored green by an undetermined species of *Euglena*. This is an important observation, indicating that some species are viable in conditions under which they could be transported aloft and to long distances and colonize situations where their origin is a puzzle.

C. Soils

The presence of *Euglena* species in soils can be readily demonstrated simply by using suitable culture media. Reference is made to the book of Sandon (1927) for generalities, but the importance of this is that certain euglenas can migrate into the soil and persist through a drying up of the pool in which they existed as trophozoites. Evans (1959), investigating the survival of freshwater algae during dry periods, found 16 species of *Euglena* in 5 small ponds, not all in any one. Some of these were found in the mud as it was drying, and *E. deses*, *E. mutabilis*, and *E. viridis* were recovered from marginal mud by culture methods. In San Francisco Bay, at low tide, no euglenas were found on mud bands heavily populated by naviculoid diatoms, whereas when these banks were covered with water

E. vermiformis appeared there. Conceivably this could be a periodicity related to tides or light intensity (Jahn, 1946) rather than dryness, but the practical result is the same—migration to a "safe" environment.

D. HIGH ALTITUDES AND DESERTS

Somewhat less-specific habitats may be those at high altitudes and in desert situations. Examination of freshwater ponds and streams in California mountains near San Diego revealed no *Euglena*, even when some of these were used by cattle. A few very high situations, up to 10,000 ft, in Colorado were also examined and no euglenas found. Durrell and Norton (1960) examined 59 samples from 41 lakes in Colorado, all over 9000 ft, and the only euglenoids found were two species of *Trachelomonas*. Lake Waiau, among the cinder cones on the top of Mauna Kea at 13,700 ft, likewise contained no *Euglena*, although other algae were present. This lake is fed by melting snow, and there is virtually no source of organic matter. Its population has presumably been wiped out repeatedly in the past by ash fall and high temperatures, but recolonization takes place each time. This lack of euglenas may have been the result of a barometric pressure factor or of great variation in temperature.

Some 30 locations in the desert around Las Vegas, Nevada, were examined in 1958. Many of these were small marshy locations around springs; some were highly alkaline, others were not. No species of *Euglena* was found, although a number of other algae and protozoa were recorded. Bad Waters, in Death Valley, some 250 ft below sea level, had a thriving population of *Gomphosphaeria aponina* but few other species. Its percentage of magnesium sulfate and other salts is very high, however, if one species of *Euglena* can exist in Great Salt Lake (Jahn, 1951) it seems reasonable to expect them in Death Valley waters. Here again are wide temperature fluctuations, daytime temperatures reaching 115°F.

E. TREE BARK OCCURRENCES

A unique occurrence of *Euglena* sp. was cited by Briscoe (1939) who cultured bark from the honey locust *(Gleditschia)* and obtained a growth which he decided developed from cysts.

VIII. Symbiosis and Parasitism

There seem to be no true instances of symbiosis in this genus. Perhaps the nearest is *E. cyclopicola* Gicklhorn, which is reported (Gojdics, 1953) to be fairly common on small copepods and other invertebrates. It also has a free-swimming stage. The principal difficulty with this species is that it is readily interpreted as *Colacium*, and Pringsheim (1956) regards it as such. If it is not, then it is certainly a connecting link to that genus.

Of interest in this respect are the euglenalike forms that one may often find in the digestive tract of tadpoles. Wenrich (1924) described *Euglenamorpha hegneri* from the intestine of *Rana pipiens*. It had three flagella, three photoreceptors, and chloroplasts. These however, are dull to almost colorless in specimens I have examined, and Wenrich says there is a colorless variety. Another genus for which I have not seen the paper, is *Hegneria*, which is said to have seven flagella. These would seem to merit much more study from a phylogenetic standpoint. Those I have examined show much variability and are therefore difficult to place. They could indicate a tendency either toward symbiosis, or toward parasitism. More information on their possible free-living conditions is needed.

There seems to be only one case of true parasitism in *Euglena*—*E. leucops* (Hall, 1931) which is colorless, although it possesses a stigma. It occurs in the flatworm *Stenostomum predatorum*. *Euglena parasitica* is not a true parasite; Pringsheim (1956) regards it as a possible *Colacium*. There seem to be no observations on it beyond the original description by Sokoloff.

It seems that this group of green and nongreen flagellates, having nutrition that encompasses soluble organic materials, would be a favorable group in which to locate symbionts and parasites. The very few reports of such indicate a paucity instead. The reader is referred to a paper by Haswell (1907) for a discussion of this matter.

IX. *Euglena* As a Food Organism

Because of its abundance, size, probable vitamin content, and occurrence in plankton and in the debris around the shallow margins of ponds, as well as a surface bloom, members of the genus might well serve as food for many of the invertebrates. Attention has been called to a paper by Senior-White on their use as food for mosquito larvae. Personal observation has shown their ingestion by ciliates (*Euplotes* and other hypotrichs), rotifers, flatworms, and other insect larvae. They must therefore be an important item in the trophic pyramid.

However, they are also resistant. Viable *Euglenae* have been recovered from copepod feces, from tadpole feces, and from mosquito larvae feces. They have also been recovered on dissection from the digestive tracts of the last two. It is not clear just what the mechanism of their resistance is, but where other protozoa and algae seem to be generally digested, a percentage of *Euglena* cells comes through in a viable condition. One possibility appears to be the production of a mucilaginous envelope by their muciferous bodies, and such an envelope has been noted in some of the cases examined.

Another possibility is that the pellicle is a very resistant structure, impervious to grinding action and digestive enzymes. Some *Euglena* apparently come through as cysts, but not all. Cysts however, are probably quite impervious, or at least resistant stages.

Euglena is sometimes parasitized and becomes a food in this manner. Mitchell (1928) found four species of *Euglena* readily attacked by an organism he decided was *Pseudosphaerita euglenae. Euglena spirogyra* was not found to be a victim of the parasite.

X. Cosmopolitan Distribution

Species of *Euglena* have been reported from every continent. Without reviewing each individual paper, but accepting occurrence as listed by Gojdics (1953), we note the following: *E. acus, E. deses, E. ehrenbergii, E. gracilis, E. limosa, E. mutabilis, E. oxyuris, E. pisciformis, E. polymorpha, E. schmitzi (geniculata), E. spirogyra,* and *E. viridis.* A complete search of the literature would probably add more species to this list, and certainly more locations for which these species have been reported.

There are some species for which a single occurrence has been reported. Examples are: *E. synchlora* (Nebraska), *E. purpurea* (Germany), *E. circularis* (Illinois), *E. minuta* (Wisconsin), *E. pascheri* (Russia), *E. subehrenbergii* (Sweden), *E. megalithos* (Latvia), and *E. fenestrata* (Russia). These species seem reasonably authenticated. The descriptions of many others are not adequate for wide acceptance. This may have been the result of lack of recognition on the part of other workers, however, since there is a smaller group interested in the taxonomy than in the ecology of the genus, and since the ecologists usually have less time and opportunity to make species determinations.

Nor is *Euglena* to be had for the asking. One has to search for it. Certain types of habitats can usually be expected to yield *Euglena,* viz., *E. mutabilis* in acid mine streams, *E. spirogyra* in soft-water, partly shaded, woodland pools. Careful sampling of such habitats may fail to turn up the expected species, however. Within 1 month after coming to the island of Oahu, Hawaii, researchers had sampled about 30 freshwater habitats. In the first 28 days only two species, *E. pisciformis* and *E. oxyuris,* had been found, despite intensive examination of collected material. Then *E. clara, E. mutabilis,* and *E. gracilis* were found in each of four samples. The reason was evident if not simple. The last four samples came from stagnant pools in a valley where surface drainage would have been from agricultural land, or land adjacent to a horse farm.

It seems probable that most but not all species of the genus are cosmopolitan, but that difficulties of colonization (vide supra) may prevent certain species from being found in isolated geographical locations. There are almost certainly species so adapted to a specialized environment as to preclude their being widespread. Examples within the genus are not at hand, but two nongreen relatives illustrate this.

One came from the Eel Pond, a small bay within the limits of Woods Hole, Massachusetts. This bay and its environs had been studied for years as a species source for classes in protozoology. Recently, from a sampling of the interface in some 6–9 ft of water, considerable numbers of a new, large, golden euglenoid *(Calkinsia aureus)* were found. The organism is very striking (Lackey, 1960) and would hardly be passed over even by a novice, yet subsequent search in many places has failed to produce others. A rather similar organism has been found a few times in the sand churned up by the surf at La Jolla, California, but not in sufficient numbers to give it validity.

The second case concerns *Pentamonas spinifera* (Lackey, 1962) which occurs very commonly in the sand at the low-tide line in front of the Oceanographic Laboratory of the University of Rhode Island Kingston, Rhode Island. This is also a very striking organism, from a particular habitat, and would certainly have been found long ago if cosmopolitan. Careful search of the sand in many other places has provided only one other occurrence—a single, unmistakable find from the sand of Drake's Island at Plymouth, England.

We readily employ the term cosmopolitan to describe the distribution of many microorganisms, but I think we all recognize that certain combinations of factors must exist for many species before they occur in sufficient numbers to be studied. Perhaps every systematist has had the frustrating experience of seeing a spectacular (to him) organism once and fleetingly. The genus *Euglena* is no exception and is a rewarding group to the ecologist who has time to study the conditions under which new or imperfectly known individuals occur. In so doing, he may offer valuable assistance to the systematist, or perhaps himself become both ecologist and systematist.

References

Bracher, R. (1929). *J. Ecol.* **17**, 36.
Bracher, R. (1937). *J. Linnean. Soc. (London), Botany* **51**, 23.
Briscoe, M. S. (1939). *Trans. Am. Microscop. Soc.* **58**, 374.
Carter, N. (1938). *Arch. Protistenk.* **90**, 1.
Durrell, L. W., and Norton, C. (1960). *Trans. Am. Microscop. Soc.* **79**, 160.

Evans, J. H. (1959). *J. Ecol.* **47**, 55.

Fraser, J. H. (1932). *J. Marine Biol. Assoc. U. K.* **18**, 69.

Fritsch, F. E. (1931). *J. Ecol.* **21**, 16.

Gojdics, M. (1953). "The Genus *Euglena*." Univ. of Wisconsin Press, Madison, Wisconsin.

Hall, S. R. (1931). *Biol. Bull.* **60**, 327.

Haswell, W. A. (1907). *Zool. Anz.* **31**, 296.

Hegner, R. W. (1926). *Biol. Bull.* **50**, 239.

Huber-Pestalozzi, G. (1955). *In* "Die Binnengewässer" (A. Theinemann, ed.), Vol. 16: Das Phytoplankton des Süsswassers. Part 4. Schweizerbart'sche Verlagbuchhandlung, Stuttgart.

Jahn, T. L. (1946). *Quart. Rev. Biol.* **21**, 246.

Jahn, T. L. (1951). *In* "Manual of Phycology," pp. 69–81. Chronica Botanica, Waltham, Massachusetts.

Kiener, W. (1944). *Proc. Nebraska Acad. Sci. 54th Ann. Meeting* **12**, 1.

Lackey, J. B. (1938). *Public Health Rept. (U.S.)* **53**, 1.

Lackey, J. B. (1939). *Lloydia* **2**, 128.

Lackey, J. B. (1940). *Ohio J. Sci.* **40**, 126.

Lackey, J. B. (1941). *Public Health Rept. (U.S.)* **54**, 740.

Lackey, J. B. (1957). *U.S. At. Energy Comm. ORNL*-**2410**, 1.

Lackey, J. B. (1960). *Trans Am. Microscop. Soc.* **79**, 105.

Lackey, J. B. (1962). *Arch. Mikrobiol.* **42**, 60.

Lackey, J. B., and Bennett, C. (1962). *Proc. Natl. Symp. Radioecology, 1st., Denver, Colorado, 1958.*

Lackey, J. B., and Morgan, G. B. (1960). *Quart. J. Florida Acad. Sci.* **23**, 287.

Leedale, G. F., Meeuse, B. J. D., and Pringsheim, E. G. (1965). *Arch. Mikrobiol.* **50**, 133.

Lund, W. J. G. (1942). *J. Ecol.* **30**, 245.

Mitchell, J. B. (1928). *Trans. Am. Microscop. Soc.* **47**, 29.

Pringsheim, E. G. (1946). *Phil. Trans. Roy. Soc. London, Ser. B* **232**, 40.

Pringsheim, E. G. (1956). *Nova Acta Leopoldina* **18**, no. 125, 1.

Sandon, H. (1927). "The Composition and Distribution of the Protozoan Fauna of the Soil." Oliver & Boyd, Edinburgh and London.

Schlichting, H. E., Jr. (1960). *Trans. Am. Microscop. Soc.* **7** , 160.

Shields, S. E., and Lackey, J. B. (1938). *J. Econ. Entomol.* **31**, 95.

Senior-White, R. (1928). *Indian J. Med. Res.* **15**, 969.

Starr, R. (1964). *Am. J. Botany* **51**, 1013.

Vorhies, C. T. (1937). *Am. Naturalist* **51**, 494.

Wenrich, D. H. (1924). *Biol. Bull.* **47**, 149.

CHAPTER 3

LOCOMOTIVE AND
MOTILE RESPONSE IN *EUGLENA**

Theodore L. Jahn and Eugene C. Bovee

I. Introductory Remarks

Although Leeuwenhoek (1674, cited in Dobell, 1932) may have first observed swimming *Euglena*, any discussion of the motile behavior of *Euglena* still requires, in large measure, a reasoning by analogy and some

* Based partly on experimental work supported by NIH grant 6462, NSF grants GB 1589 and GB 5573, and ONR contract Nonr 4756.

outright speculation. This is true partly because of the paradoxical situation that while much has been done in the past 15 years to clarify the biochemistry of *Euglena*, at least for certain strains of *E. gracilis* (see reference lists of other chapters in this treatise), there are only a few biochemical studies of the motile organelles of *Euglena* available concerning either the molecular biology or functions thereof.

There is a similar dearth of observation and experiment involving studies on the effects of ions, especially cations and their ratios, on motility and locomotion of *Euglena*. No research has been done comparable to the excellent studies of effects of ions on ciliary reversal in *Paramecium* (Oliphant, 1938, 1942; Kamada, 1940; Kamada and Kinosita, 1940; Jahn, 1961, 1962) or equivalent to the studies of Halldal (1958, 1959) on the effects of cation concentrations on the swimming and phototaxis of *Platymonas*. Further lack of literature—except for an occasional paper here and there—is the case for the effects of practically all other physical and chemical agents on the motility of *Euglena*, except for the effects of light on phototaxis and on the photoreceptor.

Moreover, the mechanics of the principal motile organelle, i.e., the flagellum, have scarcely been observed, let alone tested, except for studies by Metzner (1920) and Lowndes (1936, 1941, 1943, 1944, 1947), and some new studies by Holwill (1966a) and Jahn and Bovee (1967), admittedly because of certain difficulties in seeing and recording flagellar movements.

Somehow, *Euglena* has been generally overlooked as an object of study in cellular motility despite its well-known existence, its easy availability and culturability, and the already abundant literature (attested to by all the chapters in this treatise) on its morphology and ultrastructure, growth and division, metabolism, evolution, taxonomy, and ecology.

Therefore, we shall cite (and criticize) the pertinent literature on the motile behavior of *Euglena* wherever it is available. Where it is not, we shall try to indicate what might be expected of *Euglena* by citing works on other organisms which, by analogy, seem applicable. Where even analogy is lacking we shall speculate, at least where "educated guessing" seems to permit it, and we hope to be pardoned if our speculations are deemed wrong (by some other investigator or by future discoveries).

II. The Effects of Physical and Chemical Factors on Movements

As already mentioned, there is a serious lack of observation and experiment concerning the effects of most physical and chemical agents on the movements and locomotion of *Euglena*. This is perhaps due to the early labeling by Ehrenberg (1838) of the reddish "eyespot," or stigma, as a veritable eye, and the revelation by Wager (1900) that the stigma unilaterally shields a pale lenslike lump within the flagellar membrane and adjacent to its axoneme near the base of the flagellum. One result of these observations has been a preoccupation with and a plethora of studies involving light and the phototactic reaction that do not reveal much about the fundamental cause (see Section II,G).

A. The Shock Reaction

Engelmann (1869) noted that most cells show a shock reaction to a sudden stimulus. This usually involves an initial halt of protoplasmic movement followed by resumption of movement in a time, promptly or slowly after the removal of the stimulus, providing the shock stimulus is not lethal. The resumed locomotion is neodirectional and is often reversed (see review by Mast, 1941). Engelmann (1882) elicited this shock reaction in *Euglena* by applying a beam of light; he found the anterior end to be sensitive and the rest of the body slightly, or not at all, responsive.

Jennings (1904) carefully observed this shock reaction to light in *Euglena*, and determined that it occurs whenever a *sudden* change in intensity occurs. Similar shock reactions occur in an electric field when the current is applied (Fig. 1) (Bancroft, 1913), when the *Euglena* is suddenly immersed in heavy water (D₂O) (Mandeville *et al.*, 1964), or upon other sudden stimulus of

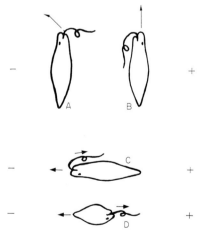

Fig. 1. Reaction of *E. viridis* in an electrical field. A, Swings toward cathode if the eyespot is on the cathodal side. B, Swims on forward at right angles to field if the eyespot is toward the anode. C, Continues to swim toward the cathode if it is already pointed that way. D, If pointed toward the anode, contracts, swings flagellum forward, and swims backward toward the cathode. All effects in a medium with added citrate. (Figures after Bancroft, 1915.)

shock strength. In its severest manifestation the shock reaction involves
sudden contraction of the body, often accompanied by the autotomy of
the flagellum, sometimes followed by convulsive metabolic movements
(Hilmbauer, 1954, and others). In lesser manifestations it involves a swerving
of the body away from (or sometimes toward) the source of stimulus while
swimming (Jennings, 1906), or in a long species with a short flagellum,
while creeping, e.g., *Euglena deses* (Mast, 1941). Although many observations
show that sudden intense stimuli of almost any physical or chemical nature
cause these shock reactions in *Euglena*, the mechanisms for both triggering
the reaction and of the reaction itself are obscure.

B. A Classification of Responses

One of the problems in discussing motile responses of organisms is the
lack of agreement on terms. Fraenkel and Gunn (1940) suggest that each
organism has a *primary orientation* in space and time, alterations thereof
invoked by environmental changes being *secondary orientations* which
direct the body and its locomotion either toward or away from the environmental stimulus. They call any such stimulus-directed movement a *taxis*,
which if bilaterally equated as a smooth movement toward or away from
the stimulus is considered a *tropotaxis*, or if similarly directed but unequated
bilaterally is termed a *telotaxis*. If such a stimulus-directed taxis involves
bending and comparative movements it is labeled a *klinotaxis*. If the response
is only a nondirected velocity change, Fraenkel and Gunn (1940) term it a
kinesis, which becomes an *orthokinesis* if only the linear rate of progress
is altered, or becomes a *klinokinesis* if the alterations are a change in direction
and rate of turning.

Some of the motile behavior of *Euglena* appears to involve *phototaxis*,
an equated response to stimulus by light. A movement away from the
stimulus is considered a *phobotaxis* by Halldal (1964), and a positive movement toward the light a *topotaxis*. Some other motile reactions of *Euglena*
appear more likely to be kineses in the Fraenkel–Gunn terminology, e.g.,
alterations of swimming rate by changes in pH or temperature (Lee, 1954 a,b).

A major problem in the use of these terms is the lack of knowledge,
not only for *Euglena* but for many and perhaps most organisms, concerning
the mechanisms on which the taxes and kineses are dependent. Most likely
each is the result of many metabolic interactions, which makes the distinctions of them and the terminology descriptive but unreliable.

C. Reactions to Chemicals

Very little is known of the specific or general motile reactions of *Euglena*
to chemicals, either to specific chemical substances or ions, or to ratios of ions.

1. *Chemotaxis*

As far as we have been able to determine, no one has found any evident chemotactic role in the motile behavior of *Euglena*. Jennings (1906) states that *Euglena* exhibits a general avoiding reaction to weak chemical solutions, but does not elucidate, and does not indicate any specific reaction to any chemical. The only allusion to a possible positive response that we can find is the usually unaccepted statement by Haase (1910) that *E. sanguinea* forms small ameboid "gametes" which fuse in pairs (presumably somehow attracted to one another) to form a "zygote."

It is also known that a gliding *Euglena* will follow old mucus trails left by itself or other gliding euglenas (Günther, 1927), but whether or not a *chemotaxis* is involved has not been investigated. The role of chemotaxis in motile behavior of *Euglena* is therefore an open question.

2. *Gases*

Protozoa that have been continuously exposed to a pure atmosphere of any one of the principal gases of the air over and/or dissolved in the water of their environment at saturation levels invariably cease moving and ultimately die, whether the gas be hydrogen, nitrogen, carbon dioxide, oxygen, or xenon (Loewy, 1950; Kuhne, 1864; Kitching, 1939; Gittleson and Sears, 1964; Sears and Gittleson, 1964). Similar tests have not been made on *Euglena*, as far as we can determine, but certain field observations suggest that such is probably true for most, and perhaps all, species of *Euglena*. In an ecological study of a polluted section of a stream in Florida, Bovee (1965) reports that all protozoa, including numerous species and individuals of *Euglena* (Sundaresan *et al.* 1965), ceased moving and died shortly after entering the polluted section of the stream if the free dissolved oxygen were very low (less than 1 ppm) and the pH of the water less than 7.0 (i.e., in the acid range). Some euglenas, although immobile at such a low O_2 tension, survived when the pH was alkaline (Bovee, unpublished notes), suggesting that some available O_2 is necessary for the motile functions. Since *E. gracilis*, either the green or bleached variety, can fix CO_2 metabolically (Cook, 1960; Levedahl, 1967; Lynch and Calvin, 1953), and presumably other species can also, survival may be possible at very low dissolved oxygen tensions by way of metabolically derived oxygen via the CO_2 fixation, although swimming may be stopped. This is also suggested by the fact that in certain species of *Euglena*, e.g., *E. mutabilis*, normally resident in and active in environments of high organic acid content, low pH, and low levels of available oxygen (Lackey, 1938, 1939; Chacharonis and Kostir, 1952; von Dach, 1943), the movement and locomotion are due to metabolic creeping, and in that species, or similar ones, e.g., *E. deses* or *E. obtusa*, under such conditions the flagellum may be absent (Bracher, 1938) or very short and relatively useless if active (Mast, 1928; Palmer, 1967).

The ratio of atmospheric gases may be more important in maintaining activity of *E. gracilis* than is the effect of a single gas. Kostir (1952) provided artificial $O_2-CO_2-N_2$ atmospheres over clonal cultures of *E. gracilis* at 21°C, and found that with 4% CO_2, zero O_2, and 96% N_2, 90% of the euglenas swam, but *very* slowly; with 4% CO_2, 11% O_2, and 85% N_2, 80% of them swam, almost as slowly; at 4% CO_2, 16% O_2, and 80% N_2, 10% swam, slower than normal; at 4% CO_2, 21% O_2, and 75% N_2, or at O_2 percentages still higher, less than 5% to very few, swam. If the O_2 were held constant at 21% (i.e., approximately that of normal air) and the CO_2 and N_2 varied, at 1% CO_2 very few swam; at 4% CO_2 only 8% swam, and at 10% CO_2, 10% swam; at 16% CO_2, however, 95% of the euglenas swam. Kostir (1952) concludes that an $O_2 : CO_2$ ratio of 4 : 1 is required for normal swimming and that either high CO_2 or low O_2 may induce swimming.

Borgers and Kitching (1956), using CO_2 mixed with N_2 on one side of a solution and pure N_2 on the other, found that *Astasia longa* (a colorless close relative of *E. gracilis*) swims toward different concentrations of CO_2, depending on the alkali reserve of the solution. When the solution had an alkali reserve of 0.214 N, it swam toward the interface bathed by 5–50% CO_2; when the reserve was 0.00248–0.00275 N, toward 5–20% CO_2; when 0.000121 N, toward 1.25% CO_2. In each case, concentrations of CO_2 higher than those cited repelled the organisms; more than 50% CO_2 always repelled the organisms; and 80% CO_2 caused circling movements, distortion of the body, anesthetization, and ultimate death. These authors reasoned that the responses were to pH changes in the solutions used, rather than a response to the gas itself.

Some evidence indicates that although toxic gases e.g., H_2O_2 (Wichterman *et al.*, 1958) immobilize and kill *Euglena* spp. as they do other protozoa, some species of *Euglena*, perhaps most, may be relatively more resistant to some toxic gases than are many other protozoa. Ludwig *et al.* (1951) report highly motile rapidly growing *Euglena* in aerated sewage, provided CO_2 is added, and Abbott (1951) reports actively motile euglenas when H_2S content is 11 ppm.

3. *Water*

Water is so necessary to life processes and so taken for granted, that its roles are often ignored in behavior and motility as well in as other biological events.

Jahn and Bovee (1964, 1965, 1967, 1968) and their co-workers have recently emphasized the importance of hydrodynamics in the description of ciliary and flagellar movements, as have also Gray (1928, 1953, 1955, 1958, 1962), Lowndes (1941, 1943, 1944), Taylor (1951, 1952a,b), Hancock (1953), Brown (1945), Machin (1958, 1963, 1965), Sleigh (1962), Pitelka and Child (1964), and most recently Holwill (1965, 1966a,b), Brokaw and Holwill (1967), Brokaw (1964, 1965), Brokaw and Wright (1963), and Rikmenspoel (1965a,b, 1966). This will be discussed as it pertains to flagellar and body movement in the swimming of *Euglena* in Section III, A, 2.

Water is also important in the movements performed by protoplasm itself (which is grossly 70–80% water) and those components of protoplasm, especially proteins, that are coated with water more-or-less bound as a quasicrystalline ice (Eisenman, 1963; Ling, 1962; Szent-Györgyi, 1960a,b). Szent-Györgyi (1960b) points out that structured water as a mediator and activator of many biological processes is "half the living machinery, and not merely medium and space filler," and further that in protoplasmic movements "half the contractile material is water; contraction is the collapse of its structure, induced by actomyosin." Steinbach and Dunham (1961, 1962) show that in terms of dry weight, the very motile, fibrous flagellum is extremely dilute, and Rikmenspoel (1965b) and Brokaw (1965) indicate that the stiffness that may be obligatory to its function may be, in part at least, the result of its turgidity as reflected by the stiffness of the flagellar membrane. Others (Gray, 1928; Hancock, 1953; Jahn and Bovee, 1964, 1965, 1967; Brokaw, 1965; Holwill, 1966b) have commented on the necessary stiffness of the flagellum if it is to be hydrodynamically efficient, without stressing the fundamental role of structured water in maintaining the requisite flagellar rigidity. The flagellum of *Euglena* cannot be an exception to these requirements.

Certainly, as for other protozoa, either a sudden increase or decrease in the proportion of water in the medium immobilizes *Euglena*. Höfler and Höfler (1952) report that *E. olivacea* and *E. deses* are immobilized by exposure to 0.4–0.6 *M* dextrose solution, but recover when put back in pond water. If the osmotic change is sudden but not great, *A. longa* avoids the change or may be attracted to it depending on the concentration (Borgers and Kitching, 1956). An osmotic change serves as a stimulus to more vigorous movement for *E. leucops* (Hall, 1931).

The chemical nature of the water itself is also important to motility. Heavy water (deuterium, or D_2O) inhibits motile behavior and metabolism of *E. gracilis* (Mandeville *et al.*, 1964), with slowing and immobilization above 50% deuterium, and shock stoppage at 90% D_2O. These authors do not indicate how the D_2O affects the *Euglena;* but other work on other cells [e.g., D_2O-treated squid axon (Stillman and Binstock, 1967)] suggests a direct slowing of metabolic processes by D_2O replacement of H_2O in the mechanisms.

Adaptation is possible, however, to some rather severe osmotic challenges. Finley (1930) was able by degrees to get *Euglena* to swim in 103% sea water; and Mandeville *et al.* (1964) ultimately adapted *E. gracilis*, by slow degrees over a 6-month period, to 99.6% deuterium. At each transfer to media more concentrated in deuterium *Euglena* loses the stigma (eyespot), rounds up, and is immobilized, although the flagellum is not structurally affected. Then it slowly recovers, if it survives. It regains a pale stigma, a measure of its spindle shape (but not entirely), its motility and swimming (to a reduced degree). These effects are said *not* to be caused by any accumulation of H_2O_2 resulting from the presence of the deuterium (Harvey, 1934). Evidently the physical structure of the deuterium is such that it cannot substitute fully for normal water in biological processes of *Euglena*, including the motile mechanism, even after long term adjustment to it. The reason is not known.

4. *Cations*

There are occasional hints in the literature, concealed under the guise of other information, that positively charged ions play a major role in motility of *Euglena*, as well as in its other metabolic processes. We can find no literature,* however, indicating that any one has investigated the role of cations either generally or specifically in movement of *Euglena*. Therefore, we can analyze their roles only by analogy and "educated" postulation as far as *Euglena* is concerned.

a. Hydrogen (Hydronium) Ions. The rapid transit and the instability of the hydronium ion (H_3O^+) and the transfer of its active component, the hydrogen ion (H^+), involving both electron and proton charge transports (Eigen and de Maeyer, 1958), are known to affect all metabolic processes of the cell, including those resulting in movements. How these effects occur, however, is still unknown (for reviews see Wiercinski, 1955; Bittar, 1964).

Alterations of pH have been reported to affect both swimming and metabolic movements of *Euglena*. Flagella generally operate efficiently within an optimal pH range (Newton and Kerridge, 1965). Borgers and Kitching (1956) give pH 5.3–6.3 as the optimal range for *A. longa*, based on studies indicating which region of the solution they swim toward and tend to congregate in, or tend to swim away from. Lee (1954a) gives pH 7.0 as that at which forward swimming is most rapid (59.7 μ/second) by *E.*

---

* *Note Added in Proof:* Wolken (1967) reports that *Euglena* in Mg^{2+}-deficient media rounds up and becomes immobile. It reacts similarly in media deficient in trace elements (Mo^{6+}, Cu^{2+}, Co^{2+}, Zn^{2+}, Mn^{2+}, Bo^{3+}). Motility is regained in the Mg^{2+} deficiency if small amounts of Mg^{2+} are added. In metal-free media to which each of the metal ions is added singly, motility was regained by only a few euglenas in Mg^{2+}- and Cu^{2+}-containing media.

gracilis, but his observation is criticized by Dryl (1959) because the pH of the medium in which they were grown was not indicated. Metabolic movement of *Euglena* spp. may be induced when the pH is changed (Alexander, 1931), but pH is not the only factor involved (Hall, 1933). If the medium is high in organic material, presumably already acidic, metaboly is augmented (Szabados, 1936) as it also is if the organic acid (acetate) is added to a highly organic (tryptone) medium (Jahn and McKibben, 1937). In acid solutions, metaboly of *E. deses* becomes accentuated and oscillatory (Kamiya, 1939); *E. mutabilis* normally lives and is very active in highly organic waters of very low pH (Lackey, 1938; von Dach, 1943). Also the usually stiff *E. acus* becomes metabolically motile if the pH is suddenly altered in the alkaline direction (Lefèvre, 1931).

Acetate as the *sole* organic (i.e., carbon) source in the growth medium, however, immobilizes *E. gracilis* at pH 4.5 and below (Danforth, 1953), and 1% acetic acid stops metaboly of *E. limosa* (Conrad, 1940).

Other reports say that alkaline pH may accentuate metaboly of some species (Alexander, 1931; Lefèvre, 1931). As Hall (1933) suggests, pH is only one of the factors involved in movements of *Euglena*, and interacts with others rather than acting directly.

b. Weak Electrolytes. The above data on the effects of pH on movements and locomotion of *Euglena*, scant as it is, is confusing partly because it has not been related to the effects of pH on the dissociation, penetration into the cell, and accumulation in the cytoplasm of weak alkalis and acids, including acetate and other organic acids and carbon dioxide in the media in which euglenas are grown. These effects are probably much greater than suggested in the early papers on growth of ciliates (Elliott, 1935a,b) and flagellates (Loefer, 1935a,b) as discussed by Jahn (1934). The undissociated electrolyte penetrates much more rapidly than the dissociated state, so that the changes of pH, which affect the extent to which weak electrolytes dissociate, also influence the penetration rate. This is true especially of the organic acids and their accumulation or the degree of their exclusion. Jacobs*(1940) and Höber (1965) give equations that predict the accumulation or exclusion of monobasic organic acids; Wilson *et al.* (1959) emended the equations to include dibasic acids, and predicted a theoretically possible

---

* The basic equation (Jacobs, 1940) is:

$$C_i = \frac{1 + 10^{\mathrm{pH}_i - \mathrm{p}K}}{1 + 10^{\mathrm{pH}_o - \mathrm{p}K}} \, C_o$$

where C_i and C_o are the total concentrations of the acid inside and outside, and $\mathrm{p}K$ is the $\mathrm{p}K$ value for the acid being dealt with. Kotyk (1962) has independently constructed another equation applicable to monobasic acids.

5

"stock-piling" to 160,000 times greater inside the cell than in the outside medium. Although the Jacobs, Kotyk, and Wilson equations have not been applied directly to the locomotion of *Euglena* or other protozoa, they suggest an explanation for the immobilizing effect of acetate on *Euglena* in some growth media (especially as the sole carbon source). It has been found by Cook and Carver (1966) that acetate activates the glyoxalate by-pass mechanism of *E. gracilis*, with added utilization of O_2 in the synthetic mechanisms at the expense of the energy production and storage paths. Levedahl (1966) shows that CO_2 fixation resulting in energy accumulation is *not* elevated in a salt and acetate medium, and growth is not accelerated; however, protein synthesis, which requires energy utilization, is increased. Perini *et al.* (1964a,b) have recently identified a c-type "cyanide by-pass" cytochrome as well as cytochrome a and cytochrome b in *Euglena*, and a succinic acid cyanide-sensitive succinoxidase system. These observations suggest that a rapid accumulation of acetate due to the pH level, predictable by the Jacobs' equation, would force *Euglena* to accumulate large quantities of acetate, and further force the use of anaerobic pathways of energy utilization in synthesis in excess over the operation of energy accumulation paths. In effect this would deplete the flow of energy to motile organelles and deplete their oxygen supply as well, bringing motion and locomotion to an abrupt halt.

c. Metallic Cations. There are few studies on flagellates concerning the effects of monovalent and polyvalent cations of metals on their movements, or on those of their flagella. There seem to be none* for *Euglena*. It is reasonable to expect, however, that data from studies on other protozoa that are ecological associates of, or taxonomically close to *Euglena*, may suggest what responses *Euglena* should make. Since some of the effects of cations result from their binding ratios to protein surfaces, especially at interfaces between the fluid medium around and within the cell, and since those ratios have general application to protoplasm (Jahn, 1962, 1966, 1967) and to many of its functions, these conclusions may be assumed to apply to *Euglena*.

An "ion antagonism" has long been supposed to exist between monovalent cations (especially K^+ and Na^+) and divalent cations (especially Ca^{2+} and Mg^{2+}), with Ba^{2+} seemingly an exception, especially as they affect the viscosity and motility of protoplasm generally (see reviews in Heilbrunn, 1958; Jahn and Bovee, 1964, 1967). These "antagonistic" effects are explained by Jahn (1962) as arising from alterations in accumulations of one or another of the cations at the protein surfaces, not only of membranes, but of all

---

* See footnote on page 52.

proteins within the protoplasm as well (Jahn, 1966, 1967), in accordance with the Gibbs–Donnan equilibrium. This is the ratio that exists between the monovalent and divalent cations attached to the protein surface and those dispersed in the aqueous phase. The ratio of K^+/Ca^{2+} at the protein surface depends upon the $K^+/\sqrt{Ca^{2+}}$ in the bulk of the aqueous phase.

Therefore, the apparent "antagonism" between K^+ and Ca^{2+} and Mg^{2+} that Halldal (1959) found for the phototactic movements of *Platymonas* indicates the involvement of the Gibbs–Donnan equilibrium in flagellate motile behavior. Halldal (1959) found that motility of *Platymonas cordiformis* absolutely requires the presence of K^+ in the medium and either Mg^{2+} or Ca^{2+} as well. He also found that within "the physiological range" all combinations of K^+ and Mg^{2+} caused a positive phototaxis, while K^+ and Ca^{2+} caused negative phototaxis *if the content of K^+ were low*, but changed again to positive if the proportion of K^+ were raised to and above a critical concentration in the medium. An inspection of the data suggests that if a family of curves were drawn for relative concentrations of the cations in the medium for *Platymonas*, such as those drawn by Jahn (1962) for the data of Kamada and Kinosita (1940) for effects on ciliary reversal of *Paramecium*, a Gibbs–Donnan relationship for the reversal of the phototactic response in *Platymonas* might be revealed.

While no similar studies have been done on *Euglena*, an old paper by Bancroft (1915) suggests a probable role for the Gibbs–Donnan equilibria in certain movements of *Euglena*. He found that citrate added to the medium caused a reversal of the direction of movement in an electric field. Since citrate is a cation chelater—especially for Ca^{2+}—it is probable that the reversal was the result of an alteration of the Gibbs–Donnan equilibrium, which was accentuated by the differential rates of ion mobilities effected by the electric current (see Jahn, 1962, 1966, 1967). Polyvalent heavy metal cations, which do not necessarily cause alterations in the direction of swimming, do tend to reduce motility of *Euglena*, and are lethal in many instances. Jirovec (1935) found ions of Cu, Zn, Ag, Hg, and Pb are detrimental to movement and survival of *E. gracilis*. De Puytorac *et al.* (1963) and Kuznicki (1963) reason that the heavy metal cations compete for active sites in the motile apparatuses of protoplasm, since Ca^{2+} is antagonistic to their anesthetic effects. Grebecki *et al.* (1956) attribute the effects to differential adsorption of cations to the surface of the cell, especially at alkaline pH levels.

Another theoretical effect that needs to be investigated concerning the effects of cations on motile behavior of *Euglena*, is the role of the zeta potential which is a partial expression of the state of the adsorption of ions at the protein (or other) surface, and of the association energies of the ions involved (see theoretical explanations by Ling, 1962; and Eisenman, 1963).

d. Other Chemicals. Very little has been published about the effects of other chemicals on the locomotion and motility of *Euglena*. The many papers on its biochemistry and growth seldom refer to any effects of metabolic inhibitors on motility. That these inhibitors do affect motility, directly or indirectly, is shown by Jira and Ottova (1950) who immobilized a variety of ciliates and flagellates, including *Astasia chattoni*, with 8% urethan, and found that urethan and sulfonamide had an additive effect in lower concentrations of urethan, the effect not being reversed by p-aminobenzoic acid (PABA). In addition, Danforth and Erve (1964) state that iodoacetate, commonly used as an inhibitor of certain phosphorylations in biochemical studies of glycolysis, inhibits flagellar movement of *E. gracilis* var. *bacillaris* (bleached strain) in concentrations so low that the glycolytic mechanism is uninterrupted.* Loefer (1967, personal communication) says that certain antibiotics that inhibit growth of *E. gracilis* var. *bacillaris* (Blumberg and Loefer, 1952; Loefer and Mattney, 1952) also reduce motility, but no specific observations suggest how.† Certain other metabolic inhibitors which inhibit motility as well as growth and division of *Euglena*, *e.g.*, arsenic (Rybinsky and Zrynkina, 1935), presumably inhibit motility because of their known interference with oxidation–reduction systems, including those of motility.

The vacuum of information concerning the effects of chemical agents on motile behavior of *Euglena* needs to be filled. It exists only because few efforts have been made to do so. The equipment, chemicals, and techniques are available, and a detailed and concerted research on the topic should reveal much of both specific and general value to studies of the motile process.

D. REACTIONS TO ELECTRIC CURRENTS

1. *Galvanotaxis*

Swimming *Euglena* normally show a tendency in an electric field to move toward the anode (Verworn, 1889; Pütter, 1900; Bancroft, 1915; Schröder,

---

* *Note Added in Proof:* Diehn and Tollin (1966) state that motility and phototaxis in *Euglena* are inhibited by chemicals that block oxidative phosphorylation (e.g., dichlorphenyl indophenol); and by some that block photophosphorylation (e.g., salicylaldoxime). Some inhibitors block phototaxis only (e.g., dichlorphenyl methylurea); In certain combinations (e.g., dichlorphenyl indophenol plus dichlorphenyl methylurea) inhibition "is prevented." These results suggest a high-energy reservoir, presumably ATP.

† *Note Added in Proof:* Goodwin (1951) found that penicillin and dihydrostreptomycin both reduce motility of *Euglena* spp.; 30,000 units/ml of culture of penicillin reduced motility below 10% in 5 hours, to zero in 18–30 hours, with no growth or motile recovery after 11 days; growth and recovery occurred in 10,000 units/ml culture after 11 days. Dihydrostreptomycin at 33,000 units/ml culture reduced motility to 25% in 5–8 hours; some recovery occurred after 18 hours exposure. At 81,300 μg/ml culture for 12 days motility and growth were recovered but chlorophyll was lost.

1927). If the fixed charges of the surface are predominantly negative, as in most other cells, euglenas should move to the anode by passive electrophoresis. In addition, they seem to be driven by the flagellum, but the mechanism by which they are oriented, anterior end foremost, has not been determined.

2. Gross Effects of Electrical Currents

Electrical current of shock strength causes *Euglena* to contract, round up, and sometimes to autotomize its flagellum, as does any other very strong environmental shock (Jennings, 1906; Schröder, 1927). Verworn (1889) found that the phototactic response was difficult to counteract in *E. viridis* with the current strength delivered by his equipment, but he considered *Euglena* probably anodally galvanotactic, and found *Trachelomonas hispida* definitely anodally galvanotactic. He observed, also, an anodal contraction in *E. viridis* (Verworn, 1890).

Ion ratios are important in the reaction. By adding citrate to the medium, Bancroft (1915) reversed the galvanotaxis of *E. viridis*, finding it then *cathodally* galvanotactic. When he reversed the current, the already cathodally oriented euglenas turned around and swam toward the new cathode. In the citrate medium, when the current was first delivered those euglenas with the anterior end toward the anode contracted strongly at the anodal end, swinging the flagellum toward the anode, and swam *backward* toward the cathode (Fig. 1).

Schröder (1927) reported that acids slow down the swimming of *E. ehrenbergii* toward the cathode, that the anodal galvanotropism is accentuated in basic medium, and that in high concentrations, $CaCl_2$ and NaCl reduced mobility in the electric field. He assumed these results were due to changes in charge at the membrane.

While these data are few, a distinct effect of the Gibbs–Donnan equilibrium is indicated, effected by changes in the ratio of K^+/Ca^{2+} in the medium (citrate is a well know Ca^{2+} chelater), and by the differential motilities of the cations in an electrical gradient ($K^+ > Ca^{2+} > Na^+$). These phenomena have been already used by Jahn (1961, 1962, 1964, 1966) to explain ciliary reversal and anodal contraction of *Paramecium* and *Spirostomum* in an electric field; they apply equally well to the observed movement of *Euglena* in an electrical field as recorded by Verworn (1890), Bancroft (1915), and Schröder (1927). Therefore, no lengthy elucidation is needed here, especially since the data are somewhat fragmentary.

Some recent studies of the effects of electricity and ions on other protozoa (Dryl, 1961a,b,c, 1964), and other studies of galvanic stimuli (Grebecki, 1963, 1964, 1965; Grebecki *et al.*, 1956; Kuznicki, 1963; Rostkowska, 1964;

Czarska, 1964), further indicate that concentrations of metallic cations at surfaces, and especially their proportions, are important factors in all motile responses of cells, as well as in their galvanotaxes.

E. REACTIONS TO MECHANICAL FORCES

1. *Local Pressure*

Whether or not a specific response to contact (thigmotaxis) exists in *Euglena* is doubtful. Its response to mechanical force is the same as its "avoidance response" to a change of light intensity (Jennings, 1906) or its reaction to an electric shock (Bancroft, 1915). There are no publications or even incidental references, as far as we know, dealing with variations of local pressure on *Euglena*. We have observed that local pressure on the body of some species of *Euglena* normally capable of metaboly will stimulate a local or general contraction of the body, depending on the amount of force exerted (Jahn and Bovee, unpublished), and that the anterior end is the more sensitive area, as it also is to other stimuli. Wager (1911) states that in dense populations of *Euglena*, motion is slowed by mutual interference and contact, but cites no specific response to contact. Pütter (1900) states that flagellates, including the testate euglenoid *Trachelomonas*, do not respond to contact, although they do to electric current.

2. *Mechanical Shock and Agitation*

No specific investigation of single or repeated mechanical shock seems to have been made for *Euglena*. Jennings (1904, 1906) makes the general statement that mechanical shock causes the "avoiding reaction" in *E. viridis*, implying a single shock, but does not indicate the strength, frequency, or duration of the shock. The subject is another open question in the motile behavior of *Euglena*.

3. *High-Frequency Sound*

Euglena demonstrates a typical shock reaction to ultrasound waves (Harvey and Loomis, 1928), essentially the same as the general shock reaction cited by Jennings (1906) for any strong stimulus, i.e., it contracts, rounds up, and sheds its flagellum. It may or may not recover, depending on the duration of the stimulus.

4. *Hydrostatic Pressure*

Gross (1965) states that a majority of *E. gracilis* are immobilized by application of a pressure of 6000 psi for 30 minutes, but retain normal shape. At 15,000 psi all are immobilized; some become rounded after 20 minutes exposure; all become rounded after 2-hours exposure. Byrne and Marsland (1965) find that *E. gracilis* resists hydrostatic pressure better than *Amoeba proteus*, and that on exposures of 7000 psi for 15–20 minutes normal swimming and metabolic movement are maintained. Metaboly is abolished, however, in 4 seconds at 10,000 psi. The cells tend to round up and stop swimming, and at 15,000 psi almost all become rounded and immobile. Flagellar undulations continue in all individuals at all pressures up to 15,000 psi. At 13,000 psi for 10 minutes, locomotion becomes sluggish, periodic, and irregular although rate and amplitude of flagellar beat are not altered; as cells become partly or completely rounded, however, locomotion stops and the euglenas settle to the bottom, although flagellar beating continues. Any alterations of flagellar beating were not detected. A few individuals were able to maintain body form, flagellar beating, swimming, and metaboly even at 15,000 psi. Any alterations in form and behavior are reversible, and within 30 seconds after release of the hydrostatic pressure the normal elongate form is regained, and swimming begins, somewhat slowly, and, after 5 minutes, shape and movements are entirely normal. Upon release of pressure, an instantaneous general contraction of cytoplasm occurs, leaving a clear fluid-filled space, temporarily, under the pellicle, similar to the massive contractions of *Amoeba* and *Arbacia* eggs following rapid decompression (Brown and Marsland, 1936; Marsland, 1964).

5. *Water Currents*

Johnson (1939) comments that *E. rubra* is positively rheotactic, but gives no supporting evidence. Schwarz (1884) decided that the centripetal swimming of *Euglena* in centrifugal forces between 0.5 and 8.5 × *g* implied a positive rheotaxis. This was denied by Aderhold (1888) who concluded that his experiments demonstrated no clear relationship between the direction of water flow and the orientation of *Euglena*. Wager (1911) correctly suggests that Schwarz's observations imply a geotaxis rather than a rheotaxis, because the water was not actually moving.

We have pointed out, however (Jahn and Bovee, 1967), that apparent rheotaxis in a stream of moving water may be based on known hydrodynamic principles applicable to swimming, and may demonstrate neither a taxis nor a kinesis, but an inevitable hydrodynamic interaction.

If the current of water were an absolutely *uniform* rectilinear one, the swimming *Euglena* (or other swimming organism) would have no means to detect such a movement, just as it cannot detect the rotation of the earth (nor can we). If the current becomes different on the two sides of the cell, or is accelerating, decelerating, or nonrectilinear, the organism might then detect the current. The experiments performed by Jennings (1904, 1906) on *Paramecium*, using water propelled from a pipette to generate currents which were obviously not rectilinearly uniform, therefore do not indicate a true rheotaxis. In a non-uniform field, an elongate organism, such as *Euglena*, headed upstream would veer toward the higher velocity of water current flow, and if headed downstream, toward the lower velocity of flow. If crosswise of the stream it would turn upstream or down stream because of the reduced weights of its opposite ends and its body geometry. The rear end being the larger and heavier would be swung downstream; and the *Euglena*, which normally swims anterior end forward, would then be swimming upstream, apparently positively rheotactic. Since the hydro-dynamic relationships with the water determine these purely physical effects, they need not involve a response in the *Euglena* and do not constitute a taxis. Any effort by the *Euglena* due to a taxis, if it exists, would be additive to or subtractive from the hydrodynamic effects. Even such complex animals as fish, which are often called rheotactic, are not, since they cannot detect uniform rectilinear motion (Gray, 1937).

Although no evidence supports rheotaxis in *Euglena*, some flagellated cells *are* rheotactic. For example, live spermatozoa *always* swim upstream in either the top or the bottom half of a parabolic gradient of velocity; but dead ones are carried in such a current with the head end pointed upstream, only if the head is in the bottom half of the gradient, and with the head pointed downstream if it is in the *upper* half of the velocity gradient (Rothschild, 1962). How the sperm manages always to swim upstream is unknown, and Rothschild (1962) says that he would appreciate suggestions.

6. Gravity and Magnetism

There are no recent studies on effects of gravity on *Euglena*.* Verworn (1889) considered "geotaxis" to be due to purely physical phenomena. In an early study, Jensen (1893a,b) contended that the flagellar end of *Euglena* theoretic-ally should be pointed downward in reaction to gravity. He explained the fact that the rear end points downward as *Euglena* settles to the bottom as the result of different hydrostatic pressures at different levels. Wager (1911) believed Jensen wrong on both assumptions, and held that the rear end is heavier, and that when flagellar action is interfered with, e.g., during crowding, the *Euglena* sinks with posterior end downward because that end

---

* *Note Added in Proof:* Wolken (1967) says that fields of 2000 × *g* interfere with enzymic reactions and chlorophyll synthesis in *Euglena*, and while he does not mention specifically the effects of magnetism on motion he suggests that magnetism has effects similar to those of high-frequency radio waves, which do affect the motion of *Euglena*.

is heavier, and the specific gravity of the *Euglena* is greater than that of the water. Schwarz (1884) believed *Euglena* to be negatively geotropic, since at 0.5–8.5 × g of force in the centrifuge it orients its anterior end, and swims, centripetally. Wager (1911) was probably correct, as he found that dead, osmium acid–fixed euglenas sink posterior end first, as do the live ones. If the posterior end is actually heavier, as indicated by Wager's results, this explains negative geotaxis, because a greater density at the posterior end will automatically orient the anterior end upwards, at least on a statistical basis. Application of this principle was first pointed out by Verworn (1889), has never been disproved, is applicable to most and probably all protozoa, and is discussed in more detail by Jahn and Bovee (1967).

No experiments or observations on the effects of magnetic fields are known for *Euglena*, such as those of Brown (1962) and Kogan and Tikhonova (1965) on *Paramecium*.

7. *Crowding and Pattern Swimming*

In crowded populations, protozoa and motile bacteria tend to swim in distinct patterns of one type or another, e.g., *E. viridis* and *E. deses* (Wager, 1911), *E. gracilis* (Robbins, 1952), algal swarmers (Nägeli, 1860; Sachs, 1876), *Chlamydomonas* sp. (Wager, 1911), *Polytomella uvella* (Gittleson, 1966; Gittleson and Jahn, 1964, 1966), *Tetrahymena pyriformis* (Loefer and Mefferd, 1952; Jahn, and Brown, 1961; Jahn *et al.*, 1962), and *Escherichia coli* (Nettleton *et al.*, 1953). It is probably a general phenomenon of small swimming cells in large populations.

The type of pattern is related to the kind of organism (Jahn and Bovee, 1967) and to the shape of the container (Gittleson and Jahn, 1964, 1966). The distinctness of the pattern is affected by anything that affects the motility of the organism (Wager, 1911; Loefer and Mefferd, 1952; Jahn and Bovee, 1967) including O_2 (air) Wager (1911), CO_2 (Wager, 1911; Jahn *et al.*, 1962; Gittleson, 1966), NH_3 (Jahn *et al.*, 1962), low temperature (Wager, 1911; Robbins, 1952), viscosity of the medium, anaerobiosis, high osmotic pressure, ultraviolet light, or metabolic poisons, e.g., parathion (Loefer and Mefferd, 1952) or bright visible light (for *Euglena*) (Wager, 1911; Robbins, 1952). *Euglena* forms triangular networks in 30 seconds to 1 minute in a flat, shallow container (Wager, 1911; Robbins, 1952), the triangles being about equilaterial, 3–5 mm on a side, or squares, about the same size (Robbins, 1952); in a horizontally placed capillary tube or vertically placed container, the patterns are conical streams with the base of the cone uppermost (Wager, 1911). The sides and angles of the triangles are empty, the triangles full, and the center of the cone is empty, the descending organisms making up the "walls" of the cone (Wager, 1911). Both Robbins (1952) and Wager (1911) show that active motion on the part of the euglenas is a requisite of the pattern formation, but do not attempt to explain why. Wager (1911) says the cause of downward movements is a mutual slowing of swimming as the euglenas swim together to form a group, descent being due to a specific gravity greater than water, and partial immobility due to interference by other organisms. He does not explain why they swim toward one another in streams initially; and considers the phenomenon a geotaxic one.

The explanation must necessarily involve hydrodynamic principles as well as the physical geometry and specific gravity of the euglenas. Jahn suggests (Jahn and Brown, 1961; Jahn and Bovee, 1967) that for *Tetrahymena* (and presumably for *Euglena* and other organisms) the calculated mean free path for the organisms is such that they will approach one another at random at a rate of 7% per second within a solid posterior angle of 90°, and an end-to-end hydrodynamic linkage will occur. A pseudoturbulent vortex ring is developed by ciliary or flagellar beating behind the body of each organism, despite a low Reynolds number (about 10^{-3}), so that an organism entering the vortex ring is tied hydrodynamically to the first and must follow it. Extraneous particles and dead organisms may join the pattern, and they behave as Wager (1911) found for dead *Euglena*.

If gravitational attraction is involved, the patterns should break up in a weightless state, and should form even more rapidly under increased gravitational force. This actually occurs for *Tetrahymena* if the cultures are made weightless during parabolic flight in an airplane for 12–15 seconds; and the patterns reform in about half the normal time upon re-entry to the gravitational field if it is then about $3 \times g$ (Jahn *et al.*, 1962)

The swimming streams of euglenas (or other microorganisms) collide, lose swimming room and motility, and sink at junctions on collision, falling faster as a group than as a single organism, and dragging each organism downward with more force than it can exert in attempting to swim upward. It can rarely leave the stream until the bottom of the vessel is reached, except at the extreme periphery of the group (Wager, 1911). W. Bradley (1965) explains that a column of descending inanimate particles, each with specific gravity greater than water (e.g., small calcite crystals), creates a columnar density current which will continue to fall, contained by a cylindrical shear zone and at rates to 50 times faster than the free fall of a single particle, as long as the density of the column is greater than that of the water, or until the bottom is reached. This theory has been used by Gittleson and Jahn (1968) to explain vertical column formation of *Polytomella*, and it seems as if the explanation can also be applied to all falling columns of swimming organisms involved in pattern swimming, including *Euglena*.

Other partial explanations of pattern swimming have been suggested (Nettleton *et al.*, 1953; Platt, 1961).

F. Reaction to Invisible Radiations

Euglena does not exhibit a shock reaction to invisible radiation unless it is so intense that it is lethal to the whole organism, or at least to that part struck if the intense ray is a microbeam (Tchakhotine, 1936a,b). The shock reaction thus stimulated is no different from that due to other forces.

1. *Radio Waves*

In radio frequency fields, elongated cells, including *E. gracilis* and *Astasia klebsi*, align parallel to the force field at low frequencies (below 8–10 Mc) and perpendicular to it at higher frequencies (above 20–25 Mc) (Teixeira-Pinto *et al.*, 1960; Griffin and Arnold, 1965; Griffin and Stowell, 1966). Dead *Euglena* always align with the force field at any frequency. Another transitional frequency above 100 Mc exists in which the live *Euglena* again align with the field (Griffin and Stowell, 1966; Griffin and Arnold, 1965). Intracellular particles, however, always align parallel to the field regardless of the orientation of the cell body (Teixeira-Pinto *et al.*, 1960; Griffin and Stowell, 1966). If the ionic strength of the medium is increased, the euglenas remain parallel to the field, up to frequencies of 200 Mc. If the *Euglena* is in the same field as a larger cell, e.g., *A. proteus*, and approaches it, the *Euglena* then spins near the surface of the *Amoeba*. Within the transitional frequency range, the strength of the voltage applied to produce the field is somehow related to the orientation of *Euglena* perpendicular to the force field; 1016 V/cm peak-to-peak voltage is needed at 11 Mc, and about half that (582 V) at 27 Mc. These phenomena are as yet unexplained.*

2. *Infrared Rays and Heat*

No observations are known to us reporting effects of the long or short infrared rays on motile behavior of *Euglena*. Heat, however, as measured by water temperature, is said by Wildeman (1893) to affect locomotion and aggregation of *E. viridis*, and reportedly it moves toward the warmer end of a differentially heated tube of fluid or wet sand, but he did not determine the optimal temperature. Swimming rate slows between 3 and 5°C (Wager, 1911; Bancroft, 1913), and movement is reduced to a quivering of the flagellum without locomotion at 2°C (Wager, 1911). Heidt (1934) says that *E. sanguinea* aggregates in groups or sheetlike layers on the surface of water

---

* *Note Added in Proof:* Wolken (1967) states that *Euglena* not only aligns with the radio frequency field, but also develops an increase of average swimming speed as the potential gradient increases. He offers no explanation.

at a temperature of 28°C, which he considers "optimal." Maximal swimming
rate given for *E. rubra* is 20 μ/second at 30°C, and sheetlike surface aggregation also occurs at similar temperatures (Johnson, 1939). Older reports
(Wager, 1911; Bancroft, 1913) cite little difference in swimming rate between
15° and 35°C, but state that above 35°C locomotion is reduced. Lee (1954b)
has definitely shown, however, that the forward swimming rate of *E. gracilis*
is heat-responsive and shows an increase from 15 μ/second at 10°C to a
peak of 84 μ/second at 30°C, followed by a drop to 38 μ/second at 40°C.
The apparent Q_{10} values near 2 for translatory velocities of *Euglena* are
correlated with flagellar beat frequencies at various temperatures, and also
apply to them, according to Holwill (1966b), who attributes both to a
change in rate of the underlying chemistry.

3. *Ultraviolet Radiations*

High-intensity ultraviolet light causes shock reaction by *Euglena*, with
contraction of the body and loss of flagellum; but no specific shock reaction
is elicted by a microbeam of ultraviolet striking the eyespot (stigma), although
the anterior end is generally the part of the body more sensitive to ultraviolet
(Tchakhotine, 1936a,b). According to Jirovec (1934), colorless strains of
E. gracilis are more sensitive to ultraviolet damage than green strains, 50%
being permanently immobilized after 2 minutes exposure, from which 100%
of the green ones recover in 45 minutes. However, prompt slowing of both
green and colorless strains of *E. gracilis* has been reported after exposure
to to 85.4 ergs/second/mm² of 2537 Å wavelength (Giese, 1938; Swann and
del Rosario, 1932).

4. *X-rays, α-, β-, and γ-Radiations*

Schaudinn (1899) first exposed protozoa, including *E. acus*, to X-rays,
in fact to the rays of one of Roentgen's first successful X-ray devices. Six
hours exposure (dosage unknown) stopped all flagellar movement of *E. acus*.
Wichterman (1955) found that 16,5000 r of X-ray caused *E. gracilis* to swim
in circles and in spurts, with frequent avoiding reactions and changes of
direction, and with ultimate slow metabolic movement, rounding, quivering
and immobilization. Doses above 16,000 r hastened these results. Recovery
of some, though few, occurred, however, even above a 55,000-r dosage,
although LD_{50} is 30,000–40,000 r.

Similar results occur upon β-radiation at 10^6 eV for *E. gracilis* (Godward,
1962), exposure to radium for *E. viridis* (Willcock, 1904) and exposure of
E. gracilis to α-rays (Swann and del Rosario, 1931). Recovery of motility
may sometimes occur, even if the dosage is ultimately lethal (Godward, 1962).

G. Reactions to Visible Light (Phototaxis)

Most of the work on motile behavior of *Euglena* has centered on its response to visible light, i.e., its phototaxis, and the cycling of it, apparently circadian. Engelmann (1882) showed distinctly that the anterior end of *Euglena* in the region of the stigma is the part most sensitive to light, especially blue light, and that *Euglena* swims toward a beam of light.

1. *Spectral Sensitivity*

Refinements of Engelmann's experiments have shown that *Euglena* (and most cells; see reviews Mast, 1941; Giese, 1964; Jahn and Bovee, 1967) is most sensitive and responsive to blue light in orienting itself (Mast, 1917; Oltmanns, 1917; Bracher, 1938; Wolken and Shin, 1958; Strother and Wolken, 1960, 1961; Wolken, 1961; Bünning and Schneiderhohn, 1956; Gössel, 1957). There has been confusion, however, because the reaction may be photopositive in some cases, photonegative in others, and aggregative in still others.

The more recent studies indicate that phototactic peaks lie in two ranges, between 410 and 420 mμ (blue-violet), and at 485–495 mμ (blue-green) (Bünning and Schneiderhohn, 1956; Wolken and Shin, 1958; Strother and Wolken, 1960, 1961; Gössel, 1957). Photokinetic peaks (i.e., affecting rate of swimming, but not direction) are at 460–465 mμ and 620–630 mμ (Wolken and Shin, 1958; Strother and Wolken, 1960, 1961). The last-mentioned two ranges of wavelength are β-carotene- and protein-absorption peaks, respectively, of both the chloroplasts and the stigma (Strother and Wolken, 1960, 1961; Batra and Tollin, 1964).

Positive phototaxis occurs maximally between 480 and 495 mμ (Mast, 1917; Bünning and Schneiderhohn, 1956); the absorptive maxima for the stigma is 460–465 mμ with a range of absorption of 400–630 mμ (Strother and Wolken, 1960, 1961; Batra and Tollin, 1964). This has led Wolken (1967) to assume direct anatomical and functional relationships between the stigma and flagellum which do not actually exist. Wager (1900) demonstrated the lack of anatomical relationship and also proposed correctly that the paraflagellar swelling is the photoreceptor and that the function of the stigma is to shade the paraflagellar swelling from light. Wolken has also assumed a *direct* transfer of energy absorbed by the stigma to the flagellum through a physical contact which does not occur. That it does not occur is evidenced by the fact that a *Euglena* bleached of its chlorophyll but retaining its stigma and photoreceptor is still positively phototactic, eliminating chlorophyll and chloroplasts in the phototaxis directly; while a *Euglena* bleached of all pigments, but retaining its photoreceptor is *negatively* phototactic, ruling out the carotenoids of the stigma as directly stimulatory in phototaxis;

and a *Euglena* lacking photoreceptor and all pigments, like *Astasia*, is no longer phototactic (Pringsheim, 1937, 1948a,b; Gössel, 1957; Vavra and Aaronson, 1962).

The paraflagellar swelling is therefore the photoreceptor or light-sensitive organelle. It has its sensitive maximum at 410 mμ (Gössel, 1957), explaining the phototactic peak at that wavelength. Positive phototaxis occurs only if the stigma, with its absorptive range of 400–630 mμ, periodically *shades* the photoreceptor.

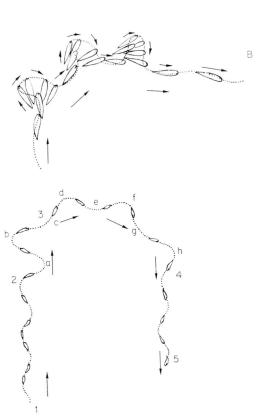

Fig. 2. Path taken by *E. viridis* if light is reversed (below), and if light is shifted 90° as a source (above). Figure above shows "pivoting" of cell as beam strikes the photosensitive region of the anterior end, and the return to normal path as it is shaded by the stigma, until oriented toward the light. A, original direction of light; B, light shifted 90°; a–h, sequential alterations in direction after reversal of light; 1–2, light from above, light reversed at 2; 3, direction altered by 90°; 4–5, direction reversed. (Adapted from Jennings, 1906.)

2. *Changes in Intensity*

How the photoreceptor functions is not known. A sudden beam of intense visible light, ultraviolet, or electricity causes the flagellum, which normally is bent where it leaves the anterior reservoir and trails at an angle of about 30° or less to the body in swimming, to be straightened at that bend and thrown forward so that the body describes a wide spiral and pivots, or is driven backward (Jennings, 1906; Bancroft, 1913; Uhlehla, 1911; Metzner, 1920) (see also Figs. 2 and 3). Or the flagellum is autotomized at that bend

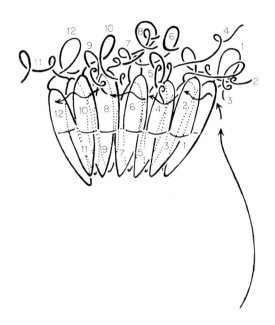

Fig. 3. "Pivoting" movement in *E. gracilis* as a reaction to an intense beam of light. Based on tracings from a high speed motion picture film by the authors. Note that the body "wobbles" as each new helix generates a "push." Positions of the body are numbered (each 40th frame of a sequence from a film taken at 340 fps); and the position of the flagellum is noted with an equivalent number. It is evident that the flagellum swings in a broadly conical pattern, from its base, as the interrupted helices are generated.

(Tchakhotine, 1936a,b; Engelmann, 1882; Verworn, 1889; Gojdics, 1953). In slowly increased radiation or electric fields, however, the *Euglena* adapts (Mast and Gover, 1922; Lozina-Lozinskii and Zear, 1963). Buder (1917) stated that the phototactic response is a function of and is proportional to the intensity of the light source; and Mast and Johnson (1932) indicate that different light intensities from more than one source may interact. The Roscoe–Bunson law does not apply (Mast, 1941).*

The photoreceptor is presumably affected by light through absorption by the photosensitive pigment that changes its structure. In the vertebrate eye it is assumed that absorption by as little as one molecule may start a chain reaction resulting in excitation of the primary neurone. An alternative theory is that the change in structure permits the photopigment (again even a single molecule) to conduct electrons across a poised oxidation–reduction system (Jahn, 1963). Either of these theories can explain a change in an electromotive force across a membrane or even across an enzyme (Geissman, 1949; Cope, 1963).

Piezoelectric effects seem to be of general occurence in biological fibers, e.g., keratins and collagens (Shamos and Lavine, 1967), and can be effective in causing changes in the growth and maintenance of bone (Bassett, 1966; Jahn, 1967). Since the structure of the flagellum is quasicrystalline, an interesting conjecture is that it is also piezoelectric. If so, a voltage change can cause it to bend, and a reverse voltage can cause it to straighten or bend in the opposite direction, and conversely, bending and straightening will produce voltages of appropriate signs. On this basis, an alternating electrical potential would accompany a wave of bending along a flagellum, and the electrical wave might be responsible for transmission of the mechanical wave. Furthermore, if a given voltage is necessary for the flagellar fibrils to bend or straighten (i.e., a threshold exists), it should be possible to explain the mechanisms of the "delayed elasticity" postulated by Machin (1963) in order to explain wave transmission in flagella. A loss of voltage by partial short-circuiting within the flagellum might make even the assumption of a threshold unnecessary.

On this basis, absorption of light causes an electrical change in the paraflagellar swelling, and also across the flagellum at its junction with swelling, and in a direction determined by the geometry of the paraflagellar swelling–flagellum relationship. The voltage changes cause a bending in the flagellum, which is elastic, and the elastic properties tend to cause a straightening and a reverse voltage. This is transmitted with resistance losses to the next section of the flagellum, which is caused to straighten and to produce another voltage change. The velocity of propagation of such a wave depends on the electrical resistance losses within the flagellum, and those energy losses can be restored by ATP, which is necessary for the bending (Brokaw, 1961). One attractive feature of this idea is that it is the only proposed mechanism

---

* *Note Added in Proof:* Wolken (1967) states that the swimming rate of *Euglena* can be practically controlled by external light conditions, with a rate of 0.11 mm/second at 2 ft-c, rising to a maximum of 0.16 mm/second at 40 ft-c (saturation intensity), decreasing slowly at intensities above 40–150 ft-c.

for "delayed elasticity," a concept which can not be explained in simple mechanical terms and which was invented to explain the mathematical equations describing propagation (Machin, 1963). In fact, in the present proposal the elasticity is no longer "delayed." It would be instantaneous, as in ordinary elasticity, and the delay would be caused by electrical resistance losses of the piezoelectric voltage in the matrix of the flagellum or possibly through a threshold effect. Another interesting feature is that it permits a direct causal linkage between the photochemical effect and the wave of flagellar bending. Furthermore, if the electrical change at the photoreceptor is great enough, i.e., if the mechanical forces developed in the flagellum are greater than its tensile strength, it may snap off; if not, it is straightened at the bend as it leaves the reservoir, the body is tilted, and the flagellum is thrown laterally or forward.

In normal swimming the periodic shading of the photoreceptor by the stigma for 2/3 to 3/4 of the spiral turn of the body prevents this piezoelectric effect except for a brief time, so that the flagellum remains bent at the reservoir opening and continues to trail along the body, with its base-to-tip undulations tending to drive the euglena toward the light source. In the photonegative response, the initial effect is to turn the body away from the light source until the body partially shades the anterior end; then the flagellum "relaxes" at the bend and the *Euglena* swims *away* from the source. These assumptions are wholly theoretical, have not been tested, and would be difficult to test. They are consistent, however, not only with the already cited observations on phototaxis, but also those of Mast (1938) that *Euglena* is photopositive in weak light, and photonegative in strong light, that the *Euglena* turns toward the light if the angle of incidence shifts 90°; and the direction taken between two beams of different intensities is intermediate between the original direction of locomotion and that of the stronger beam (Mast and Johnson, 1932).

Other effects of light intensity suggest that *E. limosa*, a crawling species of river tidal flats, burrows into the mud at 30–40 ft-c of light, and returns to the mud–water interface at 50–60 ft-c being most responsive in the range of 420–460 mμ wavelengths (Bracher, 1938) which are the major peaks of chlorophyll absorption (Strother and Wolken, 1960). The flagellum of such burrowing species of *Euglena* is usually very short and ineffectual, or only a nonfunctional stub within the reservoir (Bracher, 1919; Gojdics, 1953; Conrad, 1940; Palmer, 1967).

3. *Circadicity of Phototaxis*

Some investigators of diurnally repetitive cycles of activity ("circadian rhythms") have been much impressed with the apparent circadicity of phototaxis in *Euglena* and other algal flagellates, and have therefore used *Euglena* as an experimental organism for study of such rhythms, but often with confusing, if seemingly successful results.

Aggregation in the direction of a light source plainly involves the photo-receptor, but why this apparently is incapable of reacting during dark periods when the light is reintroduced has not been explained. Phototaxis does not occur in swimming euglenas when the photoreceptor is absent (Bünning and Schneiderhohn, 1956; Vavra and Aaronson, 1962).

The phototactic rhythm has its highest peaks and lowest lows on a 12-hour-light 12-hour-dark repetition (Pohl, 1948), with the dark inhibition period unaffected by light (Bruce, 1959; Bruce and Pittendrigh, 1957); but rhythms of 6 : 6 hours to 48 : 48 hours have been developed (Bruce, 1960) The "*Euglena* clock," as other circadian cycles, is said to be endogenous (Bünning, 1956, 1964), and is said to show a "dusk" response (Bruce, 1960) which is dependent on various light–dark periodicities. It is not, however, solely a response to light, since deuterium (D_2O) slows it (Bruce and Pittendrigh, 1960).

In continuous light at the same temperature it is temporarily eliminated, but may then be reinstituted by a change to a lower temperature (25°–18°C) (Bruce, 1960; Harker, 1960). Continuous darkness does not eliminate the rhythm unless continued for so long a period that complete bleaching results, and the euglenas become photonegative, then unresponsive as the stigma and then the photoreceptor are permanently lost. The rhythm thus lost in continuous light is regained by a "dawn" response after 10 hours of darkness, or 4–12 hours of light in "out-of-phase" circumstances (Bruce and Pittendrigh, 1958).

Similar phototactic rhythms occur in the vertical migration of tidal flat *Euglena* spp. into and out of mud of river tidal flats. These appear to coincide with tidal changes (Bracher, 1919, 1938), however, Bracher (1938) and Palmer and Round (1965) have shown the rhythm to be diurnally phototactic. The period of the phototaxis is affected by temperature if light is constant, but may persist almost unchanged in the laboratory for several weeks when both light and temperature are constant. It is ultimately eliminated, however, under those conditions (Palmer, 1967; Palmer and Round, 1965; Round and Palmer 1966). Since the motility of the species involved is that of the whole body, and the flagellum is so short as to be functionally ineffectual, more than the swimming apparatus is involved, and the phototaxis attributed to the flagellar apparatus is not the main factor. Photosynthetic mechanisms are more likely involved since these tidal *Euglena* spp. are most phototactically responsive to wavelengths in the chlorophyll absorption range, i.e., 420–460 mμ (Bracher, 1938); they also quickly reburrow even when the light is still present (Palmer, 1967).

Seemingly the "*Euglena*-clock" obeys most of the generalized "rules" for circadian rhythms (see Pittendrigh, 1960), except that it is temperature-influenced, and perhaps chemically perturbed (Bruce and Pittendrigh, 1960).

As Scherbaum and Loefer (1964) point out, however, the apparently circadian phototactic, photosynthetic, and luminescent rhythms of swimming unicellular algae run closely parallel to their diurnal growth rhythms. Leedale (1959b) found that among 22 species of euglenoid flagellates in 12 genera, including 5 species of *Euglena*, the green species had almost perfectly circadianly synchronized the mitotic cycles (see also Cook and James, 1960). The mitotic division begins within 2 hours after onset of darkness, requiring 1 hour of dark induction to trigger the division. Mitotic cycling, like phototactic cycling, is eliminated either by continuous light or continuous darkness (Cook, 1960; Cook and James, 1960; Leedale, 1959b). In minimal growth media (salts plus acetate) mitotic cycling is more variable, but with addition of proteose–peptone, or other —SH-containing nitrogenous chains, e.g., methionine, better synchrony develops (Cook and James, 1960); media extremely rich in (Leedale, 1959b) or lacking (Cook and James, 1960) organic nutrients eliminate any apparently light-induced mitotic synchrony. In *A. longa* and other light-insensitive colorless euglenoids, a "diurnal" temperature-change cycle sets a similar mitotic rhythm. A period of 11 hours at 24°C and 13 hours at 36–38°C synchronizes division of bleached *E. gracilis* Z, whether grown in continuous dark or light, with some facilitation perhaps due to continuous light (Pogo and Arce, 1964). Periods of 10 hours at 5°C and 14 hours at 25°C; or 6 hours at 28.5°C and 17 hours at 14.4°C with two 30-minute transition periods, synchronize division of *A. longa* (Padilla and James, 1960; Blum and Padilla, 1962). Availability of —SH groups also helps stabilize the cycle (Padilla, 1960).

Another relationship between growth and mitotic cycles and the assumed phototactic circadicity is revealed when it is recalled that most unicellular organisms reduce their motility to a minimum during mitotic or meiotic division, e.g., *Euglena* with rounded form and frequently loss of the flagellum (Hall and Jahn, 1929; Godjics, 1934), *Paramecium* in conjugation or division (Wichterman, 1952), and *A. proteus* in mitosis (Chalkley and Daniel, 1933), the reason for phototactic insensitivity of swimming species of *Euglena* during dark hours of a regular day and night sequence is evident. They round up and cannot swim ably even if they do not shed the flagellum, and are rendered completely immobile if the flagellum is also shed as it more often is. They do not recover their swimming mobility until both body shape and flagellum are restored. Since these mobility losses coincide with the generation time of *Euglena* during the dark hours, it is no surprise that the mobility losses, generation times, and absence of phototaxis should also coincide in time as they do.

Much more must be considered, then, in the circadicity of phototaxis than just a migration toward the light. Diversion of energy from photosynthetic mechanisms may be involved (Bruce and Pittendrigh, 1957), but so also are energy diversions through enzymic —SH-using mechanisms involved in nucleic acid and protein synthesis, enzymic reduction–oxidation potentials and reactions (Padilla, 1960), and other metabolic activities.

Cook (1960) suggests that cycling is under control of the cell, which controls its own metabolism. Sollberger (1965) puts is another way when he states that the organism *is* the clock, rather than a clock existing as a timing device localized either inside or outside the organism.

H. The Effects of Parasites on Motility

Euglena has few known parasites. The common algal parasite *Sphaerita endogena* causes no direct effect on form or motility of *E. caudata*, but causes *E. viridis* to become round and immobile (Mitchell, 1928).

I. Interactions of Physicochemical Factors in Motility

Data on the effects of physicochemical factors on motility are numerous, but also fragmentary, and are not unified by adequate theories of the mechanisms involved. Although all good data may eventually be useful when properly organized in terms of current concepts of molecular and ionic mechanisms, investigators should be encouraged to design and execute experiments with such interpretations in mind. Even if the interpretations are later proved incorrect (as many will be), they will stimulate further work and the errors will thereby become self-correcting, and progress will be more rapid than if no mechanisms had been proposed.

Obviously, any further attempt to study motile behavior of *Euglena*, or any other unicellular organism as presumably responsive directly or indirectly to any physicochemical force or another by an so-called taxis or kinesis only "clutters up" the literature with data, unless the reactions are correlated with other knowledge involving the many now known facets of its intricate biochemistry and biophysics, and also with the physical effects of hydrodynamics and thermodynamics. Much of the "clutter" needs to be reorganized, reinvestigated, and reinterpreted. Protoplasm, and its protein interfaces, must be treated as highly active ion-exchange systems in order to interpret the activities of ions, and the motile responses of the organism to them, involving Gibbs–Donnan equilibria, spin-resonance, triplet state (especially of water), proton as well as electron transfer, π-electron systems, oxidation–reduction potentials, membrane flux of, chelation and transfer of ions in ligand reactions related to their macro and counterions, and due to their differential charges and mobilities, oscillating mechanisms, piezoelectric effects, and other energy transfer machinery.

For instance, the inactivation of the flagellum by ultraviolet (Tchakhotine, 1936b) or reduction of motility by shorter wavelengths (Wichterman, 1955; Godward, 1962) may now tentatively be assumed to result from rupture of S—S linkages by the radiations (Yalow, 1959), paralyzing the protein polymers involved in motility. Inhibition of motility due to low O_2 tensions may be considered due to the interruption of energy flow by inability of cytochrome a to transfer H^+ to O^- to form water, and eliminate thereby the excess H^+, an effect also achieved by cyanides which block the H^+ to O^- linkage by incapacitating cytochrome a.

It is also possible to suggest that iodoacetate blocks motility of the flagellum (Danforth and Erve, 1964) by interfering with phosphate transfers and interactions involving —SH groups vital to the mobility of the flagellar apparatus; that acetate as a sole carbon source in nutrition of *Euglena* renders it immobile at pH 4.5 and below (Danforth, 1953) by activating an anaerobic shunt (Cook and Carver, 1966) which shuts down the flow of energy to the motile system, and that succinate as a substrate instead of acetate stimulates the functions of the entire tricarboxylic acid machine (Levedahl, 1967) and therefore accelerates a steady flow of energy therefrom into phosphate energy reservoirs (also accessible to the motile machinery). Urethan may be assumed to immobilize *Euglena* by blocking essential dehydrogenase and oxidative mechanisms in the energy flow system. Antibiotics may be expected to block motility (Loefer and Mattney, 1952) wherever they block critical enzymes related to motile activities.

The "slowdown" of motility and of the circadity of phototaxis by deuterium may perhaps be due to displacement of H_2O by D_2O in the I, II, and III water layers adjacent to and more or less bound to protein polymers, disturbing energy flow and cation transfers to and from enzymic c-positions involved in the motility of the macromolecules.

As Danforth (1966) states: "We know a great deal about the metabolic machinery within cells, and about the processes this machinery carries out, whereas we know almost nothing about the controls which regulate and integrate this machinery so that the cell as a whole functions as a harmonious homeostatic unit." Since Danforth's work has been largely with *Euglena*, his comments are especially relevant; the lack of knowledge of the interrelationship of the motile machinery with its other processes especially requires attention.

III. Swimming

THE LOCOMOTORY APPARATUS

In *Euglena*, swimming is effected only if a flagellum is present to develop the propulsive force that drives the body through the water. Flagella-less euglenas do not swim, although they may crawl or glide on a substrate (Bracher, 1919, 1938; Palmer, 1967; Palmer and Round, 1965; Round and Palmer, 1966). Since the swimming body is propelled by the flagellum, however, both body and flagellar morphologies are of necessity involved in the swimming process. Accessory to the flagellar motile apparatus, and influencing some of its roles, are the photoreceptor swelling which is adjacent to the flagellum near its base, and the stigma (eyespot) which serves to shade the photoreceptor most of the time. The structures and functions of all of these must be related to the motile activities of *Euglena*.

1. *Structure*

Although structure and function of the motile apparatus are inseparable, it is convenient to discuss the structure first and then try to relate it to the function(s). We shall discuss such structures in *Euglena* as are part of the motile apparatus. Other structure is detailed in Chapter 4.

a. The Flagellum. It has long been known that the flagellum is composed of a fibrillar "axoneme" surrounded by a membranous sheath, whether it be that of a spermatozoan (Ballowitz, 1888) or *Euglena* (Dellinger, 1909; Korschikov, 1923; Mainx, 1928). Today, flagella and cilia of all eucells are now known to be fundamentally alike in structural organization (see reviews by Fawcett, 1961; Pitelka, 1963; Pitelka and Child, 1964; Sleigh, 1962; Grimstone, 1961, 1966).

The gross morphology of the locomotory flagellum of *Euglena* has been known since the work of Fischer (1894) on *E. viridis*. He showed that the flagellum is composed of a fibrillar core (which he thought was an artifact) surrounded by a membranous sheath from which a unilateral row of delicate filaments (mastigonemes) extend. The flagellum may be very short in some species, e.g., *E. deses;* about as long as the body, e.g., *E. viridis;* or twice to three times as long as the body, e.g., *E. inflata* (Gojdics, 1953). Wager (1900) clarified the morphology of the basal attachment of the flagellum in *E. viridis*, showing it to have two roots, each anchored at the base of the anterior reservoir, one or both roots being surrounded by a paraflagellar swelling at or near the point of bifurcation, the paraflagellar swelling being adjacent to (but not adherent to) the stigma. That morphology has been confirmed for most other species (Gojdics, 1953). Other studies, especially on division of *Euglena*, show that the two "roots" are two separate flagella each with its own basal body. One is short, extending nearly to, next to, or slightly beyond the paraflagellar swelling, and adherent (in most species) to the longer, locomotory flagellum (Alexeiff, 1912; Gojdics, 1934; Krichenbauer, 1938; Hollande, 1942; Johnson, 1956). The mastigonemes were considered artifacts by some (Dellinger, 1909; Korschikov, 1923; Owen, 1947). Others considered them real fibrils (Fischer, 1894; Petersen, 1929; DeFlandre, 1934; Vlk, 1938; Hollande, 1942). DeFlandre (1934) accurately described and measured them. Awerintzew (1907) and Uhlehla (1911) described the flagellum of *Euglena* as ribbon-shaped, and Pringsheim and Hovasse (1950) attributed the ribbon-shaped appearance to the fact that the unilateral row of mastigonemes is oriented peripherally on the helically undulating flagellum.

Electron microscopy has clarified the structure of the flagellum after some initial confusion in early studies. Both the short flagellum and the longer locomotory one are of the same diameter, about 0.4–0.5 μ, for very nearly their full lengths (Brown and Cox, 1954; Leedale *et al.*, 1965a). Both flagella have the typical ring of nine peripheral bitubular fibrils, each with a unilateral pair of "arms" and two central tubular ones, the fibrils arising from a typical kinetosome which is imbedded in the peripheral cytoplasm, and is anchored by smaller fibrils to the pellicular complex of the reservoir; and the flagellum is covered by a tubular extension of the unit cytomembrane (Wolken and Palade, 1953; Ueda, 1958; Pitelka, 1963; Leedale *et al.*, 1965a).* The unilateral array of mastigonemes begins just external to the pseudostome, and extends the length of the locomotory flagellum. The mastigonemes are about 1.5–4.0 μ long and very slender, and are set about 0.1 μ apart (DeFlandre, 1934; Pitelka and Schooley, 1949, 1955, 1957). They arise just under the flagellar membrane and project through it (Leedale *et al.*, 1965a). The "segment" of the flagellum parallel to the axoneme which seemingly showed a "periodicity" of between 30 and 40 mμ (i.e., the "accessory ribbon," Houwink, 1951) is the flagellar membrane dried and flattened over the bases of the inserted mastigonemes along the one side of the flagellum. The outer surface of the membrane seems to be covered with a finely felted coat of delicate fibrils (Leedale *et al.*, 1965a; Figs. 4 and 5).

The subfibrillar organization of the tubular fibrils of the flagellar axoneme of *Euglena* spp. has not been investigated. However, since they do not differ in any significant way from those of other cilia and flagella, studies from other sources indicate how, and of what, they may be composed. In rat and other sperm flagellar fibrils, and in ciliary bitubular fibrils *(Tetrahymena pyriformis)* each tubular half is about 20 mμ in diameter, made up of about 13 fibrils, each globular subunit of which is about 35–40 Å spaced 55 Å apart, forming long filaments which are in turn side-bonded to one another (Pease, 1963; Phillips, 1966; Renaud *et al.*, 1966). Purified extracts of these suggest they have molecular weights of about 30,000 (Gibbons, 1965,

---

* *Note Added in Proof:* The nine bitubular peripheral fibrils appear to run nearly straight and parallel to one another for the full length of the flagellum (Fig. 118 in Leedale, 1967), and not in spirals as postulated by Wolken (1961, 1967) for the purpose of supporting his theoretical concept of the helical undulation of the flagellum (Wolken, 1967).

1966, 1967; Shelanski *et al.*, 1967), and sedimention constants near S 6.5, or S 30. These reaggregate to form filaments 50–60 Å in diameter (Gibbons, 1965, 1966, 1967; Renaud *et al.*, 1966; Shelanski *et al.*, 1967). All of these physicochemical characters are very nearly the same as those of the F-actin

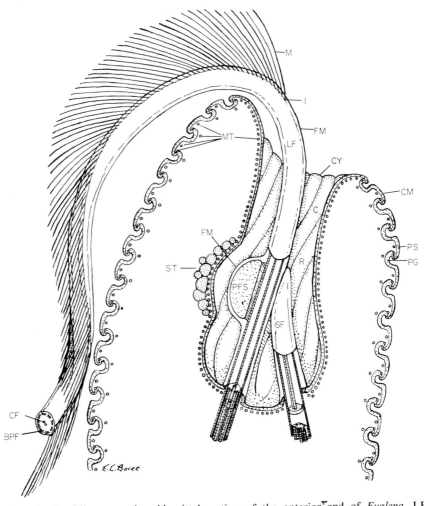

Fig. 4. Semidiagrammatic midsagittal section of the anterior end of *Euglena*. LF, Locomotory flagellum; SF, secondary flagellum; FM, flagellar membrane; CM, cell membrane; MT, microtubules; M, mastigoneme; I, insertion of mastigoneme; R, reservoir; ST, stigma. PS, pellicular strip; PG, pellicular groove; PFS, paraflagellar swelling, in two portions (protein portion, dotted, and quasicrystalline portion, lined); CF, central fiber of flagellar axoneme; BPF, bitubular peripheral fiber of the flagellar axoneme; CY, cytostome; C, canal.

strands of muscle (Karlson *et al.*, 1962; Hanson and Lowy, 1964; Panner and Honig, 1966). We may tentatively assume that the tubular fibrils are some sort of actin. The "arms" attached to the "outside" of one tubule of each peripheral bitubular unit have been extracted from *T. pyriformis*, and identified as the 14-S component of the distillate, which implies they are very similar to the myosin of vertebrate muscle (Gibbons and Rowe, 1965).

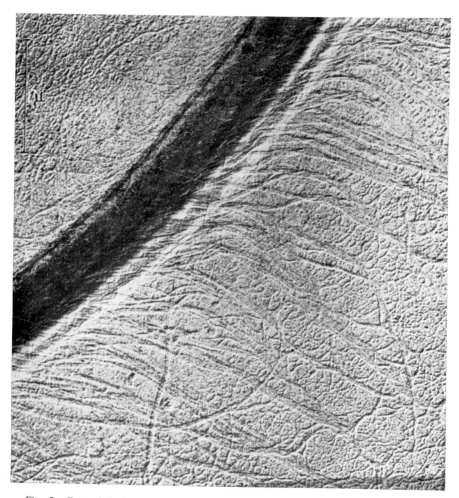

Fig. 5. Part of the locomotory flagellum, showing its coat of fibrous material and the insertion of the mastigonemes in a single row under the membrane of the flagellum (×30,000). (Leedale *et al.*, 1965a.)

b. The Paraflagellar Swelling. The morphology of this structure has also been the subject of debate. It has been said to be: (1) lenslike (Sokoloff, 1933, 1935a,b) or not (Mast, 1941); (2) lateral to (Wager, 1900) or surrounding the axoneme (Chadefaud and Provasoli, 1939; Gojdics, 1953); (3) attached to the stigma (Chadefaud and Provasoli, 1939; Wolken, 1961) or not attached (Wolken and Palade, 1953; Leedale *et al.*, 1965a); (4) endocytoplasmic (Chadefaud and Provasoli, 1939) or extracytoplasmic (Sokoloff, 1933, 1935a,b); (5) a homogeneous structure (Wager, 1900) or one composed of two distinct portions (Chadefaud and Provasoli, 1939; Leedale *et al.*, 1965a). On the basis of their electron micrographs Leedale *et al.* (1965a) consider it to be composed of an outer layer of protein and an inner layer of crystalline material lying lateral to the axoneme and adjacent to it, the swelling covered by the flagellar membrane, and the whole structure located beneath a concavity of the stigma, but not attached to the stigma by a protoplasmic bridge or secretion. (Figs. 4 and 6).

c. The Stigma (Eyespot). Like that of the photoreceptor, the morphology of the stigma has been debated. It has been considered a discrete colored plastid, more or less cup-shaped (France, 1893, and many others), a hexagonally packed layer of rods (Wolken, 1956, 1961), or a more or less loosely aggregated group of pigmented granules or globules (Gojdics, 1934; Hall and Jahn, 1929). Recent electron micrographs indicate that it is an aggregate of lipid-filled globules, each with its own membrane, clustered more or less compactly against the cell membrane of the reservoir so as to form a cup-shaped mass, partially shielding the paraflagellar photoreceptor swelling (Leedale *et al.*, 1965a; Walne and Arnott, 1966). (Figs. 4 and 6). Analysis of the reddish pigments extracted from the globules indicate they are lutein, four β-carotenes (euglenanone, echinenone, hydroxyechinone, and cryptoxanthin) and an unidentified pigment (Krinsky and Goldsmith, 1960; Batra and Tollin, 1964).*

d. The Cell Body. The primary body orientation of *Euglena* is twisted to the viewer's left, i.e., it is β-helical (Pringsheim, 1948b; Gojdics, 1953; Pochmann, 1953; Kirk and Juniper, 1964) and both the anterior end and its reservoir opening and the posterior end of the body are pointed away from the midline of the body by 10–15°. The pellicle is composed of somewhat elastic, helical, proteinaceous strips formed as more-or-less pronounced ridges, with thinner more membranous grooves between and connected to the ridges (Groupé, 1947; Pochmann, 1953; Gibbs, 1960; Pitelka, 1963; Kirk and Juniper, 1964; Mignot, 1965; Leedale, 1964; Leedale *et al.*, 1965a,b;

---

* *Note Added in Proof:* Wolken (1967) states that interpretation of the microspectrophotometry of pigments in the stigma suggests three substances, lutein, β-carotene and neoxanthine.

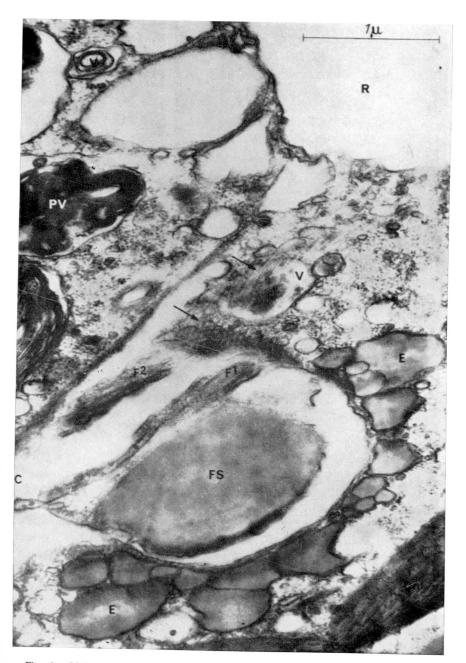

Fig. 6. Oblique longitudinal section of the transition region from reservoir to canal showing the paraflagellar swelling (FS) lying in the concavity of the eyespot (E). C, Canal; Chl, chloroplast; F¹, locomotory flagellum; F², nonemergent flagellum; PV, phospholipid vesicle; R, reservoir; V, vesicles; arrow, microtubule. (Leedale *et al.*, 1965a.)

Sommer, 1965; Arnott and Walne, 1966). The pellicle is more-or-less supple, to rigid, depending on the thickness of the pellicular ridges, and that of an underlying and sometimes interdigitated chondroid proteinaceous layer (Mignot, 1965). Each pellicular ridge has under it, usually in the section partially overhanging the groove, two or more tubules that run parallel to the ridge, and bend forward into and under the surface of the reservoir extending to the basal bodies of the flagella (Pitelka, 1963).

The body of most *Euglena* spp. is spindle-shaped, but some are long and nearly cylindrical, or long and flattened, and more-or-less secondarily twisted and ridged. Generally speaking, the shorter, broader and rounder the body, the longer the flagellum; and conversely, the longer and thinner the body, the shorter the flagellum (Gojdics, 1953), either by relative or by real measurement.

2. Functions in Swimming

Swimming *Euglena* spp., like other swimming microscopic organisms, must conform to and function with the assistance of the laws of hydrodynamics and classic physics, while also subject to the laws of the thermodynamics. Newton's third law of motion is obligatory for *Euglena*, as well as for other moving objects. The physical aspects of flagellar movement have recently been excellently treated in other pertinent reviews (Holwill, 1966b, Rikmenspoel, 1962, 1965b; Gray, 1955, 1958; Burge and Holwill, 1965) and will be only generally treated here.

a. Flagellar Movements. Briefly stated, *Euglena must continuously and actively expend energy within and along the entire length of the flagellum in order to swim* (Scourfield, 1909; Gray, 1928).

i. Hydrodynamic problems of swimming. The mass of the flagellum is small (1×10^{-7} or less) and the viscous drag of the water predictable by Stokes' law is so great on the flagellar surfaces, especially if the extended surfaces of the mastigonemes are included, that no more than half a wavelength of undulation can be generated by the flagellum if the force is applied *only* at one end, e.g., the attached base (Taylor, 1951; Gray, 1955; Machin, 1958).

The flagellar mass is also so small that if the flagellum ceases to undulate actively it cannot progress by inertia more than a few micra, a fraction of its length. It is much less likely that the flagellum of *Euglena* whipped only at one attached end could propel both itself and the relatively huge body to which it is attached.*

* Note Added in Proof: Certainly *not* "similar to that of a cowboy whipping his horse" as Wolken (1961, 1967) analogizes. He infers, however, probably correctly, that a continuous source of stimulus and energy flow must pass along the flagellum from base to tip in order to produce an undulation.

Various calculations predict, therefore, that the flagellum must actively expend energy along its entire length in order to undulate. It is suggested that it moves by means of a series of sequential bendings or one-sided contractions within segmental units of molecular size (one estimate is 80 Å, Silvester and Holwill, 1965) attached end to end, each contracting unit stimulating the onset of contraction in the unit next to it (Gray, 1955; Gray and Hancock, 1955; Machin, 1958; Brokaw, 1965; Rikmenspoel, 1965a,b; Silvester and Holwill, 1965; Holwill, 1966b). These authors have proposed that serial bendings or contractions of segments produce the travelling undulatory wave along the flagellum.

ii. Thermodynamic problems. Since the flagellum must actively and sequentially contract the serially arranged units within its structure in order to undulate, a sequential ordering of energy transferring, storing, and utilizing mechanisms is also required. The presence of actinoid and myosinoid proteins in flagella outlined in Section III,A,1,a, above, indicates that those units may be polymeric strands of actin in the bitubular peripheral fibrils which react with the myosinoid polymers of the "arms" of an adjacent bitubule (Silvester and Holwill, 1965). The presence of actin and myosin is further supported by cytostaining reactions that indicate that the electron-dense tubules of spermatozoan flagella are composed of actin, with myosin within and between the tubules (Nelson, 1962; Nelson and Plowman, 1963; Rikmenspoel, 1965a,b). Actomyosinlike ATPase reactions have been found in the kinetosomes of cilia and flagella, and distributed throughout their lengths (Levine, 1960; Rikmenspoel, 1965c). Chemical extracts also indicate the presence of both actin and myosin in flagella (Mohri, 1964; Nelson, 1966; Engelhardt and Burnasheva, 1957; Chorin-Kirsch and Mayer, 1964). Although no equivalent studies have been made on the flagella of *Euglena* it can be tentatively assumed that actomyosin reactions are involved in its flagellar organization and movements.

The energy source for flagellar motion is probably adenosine triphosphate (ATP) (Bishop, 1962; Seravin, 1961, and others). Again, no studies have been made on the direct effects of ATP on flagellar movements of *Euglena*, but studies on other flagella point to the likelihood.* Live or glycerinated flagella of *Polytoma* collected from the cell without the basal bodies (which remain in the cell) undulate and swim in ATP solutions (Brokaw, 1961; Tibbs, 1957), as do also glycerinated "models" of whole flagellates (Hoffmann-Berling, 1955). ATP is certainly produced by and is present in *Euglena*. The presence of the entire tricarboxylic acid cycle, a cytochrome system, and a TPN–DPN system (Danforth, 1953, 1966; Perini *et al.*,

---

* *Note Added in Proof:* Mahenda *et al.* (1967) report that glycerol-isolated flagella of *E. gracilis* both undulate and swim rapidly in solutions of ATP.

1964a,b; Kempner and Miller, 1965) certify its presence. How ATP reaches the flagellum and is distributed into and within it is not known.

Calculations of collision force exerted by a flagellum (3–60×10^{-8} dynes/cm; Holwill, 1965) and for its bending moment (3–5×10^{-10} dynes; Brokaw, 1965) at the viscosity of water suggest that a flagellum is relatively "stiff" with an elastic modulus of 5×10^{-8} dynes/cm (Brokaw, 1965) or more (Rikmenspoel, 1966) except when and where it bends. The amount of energy concentrated and actively used at any one point to produce the bend is tentatively calculated as about 6×10^{-6} dynes per fiber of the flagellum (Rikmenspoel, 1966). Other calculations indicate that the total capacity required to produce ATP greatly surpasses the amount needed for both locomotion and remaining metabolic functions (Carlson, 1962; Rikmenspoel, 1965a,b; Holwill, 1966b). The flagellar undulation is prodigal in its energy dissipation, but even so has an estimated mechanical efficiency of about 30% or less (Holwill, 1966b). Of the total energy dissipated, it is estimated that only 1–10% is used to propel the organism, the remainder being converted into the mechanical bending force (Brokaw 1961; Carlson, 1962). While these calculations have not been made for the flagellum of *Euglena*, its efficiency may be no greater.

iii. Flagellar undulations. Early observers were well aware that the undulations of the *Euglena* flagellum are helical, and travel from the proximal end to the distal end (e.g., Scourfield, 1909; Bancroft, 1913; Uhlehla, 1911; Metzner, 1920) (Fig. 7a,b,e,i–l). Drawings in taxonomic descriptions and monographs show the flagellum of most species as wavy or helical (e.g., Johnson, 1944; Gojdics, 1953). Lowndes (1941, 1943, 1944) verified this helical undulation of the flagellum for *E. viridis* by stroboscopic cinemicrography, and Jahn and Bovee (1967) and Holwill (1966a,b) have shown that two and slightly more helical waves are in progress along the flagellum of *E. gracilis* or *E. viridis* at any one time. (Fig. 7c,d,j–l).

The number of waves progressing along the flagellum of a *Euglena* depends partly upon the length of the flagellum. Spindle-shaped species with flagella about as long or a little longer than the body, e.g., *E. gracilis* and *E. viridis*, have about two waves progressing simultaneously; *E. inflata*, which is nearly globular and has a flagellum nearly three times the length of its body, appears to have six full wavelengths of undulation along the flagellum (Gojdics, 1953). *Euglena intermedia* and *E. deses*, with long bodies and flagella only a few micra long, may have less than one wavelength of undulation (Gojdics, 1953).

As Hollande (1942) indicates, the flagellar motion (and swimming) should be studied separately in each species of *Euglena* to discover what general statements can be made for all of them, and to discover what variations in the use of the flagellum are specific adjustments.

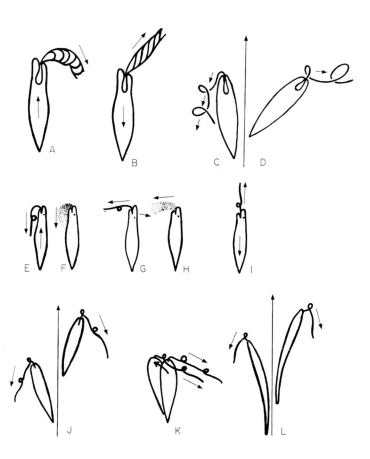

Fig. 7. Positions of the flagellum of *Euglena* in swimming. A, Forward swimming, showing helical waves from base-to-tip, flagellum bent (after Uhlehla, 1911). B, Backward swimming, with flagellum swung forward (after Uhlehla). C, Forward swimming, based on tracing from original high speed motion picture film by the authors, flagellum bent and trailing along the body. D, Turning, with flagellum extended to the side, tipping the anterior end of the body outward; based on tracing from original high speed film by the authors. E, Forward swimming, flagellum trailing. F, Movement of granules in water by flagellum during forward swimming. G, Turning, anterior end of body tipped to the side when flagellum is held laterally. H, Movement of granules when flagellum is held to side. I, Reverse swimming when flagellum is held forward. (E–I, after Bancroft, 1915) J, Positions of body and flagellum of *E. viridis* while swimming forward, in relation to the line of progress. K, Effect of each single beat of the flagellum, which pushes the body sideways slightly as each new wave originates. L, Positions of the body of *E. acus* in swimming, showing that part of the long body is inclined to the line of progress and the remainder trails along it. (J–L, after Lowndes, 1944.)

Although the waves are referred to as "helical" it should be pointed out that the beating flagellum, at least in most species of *Euglena*, is *not a true helix*. The flagellum is bent in the shape of a series of single turns of what might be parts of a helix, but there is a straight or almost straight section between the single turns (Fig. 7c,d and Fig. 8). Therefore, the wave is really an *interrupted helix*.*

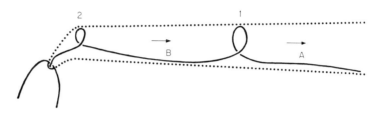

Fig. 8.   Semidiagrammatic sketch based on tracing from an original high-speed motion picture film (340 fps) showing the interrupted helices along the flagellum with the relatively straight segments between each travelling helix. Note that each helix increases in amplitude as it progresses from the base to the tip of the flagellum. Mastigonemes are not shown since their positions cannot be clearly seen with light microscopy. 1, 2, Successive helices; A, B, relatively straight segments.

Scourfield (1909) observed that while the flagellum undulates in a spiral fashion, *it does not rotate*, and that the spiral waves run along the flagellum from base to free tip, exerting force to propel the body, and also causing the body to rotate. This concept has been mathematically extended by a number of researchers, and found to fit the hydrodynamic requirements for flagellar movement. There are more than enough fibrils and subfibrils in the flagellum of *Euglena* to provide for a series of irrotational helical waves generated along it (a minimum of three or four fibrils is required, see Gray, 1928, 1955; Jahn and Landman, 1965).

Lowndes (1941) found for *E. viridis* that the flagellum is 128 μ long, twice the length of the body, and generates 12.7 beats of the flagellum per second (Fig. 9B), with never more than two waves moving along the flagellum simultaneously (Fig. 7j,k). The flagellum is bent backward as it leaves the anterior reservoir and trails almost directly backward, with the body tipped at an angle of about 30° to the linear path of progress (Fig. 9A,C). The anterior tip of the body describes a helical path about 65 μ in diameter and the rear end stays nearly on the linear path. The flagellum causes the body

* *Note Added in Proof:* Leedale (1967) refers to the flagellum of *Euglena* as being "continually mobile" from base to tip, implying an uninterrupted series of undulations along it. His photographs, however, especially Fig. 132 on page 149 of his book, clearly show that two helical waves are present, moving along the flagellum with almost straight sections of flagellum between, i.e., interrupted helical waves.

to rotate in the direction opposite the direction of the helical waves along the flagellum, and the flagellar waves push against the water, toward its tip, thereby propelling the flagellum in the direction of its base, dragging the cell body after it (Wager, 1900; Scourfield, 1909; Lowndes, 1941; Jahn and Bovee, 1967; Holwill, 1965, 1966a,b). The flagellum of *E. viridis* thereby provides both a linear component of force and a rotary one, i.e., torque.

Although Lowndes (1941, 1943, 1947) thought the flagellum of *E. viridis* (and all other flagella) to be capable only of base-to-tip helical undulations, others disagree. Leedale *et al.* (1965a,b) say the flagellum of *E. spirogyra* is moved in a variety of ways during slow progress and during changes of direction, but they do not describe them otherwise; and they state that in active swimming the flagellum projects forward like a spinning lasso. This suggests that in longer-bodied *Euglena* spp. (e.g., *E. spirogyra*) the flagellum may provide a tractile component by swinging in a more-or-less conical path, as has been demonstrated with models by Lowndes (1943) and Brown (1945). A possible variety of movement is also supported by observations that other flagellated cells perform a variety of flagellar movements (Krijgsman, 1925; Bovee, 1964; Bovee *et al.*, 1963; Jahn *et al.*, 1962, 1964; for reviews see Jahn and Bovee, 1964, 1965, 1967; Holwill, 1965, 1966b).

One movement of the flagellum of *Euglena* which has rarely been observed and remains completely unexplained, is the retraction and extension of the flagellum into and from the reservoir canal with a straight back-and-forth motion by *E. fusca* (Gojdics, 1953) and retraction by *E. gracilis* (Chadefaud and Provasoli, 1939). A similar retraction and extension of the posterior flagellum has been cinephotomicrographed for *Ceratium* (Jahn, unpublished) and also remains unexplained.

Whether the surface of the *Euglena* flagellum plays a hydrodynamic role is not known. That it may is suggested by the fact that its surface is relatively rough, of felted fibrils (Leedale *et al.*, 1965a), and by the calculations of Taylor (1952a), which indicate that rough cylinders perform differently than smooth ones in relation to the drag of the water.

The role of the unilateral row of mastigonemes in swimming is also unexplained. Fischer (1894) thought them to be independently motile. Much more recent observations of living material with better microscopes of recent design suggest they are flexible, elastic, and trail, rather than being actively motile (Chen, 1950; Leedale *et al.*, 1965a). Since they are flexible rather than stiff, they can not function in the same manner as those of *Ochromonas* (Jahn *et al.*, 1964), but they do increase the total flagellar surface and therefore increase the total viscous drag of the flagellum. This should increase the effectiveness of the flagellar action and thereby increase the total efficiency of propulsion. If the beat of the flagellum were in the form of a true helix the mastigonemes, being flexible, would tend to be wound around the flagellum, thereby becoming relatively ineffective as far as increasing flagellar drag is concerned. Since the wave form is that of an interrupted helix, however, the mastigonemes have an opportunity to unwind between successive turns. The only requirement for unwinding is a certain degree of elasticity in the mastigonemes; since they are usually found to be gently curved in dried preparations, this degree of elasticity seems highly probable. The electron micrographs by D. Bradley (1965, 1966) suggest that mastigonemes are structurally complex, of two or three subfibrils bound by a winding, fibrous material. This also suggests elasticity.

b. The Photoreceptor or Paraflagellar Swelling. This organelle, although within the flagellar membrane, and adjacent to the fibrils of the axoneme, is functionally distinct from, although accessory to, the flagellar undulatory mechanism. This is evident, since a green strain of *E. gracilis*, bleached in continued darkness, permanently loses its chloroplasts, stigma, and photoreceptor, but retains full capacity to swim by means of flagellar undulations (Pringsheim, 1948a). If it loses chloroplasts, but retains stigma and photoreceptor, it remains photopositive; if it loses both chloroplasts and stigma and retains the photoreceptor it becomes photonegative; if it loses all three, it is no longer phototactic (Bünning and Schneiderhohn, 1956; Gössel, 1957; Vavra and Aaronson, 1962). The photoreceptor is plainly responsive to light, apparently reacting to it by somehow transferring light energy to the flagellum. A suggested means (i.e., piezoelectric effect) of its action is considered in Section II,G,2, under phototaxis.

c. The Stigma. The stigma is plainly a shading device for the photoreceptor, despite the contention of Wolken (1961) that the stigma absorbs energy and delivers it to the photoreceptor. The photoreceptor absorbs and reacts to rays of 415 mμ (Gössel, 1957). The stigma absorbs mostly rays at 460–465 mμ (Wolken and Shin, 1958), but absorbs noticeably in the range 400–500 mμ, and somewhat to 630 mμ. There is no physical connection between the stigma and the eyespot, except as may be indicated by an occasional old report (Chadefaud and Provasoli, 1939), practically all other morphological studies by whatever means indicate only a proximity, but no physical connection (Leedale *et al.*, 1965a). This is also discussed in Section II,G,2, under phototaxis.

d. The Body. Dangeard (1890) proposed that the shape of *Euglena* spp., especially the irregularities and striations of the pellicle, interacted with flagellar movements to produce the rotation of the body and the helical pathway of advance; much more recently, Pochmann (1953) has assumed that the degree of torsion of the body, evidenced by its pellicular structure, affects its path of swimming. Holwill (1966b) says this may not be the case, however, since he finds that *E. tripteris*, a ridged and twisted species, rotates in a direction such as to oppose its forward movement if its torsional form were a factor. *Euglena* spp. are large enough that the hydrodynamics of viscous flow of water past the body may not be treated adequately by using equations applicable only to a sphere (e.g., Jenson, 1959; Taylor, 1951), and Holwill (1966a) treats it as a cylinder. Physical evidence that the hydrodynamic parameters vary with the shape of the body, and therefore demand varied orientations and parameters of force from and its directional application by the flagellum, is obtained from observations that when *E. gracilis* tends to change from its normal spindle shape to a rounded form in response to certain physical and chemical conditions, its motility is reduced in proportion to the degree it becomes rounded, and it can swim little, if at all, when nearly spherical even though its flagellar movements do not appear to be affected (Byrne and Marsland, 1965; Mandeville *et al.*, 1964). In rounding up, the flagellum is projected for more of its length laterally rather than posteriorly, and thereby exerts a greater amount of its force into the rotary component, and less in to the linear component of driving force. However, the normally rounded species, *E. inflata*, has a very long flagellum, more than three times the length of its body, and swims effectively, its flagellum being able to trail sufficiently straight and far enough behind, at least its distal portion, to provide an adequate linear component of drive.

Lowndes (1943) astutely noted that the flattening and lengthening of the body affect the forward swimming of euglenoids. He calculated that organisms with a short flagellum held laterally developed mostly a rotary component of drive with the flagellum. Experimentally, he showed with models that a planar or cylindrical body revolving about a conical path in the water propels itself forward (Lowndes, 1943). He assumed, therefore, that the forward component of force developed by a flagellum swung in a broad cone or circle by undulations, also driving the body to which it is attached in a conical gyration, thereby developed a *forward* component of force in the body (Fig. 9A). Brown (1945) demonstrated experimentally that a relatively thin cylindrical body conically rotated ahead of a larger body to which it is basically attached develops a forward component of force as well as a rotary one, and pulls the larger body forward (in one experiment the larger body was his *own* body in a swimming pool propelled by the conical gyrations of one arm held forward).* The latter indicates how a long,

---

* *Note Added in Proof:* Wolken (1967) likens the motion imparted to the cell body by the flagellar undulations to the movement of a single-bladed propellor screwing its way through the water, basing his analogy on the observations of Lowndes (1943) and Brown (1945).

thin, somewhat flattened *Euglena* can swim and gyrate with a relatively short flagellum. Lowndes (1944) suggests that the shorter a flagellum is, the more stable it is, and the greater its rate of undulation; he assumes it then consequently has a shorter wavelength and amplitude of undulation. The latter

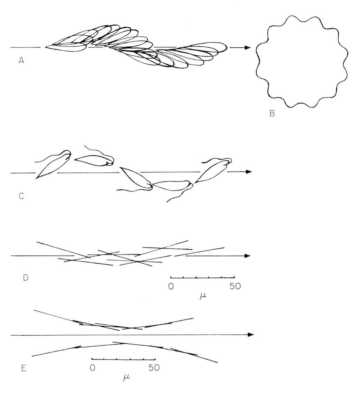

Fig. 9. Path of translocation of *E. viridis* showing spiral path about the linear direction. A, After Lowndes (1944). B, Path of the anterior end of the body seen from the front, showing "wobble" with each new travelling wave along the flagellum; after Lowndes (1944). C, after Jennings (1906). D, After Holwill (1966a), showing as lines the position of that surface of the body which is next to the line of progress. E, Positions of the body as lines, shown as if no translocation had occurred, indicating that the path of progress is as if the *Euglena* moved along an hyperbola cast about the line of progress. After Holwill (1966a).

two assumptions need not be and may not be true; and if the flagellum is too short to support less than one wavelength, it will gyrate in a conical pathway, and perhaps still be able to pull the body along. Hydrodynamic calculations have not yet been made that fully explain these phenomena;

but Holwill (1966a) presents formulas which ostensibly suggest that *E. viridis* generates only a slight forward component, not enough to account for the total or even a major portion of the forward speed; and he contends that the models used by Lowndes (1943) and Brown (1945) work because of the high Reynolds' number (10^3) and would not work at the low Reynolds' number for *Euglena* (10^{-3}) (Fig. 9,D,E). It is evident, however, that the body shape plays a more-or-less effective role in the swimming of *Euglena* spp., and the type of calculations made by Holwill (1966a) should be made for other larger, twisted, or flattened species.

   e. *Stimulus; The "Neuroid" Problem.* The idea that even single cells must have a "nervous system" is an old one generated by Ehrenberg (1838), and reopened by Sharp (1914) by the assumption on morphological grounds that subpellicular fibrils of *Paramecium* (and other ciliates) conduct impulses. The experiments of Taylor (1920) on *Euplotes*, now somewhat controverted by more careful recent work (Okajima and Kinosita, 1966), are usually cited as the "clinching" argument. Even Taylor had his doubts, however (see his review, Taylor, 1941), and so have most recent reviewers of the "evidence" (Worley, 1934; Pitelka, 1963; Pitelka and Child, 1964; Grimstone 1961, 1966; Sleigh, 1962; Jahn and Bovee, 1964, 1965, 1967).

   Some descriptions of *Euglena* morphology claim the presence of a filament ("rhizoplast") connecting the basal body ("blepharoplast" or "kinetosome") of the locomotory flagellum to either the nuclear membrane or to a centriole itself attached to the nuclear membrane; and therefore they called the nucleus–rhizoplast–flagellum a "neuromotor" apparatus (e.g., Baker, 1926; Ratcliffe, 1927; Hall and Jahn, 1929). Others denied the presence of any such "rhizoplast" (Wenrich, 1924; Lackey, 1934; Krichenbauer, 1938). Careful recent work with anoptral phase-contrast and electron microscopy does not show a rhizoplast to be present (Leedale, 1958, 1959a; Leedale *et al*, 1965a). In addition, the presence of the nucleus does not directly influence flagellar movements, since anucleate daughters from some fissions of *Euglena* swim normally until they die (Leedale, 1959b; Jahn and McKibben, unpublished).

   Wolken (1961), ignoring the observation that the stigma is *not* attached to the photoreceptor, and that neither is attached to the basal body of the flagellum (Wolken and Palade, 1953), calls the stigma–photoreceptor–flagellum complex "intimately connected," diagrams it that way (his Fig. 72, p. 140), and calls the "eyespot plus flagellum" the most "elementary nervous system." This is incorrect, both as to morphology and function, since the stigma only shades the photoreceptor, and does not originate nor coordinate the flagellar undulation (see Section II,G,2, phototaxis).

Lowndes (1943, 1947), believing that a flagellar undulation always begins at the attached base and progresses to the free tip, insists that an impulse is generated in the cell mass, is transmitted to the basal body of the flagellum, and is carried along the axial core by chemical reactions, with accessory aid from the surface membrane. This idea was "updated," with the revelation of the number of bitubular fibrils in the cilium and flagellum, as a kind of "commutator" device operating from basal body of the flagellum via either peripheral or central fibrils or both (Brown, 1945; Hodge, 1949; Astbury *et al.*, 1955; Bradfield, 1955; Gray, 1955). As Holwill (1966b) indicates, the "commutator" theories do not explain how a flagellum can theoretically generate several types of waves sometimes going in *opposite* directions (Machin, 1958), or observations that in many cases waves of undulation normally traverse the flagellum from its free tip to its attached base (Bovee, 1964; Jahn *et al.*, 1962; Walker and Walker, 1963; Holwill, 1965, and others), or the fact that waves may start nearly simultaneously at both tip and base of the flagellum and cancel out (Holwill, 1965), or that a flagellum held motionless by pressure midway of its length undulates from that point both toward the base and toward the tip (Holwill, 1965).

Lately there has been a tendency to consider the surface membrane of the flagellum the port of entry for stimulus (Holwill, 1965, 1966a,b; Parducz, 1957; Seravin, 1961, 1962; Jahn, 1964) and to consider the flagellum or cilium a mechanical receptor (Postma, 1959; Jahn, 1964; Jahn and Bovee, 1964, 1965, 1967), a concept perhaps first introduced by Verworn (1899). Holwill (1966b) contends that existence of some coordinating control system is necessary for the initiation in the flagellum of an undulatory wave (with the membrane being his candidate), and that its propogation is a mechanical process. Cook (1960) indirectly suggests such a control process by stating that the cell controls its metabolic activities (see Section II,E on interactions of forces). Where such a coordinative, initiative system morphologically resides or is morphologically distributed, and how it operates is only theoretical and still a mystery. That it is a kind of "nervous system" is highly unlikely.

One experimentally unexplored possibility is that a flagellum (or cilium) is a self-powered mechanical oscillator which when once set in operation continues to oscillate (i.e., undulate) with the orientation (i.e., direction) determined by that of the energy flow set off by and at the site of the stimulus. A piezoelectric effect might be involved in both the stimulus and the ATPase activities of fibrillar proteins and in the so-called "delayed-elasticity" of the sequentially arranged subunits of the filaments (Machin, 1958, 1963) theoretically required to set off stimulating energy flow from one contractile unit to another in the oscillating system.

IV. Contractile Body Movements (Metaboly*)

Most species of *Euglena* are able to bend or contract the body to a greater or lesser degree. These contractions vary from a slight unilateral bending, or slight twisting of the body, as in *E. tripteris*, or anterior–posterior contractions which round up the body, e.g., *E. proxima* (Dangeard, 1902) or *E. viridis* (Khawkine, 1887), or "conicalize" the body, e.g., *E. granulata* (Arnott and Walne, 1966), to the extension and retraction of ameboid lobes, e.g., *E. obtusa* (Gojdics, 1953). Or there may be waves of contraction from one end to the other, e.g., *E. deses* (Kamiya, 1939), *E. mutabilis* (Hollande, 1942; Hein, 1953), or *E. gracilis* (Chadefaud and Provasoli, 1939), or a variety of movements encompassing most of those already mentioned, e.g., *Astasia dangeardi* (Lackey, 1934), *Astasia haematodes* (Lockwood, 1884), *E. geniculata* (Dangeard, 1902), or *E. limosa* (Conrad, 1940).

Pringsheim (1948b) states that metaboly "is complex and assumes manifold forms." He lists some of these as: (1) a slight curvature of the body; (2) unilateral bulging; (3) end-to-end contraction with increase of width; (4) local bulging and distension; and (5) local "pseudopodial surface protuberances." For *E. granulata*, Arnott and Walne (1966) cite three types of metaboly: (1) simple surface fluctuations; (2) anterior and posterior conicalization; and (3) axial deformation. They consider all forms of metaboly completely "reversible."

This variety of movements suggests some principal variation in organization of the body to which these movements may be attributed and/or by which they may be restricted. The pellicle, the protoplasm, and any fibers that might exist in either have been considered most likely responsible.

A. The Role of the Pellicle

Stein (1878) thought the striations of the pellicle, visible in practically all *Euglena* spp., were contractile myonemes and were therefore responsible for metaboly. Khawkine (1887) described both longitudinal and circular myonemes in the pellicle of *E. viridis*. Günther (1927) and Pringsheim and Hovasse (1950) argue that the distinctness of pellicular striation does not correlate with the degree of metaboly. Pronounced striae are often found on *Euglena* spp. with rigid pellicles, e.g., *E. tripteris*, but also on metabolic

---

* This term, coined by Perty (1852) is not satisfactorily descriptive in the modern sense, but has become so generally used that is difficult to suggest a replacement term. The use of *euglenoid movement*, suggested by Jahn (1934) is more specifically applicable in a taxonomic sense, but no more accurate otherwise. There are no other terms available which are more accurate, and it is hardly useful here to coin another one (see also Leedale *et al.*, 1965a,b).

species, e.g., *E. granulata*. Conversely, smooth, lightly striated pellicles occur on rigid species, e.g., *E. acus*, and in highly metabolic ones, *e.g.*, *E. mutabilis*. Wager (1900) and Günther (1927) suggest that the contractile elements are within the cytoplasm, the type of metaboly being dependent on the relative flexibility and orientation of pellicular striations. Pringsheim (1948b) also suggests that the contractile mechanisms are cytoplasmic and essentially ameboid, but are limited to that degree permitted by the stiffness of the pellicle; and Gojdics (1953) states that the bending of the keeled species with little flexibility may mainly change the degree of body torsion, as is also the case for spindle-shaped species that appear only to contract.

Chadefaud (1937) reinvoked the idea that there may be both circular and longitudinal myonemes, but in the cytoplasm of the *Euglena* rather than in the pellicle, assuming that the elasticity of the pellicle passively lengthens the body after the myonemes have contracted it; but he gave no evidence for the locale of the myonemes.

Recent electron micrographs support the proposal that the degree to which the pellicle is flexible largely determines the degree of metabolic movement, especially the electron micrographs of Reger and Beams (1954) which indicate that the pellicle is bilaminar, supporting similar older contentions of more than one layer, one of which is proteinaceous, perhaps a keratin or elastin (Hamburger, 1911; Leedale, 1964), and the other not (Klebs, 1885; Hamburger, 1911; Chadefaud, 1937; Pigon, 1947). The recent electron micrographs of Leedale *et al.* (1965a) and Mignot (1965, 1966) show that the pellicle is a continuous peripheral layer *within* the true cell membrane, and is a cytoplasmic structure probably composed of fibers in a matrix (see Chapter 4; and Section III,A,1,d) Two or more microtubules 20–50 mμ in diameter parallel each ridge of a striation. A mucoid material is secreted from mucocysts in the cytoplasm adjacent to the pellicle through pores in the latter. The mucoid material forms an *external* layer over the cell membrane, and also perhaps serves as a lubricant between grooves and ridges of the striations during contractions. The pellicular layer is thick in poorly metabolic *Euglena* spp., often with accessory dentate interdigitations; and the pellicle of highly metabolic species is thin (Figs. 10, 11, and 12).

In general, present evidence supports a passively elastic function of the pellicular strips in metabolic movements, assisting the regaining of the usual contours of the body.

B. FUNCTIONS OF THE MICROTUBULES

In all *Euglena* spp. so far examined, two or more microtubules 20–50 mμ in diameter are present under each pellicular strip, usually adjacent to the pellicular ridge. Although these tubules have been sometimes assumed to

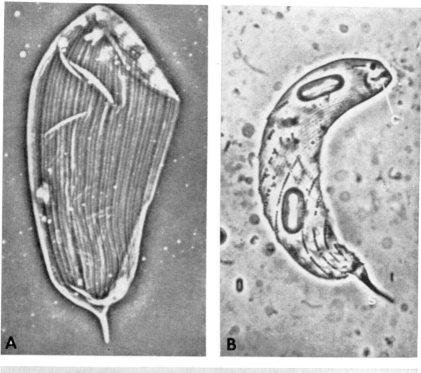

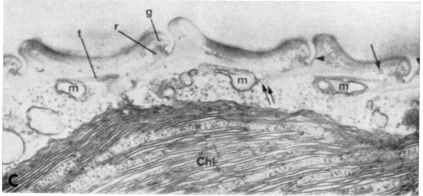

Fig. 10. Pellicle structure. A, Pellicle opened along one striation and flattened to show the internal surface; the striations are in pairs (anoptral phase, ×1000). B, Cell with most of the contents removed; the lowermost surface of the pellicle is focused to show the striations and a few rows of pellicular warts; C, canal of reservoir, S, posterior spine; also visible is the pellicular lining of the reservoir, and the posterior spine (phase contrast, ×1000). C. A thin section cut transversely to several pellicular strips, showing components: g, groove; r, ridge; t, tooth; m, muciferous bodies; Chl, section through chloroplast; arrowhead, cell membrane; arrows, microtubules (electron micrograph, ×40,000). (Leedale *et al.*, 1965a.)

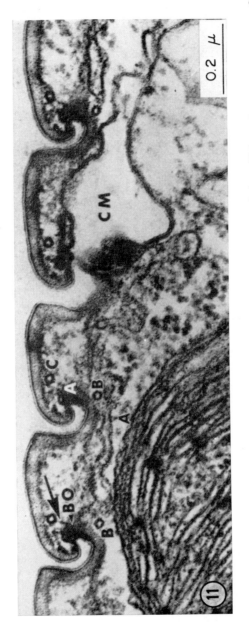

Fig. 11. Pellicle of *E. stellata* (*E. viridis*). Shows the thickness of the internal strata of the pellicle, and apparent transverse fibrils (arrow). A, B, C, Group of three fibrils; CM, muciferous body, the lining of which is a membrane continuous with the cell surface membrane. (From Mignot, 1965.)

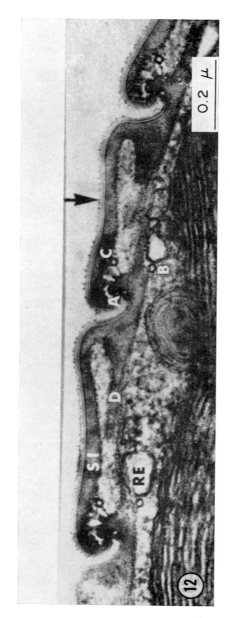

Fig. 12. Pellicle of *E. acus*. The internal strata of the pellicle (SI) turn to form teeth (D) under each striation. The external mucous coat is visible. The endoplasmic reticulum (RE) and the fiber (B) are evident for each fold. (From Mignot, 1965.)

be associated with metabolic movement (de Haller, 1959; Gibbs, 1960; Pitelka, 1963), perhaps as the contractile units, this assumption is clouded by the fact that some rigid euglenoids have more of them than some highly metabolic ones, and vice versa. Mignot (1965, 1966) considers the microtubules only cytoskeletal. Leedale *et al.* (1965a) consider them more likely transportive tubes.

C. ROLE OF THE CYTOPLASM

Arnott and Walne (1966, 1967) assume that the microtubules are functional in the various types of metaboly through interaction with the quasimuscular layer of fibrous proteins which are found underlying the pellicle. They therefore consider "conicalization" to be the result of general contraction of the quasimuscular sheet interacting with the microtubules, and local surface fluctuations and unilateral bending to be the result of more localized interactions.

The presence of both such structurally actinoid tubules and myosinoid fibrils suggests some sort of musclelike action, perhaps a sliding one similar to that currently considered probable in muscle (Huxley and Hanson, 1955, and many others). This is further supported by the localization of acid phosphatases in *E. gracilis* at the pellicle, and especially near the microtubules where they parallel each groove and overhanging ridge of the pellicle (Sommer and Blum, 1965). The abundance of fibrils now known to exist in the cytoplasm certainly would permit, by varying degrees of local to general interaction, any of the great variety of metabolic movements described for the various *Euglena* spp., assuming the known variation in the flexibility of the pellicular layers of cytoplasm.

The pellicular layer of cytoplasm is apparently composed partly of a labile matrix since the normally rigid nonmetabolic *E. acus* stops swimming and begins metaboly on being placed in an alkaline medium (LeFévre, 1931), or in one with a high organic acid content (Szabados, 1936). Similar apparent discrepancies of argument also appear for what supposedly stimulates metaboly in other species. Changes in temperature (Hall, 1931), following the Arrhenius rule, to 31°C at least (Kamiya, 1939), are effective. Changes in pH may be (Alexander, 1931), but not alone (Hall, 1933). Organic media are most often cited as stimulating metaboly in the majority of species (Jahn and McKibben, 1937; Kamiya, 1931; Szabados, 1936); and some species that normally live in organically acid waters are highly metabolic, e.g., *E. mutabilis* (Hein, 1953). Also, many species become "metabolic" if suddenly immersed in such waters (Bovee, unpublished). *Euglena limosa*, which lives on the organic interface of tidal river flats, however, is stimulated to metaboly by 1% K_2CO_3 (Conrad, 1940). Other stimulants to metaboly for various species include light, temperature, osmotic changes (Hilmbauer, 1954) in either direction (Hall, 1931); a 1/10,000 solution of brilliant cresyl blue at pH 8.0 (Diskus, 1955); changes in light intensity (Bracher, 1938; Palmer and Round, 1965), mitotic activity (Jahn and McKibben, 1937), or a force causing autotomy of the flagellum (Hilmbauer, 1954). All these data seem very confusing, but upon inspective analysis suggest that sudden changes of membrane potentials may cause release of ions internally, especially Ca^{2+}, causing contractions, and also triggering a dissolution of the pellicular matrix, but not its fibrous component, resulting in a greater degree of pellicular flexibility. This is even more likely to occur than previously thought, since evidence now shows that the pellicle is cytoplasmic and internal in origin and location.

V. Gliding Movements

Various *Euglena* spp. with short flagella, or individuals that have autotomized the locomotory flagellum have been said to creep, crawl, or glide on the substrate (Mast, 1917; Günther, 1927; Conrad, 1940; Hein, 1953; Diskus, 1955; Palmer, 1967) over a secreted path of mucus (Günther, 1927; Conrad, 1940; Diskus, 1955).

The mucus is secreted as filaments paralleling the grooves of the pellicle (Diskus, 1955; Leedale, 1964), probably from "mucocysts" in rows under the pellicular grooves, and through pores between certain pellicular strips, sometimes accumulating with extraneous materials as "warts" externally about the pores (Leedale, 1964, 1967). Acid phosphatases in the endoplasmic reticulum and around vesicles adjacent to the pellicle (Sommer and Blum, 1965) suggest such a secretory activity.

Gliding euglenas often follows old slime trails laid down by other euglenas (Günther, 1927), but may secrete new ones. How the mucus provides traction is not known. One theory suggests that a pileup of mucus at the rear pushes the *Euglena* forward (Günther, 1927). This does not explain how the mucus is driven to accumulate at the rear. Also, calculations for other algal organisms that similarly glide over self-secreted mucus trails, e.g., diatoms, suggest that a gliding algal cell would have to extrude a volume of mucus greater than its entire cytoplasmic content at least every 30 seconds in order to pile up enough mucus posterially to push it ahead at known rates (Holton and Freeman, 1965).

Some active mechanism in the cell other than secretion alone is involved to produce the gliding. Jarosch (1964, and other papers) has revived an old theory of Max Schultze (1865) that suggests that an interface-active energy-using mechanism reacts with the mucus trail at the surface of the cell, and

Jarosch (1964) postulates a sort of worm-gear drive involving multiple rotating protein helices as the mechanism. Drum and Hopkins (1966) suggest that cytoplasmic subfibrils resembling smooth muscle fibrils exert rhythmic contractions that propel the cell along the mucus track. Others suggest pellicular undulations (Günther, 1927), peristaltic movements (Chadefaud, 1937), swaying movements (Mast, 1917), gliding forward with the body twisted as a helix (Hein, 1953), or rotating with the rear end attached (Hein, 1953).

The helical gliding that occurs in most *Euglena* spp. is probably caused by the helical torsion of body and pellicle. A similar helical gliding is also performed by sporozoan trophozoites, and has been recently analyzed by means of cinematograph records by Jahn and Bovee (1968). They observed that mucus is propelled along the body, apparently by a force active between the mucus layer and the cell surface. This force is sufficient to propel the body along the mucus adhering to the substrate, and is simultaneously able to pick up an adherent blood cell anteriorly, transport it along the helix of the body and drop it off the rear end without interrupting the forward gliding of the body along the substrate. How the tractile force is exerted between the cell surface and the extruded mucus is unknown, but observations suggest it occurs; and if not as Jarosch (1964) suggests, perhaps by some "ratchet action" of shifting polymeric linkages mediated by the ATPases known to exist in living membranes.

VI. Recapitulation and Summary

In general, it may be assumed that the swimming of *Euglena* by flagellar motion operates by mechanical and thermodynamic mechanisms such as are present in other flagella, probably actomyosin–ATPase systems involving interactions between flagellar fibrils, producing the undulations of the flagellum and thereby the motive force. Torque developed in the flagellum tips the body and gyrates it, causing part of the forward component of drive to be developed by the planar surfaces of the body. The motive force developed is applied against the water in conformity with hydrodynamic principles and the laws of mass in motion. Environmental forces, acting upon the cytoplasmic response mechanisms, may initiate and vary the oscillatory mechanisms within the flagellum that produce its undulations; but little is known as to how these forces act.

Internal movements of cytoplasmic nature in *Euglena*, resulting in changes of its external contours (metaboly), have not been critically investigated, and need to be, to determine what sort of cytoplasmic movements are involved; and the mechanism of creeping and gliding is still a mystery.

We are now back where we began in the opening portion of this chapter, and must again point out the dearth of critical investigations on movement in *Euglena*. An intensive investigation, while probably yielding confirmatory evidence of generalities of flagellar and cytoplasmic movements, will also reveal what unique adaptations *Euglena* spp. may have.

As noted by Barker (1943) more than two decades past, the need is still great for studies on *Euglena;* more studies are needed on the roles of thermodynamics, protein chemistry, and biophysics, in all of the cellular organelle systems as related to motion, as well as in the actions of the motile mechanisms themselves, and the relationships thereof to hydrodynamic principles. As DeFlandre (1934) suggests, motion should be studied in as many species of *Euglena* as possible.

Much more equipment, and many more techniques are now available for such studies than ever before, and it is unfortunate that *Euglena* has been so much neglected as a tool for the study of motility. It has many unique advantages as an organism for such studies.

References

Abbott, W. E. (1951). *Sewage Ind. Wastes* **23**, 1310.
Aderhold, R. (1888). *Jena. Z. Med. Naturwiss.* **22**, 310.
Alexander, G. (1931). *Biol. Bull.* **61**, 165.
Alexeiff, A. (1912). *Arch. Zool. Exptl. Gen.* [5], **10**, 66.
Arnott, H. J., and Walne, P. L. (1966). *J. Phycol.* **2**, Suppl., (Abstr.).
Arnott, H. J., and Walne, P. L. (1968). *Protoplasma* (in press).
Astbury, W. T., Beighton, E., and Weibull, C. (1955). *Symp. Soc. Exptl. Biol.* **9**, 282.
Awerintzew, S. (1907). *Zool. Anz.* **31**, 834.
Baker, W. B. (1926). *Biol. Bull.* **51**, 321.
Barker, D. (1943). *New Phytologist* **42**, 49.
Ballowitz, E. (1888). *Arch. Mikroskop. Anat. Entwicklungsmech.* **32**, 401.
Bancroft, F. W. (1913). *J. Exptl. Zool.* **15**, 383.
Bancroft, F. W. (1915). *Studies Rockefeller Inst. Med. Res.* **20**, 384.
Bassett, C. A. L. (1966). *In* "Electromechanical Factors Regulating Bone Architecture" (H. Reisch, H. J. J. Blackwood, and M. Owen, eds.), pp. 78–89. Springer, Berlin.
Batra, P. P., and Tollin, G. (1964). *Biochim. Biophys. Acta* **79**, 371–378.
Bishop, D. W. (1962). *In* "Spermatozoan Motility" (D. W. Bishop, ed.), pp. 251. Am. Assoc. Advance Sci., Washington, D. C.
Bittar, E. E. (1964). "Cell pH," 129 pp. Butterworth, London and Washington, D. C.
Blum, J. J., and Padilla, G. M. (1962). *Exptl. Cell Res.* **28**, 512.
Blumberg, A. J., and Loefer, J. B. (1952). *Physiol. Zool.* **25**, 276.
Borgers, J. A., and Kitching, J. A. (1956). *Proc. Roy. Soc.* **B144**, 507.
Bovee, E. C. (1964). *In* "Primitive Motile Systems in Cell Biology" (R. D. Allen and N. Kamiya, eds.), p. 189. Academic Press, New York.
Bovee, E. C. (1965). *Hydrobiologia* **25**, 69.
Bovee, E. C., Jahn, T. L., Fonseca, J. R., and Landman, M. (1963). *Abstr. Proc. Biophys. Soc., 7th Ann. Meeting, New York, 1963* MD2.

Bracher, R. (1919). *Ann. Botany* **33**, 93.

Bracher, R. (1938). *J. Linnean Soc. (London) Botany* **51**, 23.

Bradfield, J. R. G. (1955). *Symp. Soc. Exptl. Biol.* **9**, 306.

Bradley, D. E. (1965). *Excerpta Med. Intern. Congr. Ser.* **91**, 262.

Bradley, D. E. (1966). *Exptl. Cell Res.* **41**, 162.

Bradley, W. H. (1965). *Science* **150**, 1423.

Brokaw, C. J. (1961). *Exptl. Cell Res.*, **22**, 151.

Brokaw, C. J. (1964). *J. Cell Biol.* **25**, 15A.

Brokaw, C. J. (1965). *Abstr. Biophys. Soc.*, *9th Ann. Meeting San Francisco, 1965* p. 148.

Brokaw, C. J., and Holwill, M. E. J. (1967). *Abstr. Biophys. Soc.*, *11th Ann. Meeting, Houston, Texas, 1967* p. 116.

Brokaw, C. J., and Wright, L. (1963). *Science* **142**, 1169.

Brown, D. E. S., and Marsland, D. A. (1936). *J. Cellular Comp. Physiol.* **8**(2), 159.

Brown, F. A., Jr. (1962). *Biol. Bull.* **123**, 264.

Brown, H. P. (1945). *Ohio J. Sci.* **45**, 247.

Brown, H. P., and Cox, A. (1954). *Am. Midland Naturalist* **52**, 106.

Bruce, V. G. (1959). *J. Protozool.* **6** (Suppl.), 11.

Bruce, V. G. (1960). *Cold Spring Harbor Symp. Quant. Biol.* **25**, 229.

Bruce, V. G., and Pittendrigh, C. S. (1957). *Proc. Natl. Acad. Sci. U.S.* **42**, 676.

Bruce, V. G., and Pittendrigh, C. S. (1958). *Am. Naturalist* **92**, 295.

Bruce, V. G. and Pittendrigh, C. S. (1960). *J. Cellular Comp. Physiol.* **56**, 25.

Buder, J. (1917). *Jahrb. Wiss Botan.* **58**, 105.

Bünning, E. (1956). *In* "Handbuch der Pflanzenphysiologie" (W. Ruhland, ed.), Vol. 11. p. 878. Springer, Berlin.

Bünning, E. (1964). "The Physiological Clock; Endogenous Duirnal Rhythms and Biological Chronometry," 2nd ed. Academic Press, New York.

Bünning, E., and Schneiderhohn, G. (1956). *Arch. Mikrobiol.* **24**, 80.

Burge, R. E., and Holwill, M. E. J. (1965). *Symp. Soc. Gen. Microbiol.* **15**, 250.

Byrne, J., and Marsland, D. (1965). *J. Cellular Comp. Physiol.* **65**, 277.

Carlson, F. D. (1962). *In* "Spermatozoan Motility" (D. W. Bishop, ed.), p. 137. Am. Assoc. Advance Sci., Washington, D. C.

Chacharonis, P., and Kostir, W. J. (1952). *Proc. Soc. Protozoologists* **3**, 5.

Chadefaud, M. (1937). *Botaniste* **28**, 86.

Chadefaud, M., and Provasoli, L. (1939). *Arch. Zool. Exptl. Gen. Notes Rev.* **80**, 55.

Chalkley, H. W., and Daniel, G. E. (1933). *Physiol. Zool.* **6**, 592.

Chen, Y. T. (1950). *Proc. Conf. Electronmicroscop.*, *Delft, 1949* p. 156. Hoogland, Delft.

Chorin-Kirsh, I., and Mayer, A. M. (1964). *Plant Cell Physiol. (Tokyo)* **5**, 441.

Conrad, W. (1940). *Bull. Musée Roy. Hist. Nat. Belg.* **16**, 1.

Cook, J. R. (1960). Ph. D. Thesis, UCLA Library, Los Angeles, California.

Cook, J. R., and Carver, M. (1966). *Plant Cell Physiol. (Tokyo)* **7**, 377.

Cook, J. R., and James, T. W. (1960). *Exptl. Cell Res.* **21**, 583.

Cope, F. W. (1963). *Arch. Biochem. Biophys.*, **103**, 352,

Czarska, L. (1964). *Acta Protozool.* **2**, 287.

Danforth, W. (1953). *Arch. Biochem. Biophys.* **46**, 164.

Danforth, W. (1966). *In* "Research in Protozoology" (T. T. Chen, ed.), Vol. I, p. 199. Pergamon, Oxford.

Danforth, W., and Erve, P. (1964). Personal communication.

Dangeard, P. A. (1890). *Botaniste* **1**, 1.

Dangeard, P. A. (1902). *Botaniste* **8**, 97.

DeFlandre, G. (1934). *Ann. Protistol.* **4**, 31.

de Haller, G. (1959). *Arch. Sci.* **12**, 309.
Dellinger, O. P. (1909). *J. Morphol. Physiol.* **20**, 171.
de Puytorac, P., Andrivon, C., and Serre, F. (1963). *J. Protozool.* **10**. 10.
Diehn, B. and Tollin, G. (1967). *Arch. Biochem. Biophys.*, **121**, 169.
Diskus, A. (1955). *Protoplasma* **45**, 460.
Dobell, C. (1960). Reprint of 1932. "Antony van Leeuwenhoek and His Little Animals," 435 pp. Dover, New York.
Drum, R. W., and Hopkins, J. T. (1966). *Protoplasma* **62**, 1.
Dryl, S. (1959). *Acta Biol. Exptl.* **19**, 83.
Dryl, S. (1961a). *Bull. Acad. Polon. Sci. Ser. Sci. Biol.* **9**, 71.
Dryl, S. (1961b). *Acta Biol. Exptl.* **21**, 75.
Dryl, S. (1961c). *J. Protozool.* **8**, Suppl. 16.
Dryl, S. (1964). *J. Protozool.* **11**, Suppl., 30.
Ehrenberg, C. G. (1838). "Die Infusiorenthierchen als vollkommene Organismen," 547 pp. Felix, Leipzig.
Eigen, M., and de Maeyer, L. (1958). *Proc. Roy. Soc.* **A247**, 505.
Eisenman, G. (1963). *Bol. Inst. Estud. Med. Biol. (Mex.)* **21**, 155.
Elliott, A. M. (1935a). *Arch. Protistenk.* **84**, 156.
Elliott, A. M. (1935b). *Arch. Protistenk.* **84**, 472.
Engelhardt, V. A., and Burnasheva, S. A. (1957). *Biokhimiya* **22**, 554.
Engelmann, T. W. (1869). *Arch. ges. Physiol.* **2**, 307.
Engelmann, T. W. (1882). *Arch. ges. Physiol.* **29**, 387.
Fawcett, D. (1961). *In* "The Cell" (J. Brachet and A. E. Mirsky, eds.), Vol. II, p. 218. Academic Press, New York.
Finley, H. E. (1930). *Ecology* **11**, 337.
Fischer, A. (1894). *Jahrb. Wiss. Botan.* **26**, 187.
Fraenkel, G. S., and Gunn, D. L. (1940). "Kineses, Taxes and Compass Reactions," 352 pp. Oxford Univ. Press, London and New York.
Francé, R. (1893). *Z. Wiss. Zool.* **56**, 38.
Geissman, T. A. (1949). *Quart. Rev. Biol.* **24**, 309.
Gibbons, I. R. (1965). *J. Cell Biol.* **27**, 33A.
Gibbons, I. R. (1966). *J. Biol. Chem.* **241**, 5590.
Gibbons, I. R. (1967). *In* "Molecular Organization and Biological Function" (J. M. Allen, ed.), pp. 221–237. Harper, New York.
Gibbons, I. R., and Rowe, A. J. (1965). *Science* **149**, 424.
Gibbs, S. P. (1960). *J. Ultrastruct. Res.* **4**, 127.
Giese, A. C. (1938). *J. Cellular Comp. Physiol.* **12**, 129.
Giese, A. C., ed. (1964). "Photophysiology," Vol. I, 377 pp. Academic Press, New York.
Gittleson, S. M. (1966). Ph.D. Thesis, UCLA Library, Los Angeles, California.
Gittleson, S. M., and Jahn, T. L. (1964). *J. Protozool.* **11**, Suppl., 13.
Gittleson, S. M., and Jahn, T. L. (1966). *J. Protozool.* **13**, Suppl., 24.
Gittleson, S. M. and Jahn, T. L. (1968) *Expt. Cell Res.* (in press)
Gittleson, S. M., and Sears, D. F. (1964). *J. Protozool.* **11**, 191.
Godward, M. B. E. (1962). *In* "Physiology and Biochemistry of Algae" (R. A. Lewin, ed.), p. 551. Academic Press, New York.
Gojdics, M. (1934). *Trans. Am. Microscop. Soc.* **53**, 299.
Gojdics, M. (1953). "The Genus *Euglena*," 268 pp. Univ. of Wisconsin Press, Madison, Wisconsin.
Goodwin, C. M. (1951). *Proc. Iowa Acad. Sci.*, **58**, 451.
Gössel, J. (1957). *Arch. Mikrobiol.* **27**, 288.

Gray, J. (1928). "Ciliary Movement." 162 pp. Cambridge Univ. Press, London and New York.
Gray, J. (1937). *J. Exptl. Biol.* **14**, 95.
Gray, J. (1953). *Quart. J. Microscop. Sci.* **94**, 551.
Gray, J. (1955). *J. Exptl. Biol.* **32**, 775.
Gray, J. (1958). *J. Exptl. Biol.* **35**, 96.
Gray, J. (1962). *In* "Spermatozoan Motility" (D. W. Bishop, ed.), pp. 1–12. Am. Assoc. Advance Sci. Washington, D.C.
Gray, J., and Hancock, G. J. (1955). *J. Exptl. Biol.* **32**, 504–514.
Grebecki, A. (1963). *Acta Protozool.* **1**, 99.
Grebecki, A. (1964). *Acta Protozool.* **2**, 69.
Grebecki, A. (1965). *Excerpta Med. Int. Congr. Ser.* **91**, 242.
Grebecki, A., Kužnicki, L., and Kinostrowski, W. (1956). *Folia Biol.* **3**, 127.
Griffin, J. L., and Arnold, E. A. (1965). *J. Cell Biol.* **27**, 117A.
Griffin, J. L., and Stowell, R. E. (1966). *Exptl. Cell Res.* **44**, 684.
Grimstone, A. V. (1961). *Proc. Linnean Soc. (London)* **174**, 49.
Grimstone, A. V. (1966). *Ann. Rev. Microbiol.* **20**, 131.
Gross, J. A. (1965). *Science* **147**, 741.
Groupé, V. (1947). *Proc. Soc. Exptl. Biol. Med.* **64**, 401.
Günther, F. (1927). *Arch. Protistenk.* **60**, 511.
Haase, G. (1910). *Arch. Protistenk.* **20**, 47.
Hall, R. P. (1933). *Arch. Protistenk.* **79**, 239.
Hall, R. P., and Jahn, T. L. (1929). *Trans. Am. Microscop. Soc.* **48**, 388.
Hall, S. R. (1931). *Biol. Bull.* **60**, 327.
Halldal, P. (1958). *Physiol. Plantarum* **11**, 118.
Halldal, P. (1959). *Physiol. Plantarum* **13**, 742.
Halldal, P. (1964). *In* "Biochemistry and Physiology of Protozoa" (S. H. Hutner, ed.), Vol. III, p. 277. Academic Press, New York.
Hamburger, C. (1911). *Sitzber. Akad. Wiss. Heidelberg, Math- Naturw. Kl.* **4**, 1.
Hancock, G. J. (1953). *Proc. Roy. Soc.* **A217**, 96.
Hanson, J., and Lowy, J. (1964). *Proc. Roy. Soc.* **B160**, 449.
Harker, J. E. (1960). "The Physiology of Diurnal, Rhythms," 114 pp. Cambridge Univ. Press, London and New York.
Harvey, E. N. (1934). *Biol. Bull.* **66**, 91.
Harvey, E. N., and Loomis, A. L. (1928). *Nature* **121**, 622.
Heidt, K. (1934). *Ber. Deut. Botan. Ges.* **52**, 607.
Heilbrunn, L. V. (1958). *Protoplasmatologia* **2**, 1.
Hein, G. (1953). *Arch. Hydrobiol.* **47**, 516.
Hilmbauer, K. (1954). *Protoplasma* **43**, 192.
Höber, K. (1965). "Physical Chemistry of Cells and Tissues," 676 pp. Blakiston, Philadelphia, Pennsylvania.
Hodge, A. J. (1949). *Australian J. Sci. Res. Ser. B* **2**, 368.
Hoffmann-Berling, H. (1955). *Biochim. Biophys. Acta* **16**, 146.
Höfler, K. and Höfler, L. (1952). *Protoplasma* **41**, 76.
Hollande, A. (1942). *Arch. Zool. Exptl. Gen.* **83**, 1.
Holton, R. W., and Freeman, A. W. (1965). *Am. J. Botany* **52**, 640 (Abst.).
Holwill, M. E. J. (1965). *Excerpta Med. Intern. Congr. Ser.* **91**, 109.
Holwill, M. E. J. (1966a). *J. Exptl. Biol.* **44**, 579.
Holwill, M. E. J. (1966b). *Physiol. Rev.* **46**, 696.
Houwink, A. L. (1951). *Koninkl. Ned. Akad. Wetenschap. Proc. Ser. C* **54**, 32.
Huxley, H. E., and Hanson, J. (1955). *Symp. Soc. Exptl. Biol.* **9**, 228.

Jacobs, M. H. (1940). *Cold Spring Harbor Symp. Quant. Biol.* **8**, 30.
Jahn, T. L. (1934). *Cold Spring Harbor Symp. Quant. Biol.* **2**, 167.
Jahn, T. L. (1936). Unpublished notes.
Jahn, T. L. (1961). *J. Protozool.* **8**, 369.
Jahn, T. L. (1962). *J. Cellular Comp. Physiol.* **60**, 217.
Jahn, T. L. (1963). *Vision Res.* **3**, 25.
Jahn, T. L. (1964). *Abstr. Biophys. Soc., 8th Ann. Meet. Chicago, Illinois, 1964* FE13.
Jahn, T. L. (1966). *J. Cellular Comp. Physiol.* **68**, 135.
Jahn, T. L. (1967). "A possible mechanism for the effect of electrical potentials on apatite formation in bone", *Am. Zool.*, **7**, 189. (1968). *Clin. Orthopaed. Related Res.*, **56**, 261,
Jahn, T. L., and Bovee, E. C. (1964). *In* "Biochemistry and Physiology of Protozoa" (S. H. Hutner, ed.), Vol. III, pp. 62–129. Academic Press, New York.
Jahn, T. L., and Bovee, E. C. (1965). *Ann. Rev. Microbiol.* **19**, 21.
Jahn, T. L., and Bovee, E. C. (1967). *In* "Research in Protozoology" (T. T. Chen, ed.), Vol. I, pp. 39–198. Macmillan (Pergamon), New York.
Jahn, T. L., and Bovee, E. C. (1968). *In* "Infectious Blood Diseases of Man and Animals" (D. Weinman and M. Ristic, eds.), Vol. I, pp. 303-436. Academic, New York.
Jahn, T. L. and Brown, M. (1961). *Am. Zool.*, **1**, 454.
Jahn, T. L., and Landman, M. D. (1965). *Trans. Am. Microscop. Soc.* **84**(3), 395.
Jahn, T. L., and McKibben, W. R. (1937). *Trans. Am. Microscop. Soc.* **56**, 48.
Jahn, T. L., Brown, M., and Winet, H. (1962). *Excerpta Med. Intern. Congr. Ser.* **48**, 638.
Jahn, T. L., Landman, M. D., and Fonseca, J. R. (1964). *J. Protozool.* **11**, 291.
Jarosch, R. (1964). *In* "Primitive Motile Systems in Cell Biology" (R. D. Allen and N. Kamiya, eds.), pp. 599–622. Academic Press, New York.
Jennings, H. S. (1904). *Carnegie Inst. Wash. Publ.* **16**, 129.
Jennings, H. S. (1906). "The Behavior of Lower Organisms," 366 pp. Reprinted (1964). Indiana Univ. Press, Bloomington, Indiana.
Jensen, P. (1893a). *Arch. Ges. Physiol.* **53**, 428.
Jensen, P. (1893b). *Arch. Ges. Physiol.* **54**, 537.
Jenson, V. G. (1959). *Proc. Roy. Soc.* **A249**, 346.
Jira, J., and Ottova, L. (1950). *Biol. Listy* **31**, 82.
Jirovec, O. (1934). *Protoplasma* **21**, 587.
Jirovec, O. (1935). *Cseck. Zool. Spolec.* **2**, 101.
Johnson, L. P. (1939). *Trans. Am. Microscop. Soc.* **58**, 42.
Johnson, L. P. (1944). *Trans. Am. Microscop. Soc.* **63**, 97.
Johnson, L. P. (1956). *Trans. Am. Microscop. Soc.* **75**, 271.
Kamada, T. (1940). *Proc. Imp. Acad. (Tokyo)* **16**, 24.
Kamada, T., and Kinosita, H. (1940). *Proc. Imp. Acad. (Tokyo)* **16**, 125.
Kamiya, N. (1939). *Ber. Deut Botan Ges.* **57**, 231.
Karlsson, U., Mommaerts, W. F. H. M., and Sjöstrand, F. S. (1962). *Abstr. Am. Soc. Cell Biol., 2nd Meeting, San Francisco, 1962* p. 88.
Kempner, E. S., and Miller, J. H. (1965). *Biochemistry* **4**, 2735.
Khawkine, W. (1887). *Ann. Sci. Nat. Botan. Biol. Vegetale* [7] **1**, 319.
Kirk, J. T. O., and Juniper, B. E. (1964). *J. Roy. Microscop. Soc.* **82**, 205.
Kitching, J. A. (1939). *J. Cellular Comp. Physiol.* **14**, 219.
Klebs, G. (1885). *Untersuch. Botan Inst., Tubingen* **1**, 1.
Kogan, A. B., and Tikhonova, N. A. (1965). *Biophysics* **10**, 322.
Korschikov, A. (1923). *Arch. Soc. Russe Protistol.* **2**, 195.
Kostir, W. J. (1952). *Proc. Soc. Protozoologists* **3**, 5.
Kotyk, A. (1962). *Folia Microbiol.* **7**, 109.

Krichenbauer, H. (1938). *Arch. Protistenk.* **90**, 88.

Krijgsman, B. J. (1925). *Arch. Protistenk.* **52**, 478.

Krinsky, N. I., and Goldsmith, T. H. (1960). *Arch. Biochem. Biophys.* **91**, 271.

Kuhne, W. (1864). "Untersuchungen über das Protoplasma und die Contractilitat." Engelmann, Leipzig.

Kuznicki, L. (1963). *Acta Protozool.* **1**, 301.

Lackey, J. B. (1934). *Biol. Bull.* **67**, 145.

Lackey, J. B. (1938). *Public Health. Rept. (U.S.)* **53**, 1499.

Lackey, J. B. (1939). *Public Health. Rept. (U.S.)* **54**, 740.

Lee, J. W. (1954a). *Physiol. Zool.* **27**, 272.

Lee, J. W. (1954b). *Physiol. Zool.* **27**, 275.

Leedale, G. F. (1958). *Arch. Mikrobiol.* **32**, 32.

Leedale, G. F. (1959a). *J. Protozool.* **6**, Suppl., 16.

Leedale, G. F. (1959b). *Biol. Bull.* **116**, 162.

Leedale, G. F. (1964). *Brit. Phycologists Bull.* **2**, 291.

Leedale, G. F. (1967). "The Euglenoid Flagellates", Prentice-Hall, Englewood Cliffs, New Jersey.

Leedale, G. F., Meeuse, B. J. D., and Pringsheim, E. G. (1965a). *Arch. Mikrobiol.* **50**, 68.

Leedale, G. F., Meeuse, B. J. D., and Pringsheim, E. G. (1965b). *Arch. Mikrobiol.* **50**, 133.

Lefèvre, M. (1931). *Trav. crypt. dédiés à L. Mangin* p. 343. Paris.

Levedahl, B. H. (1966). *Exptl. Cell Res.* **44**, 393.

Levedahl, B. H. (1967). *Expt. Cell Res.* **48**, 125.

Levine, L. (1960). *Anat. Record* **138**, 364.

Ling, G. N. (1962). "A Physical Theory of the Living State: The Association-Induction Hypothesis," 660 pp. Blaisdell, New York.

Lockwood, S. (1884). *J. Micrographie* **8**, 220–221.

Loefer, J. B. (1935a). *Arch. Protistenk.* **84**, 456.

Loefer, J. B. (1935b). *Arch. Protistenk.* **85**, 209.

Loefer, J. B. (1967). Personal communication.

Loefer, J. B., and Mattney, T. S. (1952). *Physiol. Zool.* **25**, 276.

Loefer, J. B., and Mefferd, R. B., Jr (1952). *Am. Naturalist* **86**, 325.

Loewy, A. G. (1950). *J. Cellular Comp. Physiol.* **35**, 151.

Lowndes, A. G. (1936). *Nature* **155**, 210.

Lowndes, A. G. (1941). *Proc. Zool. Soc. London* **A111**, 111.

Lowndes, A. G. (1943). *Proc. Zool. Soc. London* **A113**, 99.

Lowndes, A. G. (1944). *Proc. Zool. Soc. London* **A114**, 325.

Lowndes, A. G. (1947). *Sci. Progr.* **35**, 61.

Lozina-Lozinskii, L. K., and Zear, E. I. (1963). *Tsitologiya* **5**, 263.

Ludwig, H. F., Oswald, W. J., Gotaas, H. B., and Lynch, V. H. (1951). *Sewage Ind. Wastes* **23**, 1337.

Lynch, V. H., and Calvin, M. (1953). *Ann. N. Y. Acad. Sci.* **56**, 890.

Machin, K. E. (1958). *J. Exptl. Biol.*, **35**, 796.

Machin, K. E. (1963). *Proc. Roy. Soc.* **B158**, 88.

Machin, K. E. (1965). *Excerpta Med. Intern. Congr. Ser.* **91**, 111.

Mahenda, Z., Vora, R., Wolken, J. J. and Ahn, K. S. (1967). *J. Protozool.*, **14** Suppl., 17.

Mainx, F. (1928). *Arch. Protistenk* **60**, 305.

Mandeville, S. E., Crespi, H. L., and Katz, J. J. (1964). *Science* **146**, 769.

Marsland, D. A. (1964). *In* "Primitive Motile Systems in Cell Biology" (R. D. Allen and N. Kamiya, eds.), p. 173. Academic Press, New York.

Mast, S. O. (1917). *J. Exptl. Zool.* **22**, 471.
Mast, S. O. (1928). *Arch. Protistenk.* **60**, 197.
Mast, S. O. (1938). *Biol. Rev.* **13**, 186.
Mast, S. O. (1941). *In* "Protozoa in Biological Research" (G. N. Calkins and F. M. Summers, eds.), p. 271. Columbia Univ. Press, New York.
Mast, S. O., and Gover, M. (1922). *Biol. Bull.* **43**, 203.
Mast, S. O., and Johnson, L. P. (1932). *Z. Vergleich. Physiol.* **16**, 252.
Metzner, P. (1920). *Biol. Zentr.* **40**, 49.
Mignot, J.-P. (1965). *Protistologica* **1**, 5.
Mignot, J.-P. (1966). *Protistologica* **2**, 51.
Mitchell, J. B. (1928). *Trans. Am. Microscop. Soc.* **47**, 29.
Mohri, H. (1964). *Biol. Bull.* **127**, 181.
Nägeli, C. (1860). *Beitr. Wiss. Botan.* **2**.
Nelson, L. (1962). *Biol. Bull.* **123**, 468.
Nelson, L. (1966). *Biol. Bull.* **130**, 378.
Nelson, L., and Plowman, K. (1963). *Abstr. Proc. Biophys. Soc., 7th Ann. Meeting, New York, 1963* MD4.
Nettleton, R. M., Jr., Mefferd, R. B., Jr., and Loefer, J. B. (1953). *Am. Naturalist*, **87**, 117.
Newton, B. A., and Kerridge, D. (1965). *Symp. Soc. Gen. Microbiol.* **15**, 220.
Okajima, A., and Kinosita, H. (1966). *Comp. Biochem. Physiol.* **19**, 155.
Oliphant, J. F. (1938). *Physiol. Zool.* **11**, 19.
Oliphant, J. F. (1942). *Physiol. Zool.* **15**, 443.
Oltmanns, F. (1917). *Z. Botan.* **9**, 257.
Owen, H. M. (1947). *Trans. Am. Microscop. Soc.* **66**, 50.
Padilla, G. M. (1960). Ph.D. Thesis, UCLA Library, Los Angeles, California.
Padilla, G. M., and James, T. W. (1960). *Exptl. Cell. Res.* **20**, 401.
Palmer, J. D. (1967). *Nat. Hist.* **76(2)**, 60.
Palmer, J. D., and Round, F. E. (1965). *J. Marine Biol. Assoc. U.K.* **45**, 567.
Panner, C. J., and Honig, C. R. (1966). *Federation Proc.* **25**, 331.
Parducz, B. (1957). *Acta Biol. Acad. Sci. Hung.* **8**, 219.
Pease, D. C. (1963). *J. Cell. Biol.* **18**, 313.
Perini, F., Kamen, M. D., and Schiff, J. A. (1964a). *Biochim. Biophys. Acta* **88**, 74.
Perini, F., Schiff, J. A., and Kamen, M. D. (1964b). *Biochim. Biophys. Acta* **88**, 91.
Perty, M. (1852). "Zur Kenntnis kleinster Lebensformen nach Bau, Funktionen, Systematik, mit Spezialverzeichnis in der Schweiz beobachtet." Verlag von Jent und Reineor. Bern, Switzerland.
Petersen, J. B. (1929). *Botan Tidsskr.* **40**, 373.
Phillips, D. M. (1966). *J. Cell. Biol.*, **31**, 635.
Pigon, A. (1947). *Bull. Intern. Acad. Polon. Sci. Ser. B* **2**, 1946. 111.
Pitelka, D. R. (1963). "Electron-microscopic Structure of Protozoa," 267 pp. Pergamon, Oxford.
Pitelka, D. R., and Child, F. M. (1964). *In* "Biochemistry and Physiology of Protozoa" (S. H. Hutner, ed.), Vol. III, p. 131. Academic Press, New York.
Pitelka, D. R., and Schooley, C. N. (1949). *Univ. Calif. (Berkeley) Publ. Zool.* **53**, 377.
Pitelka, D. R., and Schooley, C. N. (1955). *Univ. Calif. (Berkeley) Publ. Zool.* **61**, 79.
Pitelka, D. R., and Schooley, C. N. (1957). *J. Protozool.* **4**, Suppl., 10.
Pittendrigh, C. S. (1960). *Cold Spring Harbor Symp. Quant. Biol.* **25**, 159.
Platt, J. R. (1961). *Science* **133**, 1766.
Pochmann, A. (1953). *Planta* **42**, 478.

Pogo, A. O., and Arce, A. (1964). *Exptl. Cell Res.* **36**, 390.

Pohl, R. (1948). *Z. Naturforsch.* **3b**, 367.

Postma, N. (1959). *Proc. Intern. Congr. Zool.* **15**, 523.

Pringsheim, E. G. (1937). *Cytologia (Tokyo)* Fujii-Jubilaumsband, pp. 234–255.

Pringsheim, E. G. (1948a). *New Phytologist.* **47**, 52.

Pringsheim, E. G. (1948b). *Biol. Rev.* **23**, 46.

Pringsheim, E. G., and Hovasse, R. (1950). *Arch. Zool. Exptl. Gen.* **86**, 499.

Pütter, A. (1900). *Arch. Anat. Physiol., Physiol Abt. Suppl.* **1900**, 243.

Ratcliffe, H. (1927). *Biol. Bull.* **53**, 109.

Reger, J. F., and Beams, H. W. (1954). *Proc. Iowa Acad. Sci.* **61**, 593.

Renaud, F. L., Rowe, A. J. & Gibbons, I. R. (1966). *J. Cell Biol.* **33**, 92A.

Rikmenspoel, R. (1962). *In* "Spermatozoan Motility" (D. W. Bishop, ed.), p. 31. Am. Assoc. Advance. Sci., Washington, D. C.

Rikmenspoel, R. (1965a). *Abstr. Biophys. Soc., 9th Ann. Meeting, San Francisco, 1965* p. 149.

Rikmenspoel, R. (1965b). *Biophys. J.* **5**, 365.

Rikmenspoel, R. (1965c). *Exptl. Cell Res.* **37**, 312.

Rikmenspoel, R. (1966). *Abstr. Biophys. Soc., 10th Ann. Meeting, Boston, Massachusetts, 1966* p. 119.

Robbins, W. J. (1952). *Bull. Torrey Botan. Club* **79**, 107.

Rostokowska, J. (1964). *Acta Protozool.* **2**, 91.

Rothschild, L. (1962). *In* "Spermatozoan Motility," (D. W. Bishop, ed.) pp. 13–29, Am. Assoc. Advance. Sci., Washington, D. C.

Round, F. E., and Palmer, J. D. (1966). *J. Marine Biol. Assoc. U.K.* **46**, 191.

Rybinsky, S. B., and Zyrnkina, L. M. (1935). *Arch. Protistenk.* **85**, 334.

Sachs, J. (1876). *Flora* **34**.

Schaudinn, F. (1899). *Arch. Ges. Physiol.* **77**, 29–44.

Scherbaum, O. H., and Loefer, J. B. (1964). *In* "Biochemistry and Physiology of Protozoa" (S. H. Hutner, ed.), Vol. III, p. 9. Academic Press, New York.

Schröder, V. N. (1927). *Proc. Congr. Zool. Anat., 2nd, USSR.* pp. 167–168.

Schultze, M. (1865). *Arch. Mikroskop. Anat.* **1**, 376.

Schwarz, F. (1884). *Ber. Deut. Botan. Ges.* **2**, 51.

Scourfield, D. J. (1909). *J. Quekett Microscop Club* [2] **10**, 357–366, 425.

Sears, D. F., and Gittleson, S. M. (1964). *J. Protozool.* **11**, 538.

Seravin, L. N. (1961). *Biokhimiya* **26**, 160.

Seravin, L. N. (1962). *Tsitologiya* **4**, 545.

Shamos, M. H., and Lavine, L. S. (1967). *Nature* **213**, 267.

Sharp, R. G. (1914). *Univ. Calif. (Berkeley) Publ. Zool.* **13**, 43.

Shelanski, M., Weisenberg, R., and Taylor, E. W. (1967). *Abstr. Biophys. Soc., 11th Ann. Meeting. Houston, Texas, 1967* p. 121.

Silvester, N. R., and Holwill, M. E. J. (1965). *Nature* **205**, 665.

Sleigh, M. A. (1962). "The Biology of Cilia and Flagella," 242 pp. Pergamon, Oxford.

Sokoloff, D. (1933). *Ann. Inst. Biol. (Mex.)* **4**, 147.

Sokoloff, D. (1935a). *Ann. Inst. Biol. (Mex.)* **6**, 71.

Sokoloff, D. (1935b). *Ann. Inst. Biol. (Mex.)* **6**, 189.

Sollberger, A. (1965). "Biological Rhythm Research." Elsevier, Amsterdam. (46 pp.)

Sommer, J. R. (1965). *J. Cell Biol.* **24**, 253.

Sommer, J. R., and Blum, J. J. (1965). *J. Cell Biol.* **24**, 235.

Stein, F. Ritter von (1878). "Der Organismus der Infusionsthiere...," Section III: Die Naturgeschichte der Flagellaten oder Geisselinfusorien, 154 pp. Engelmann, Leipzig.

Steinbach, H. B., and Dunham, P. B. (1961). *Biol. Bull.* **120**, 411.

Steinbach, H. B., and Dunham, P. B. (1962). *In* "Spermatozoan Motility" (D. W. Bishop, ed.) pp. 55–57. Am. Assoc. Advance Sci., Washington, D. C.

Stillman, I. M., and Binstock, L. (1967). *Abstr. Biophys. Soc., 11th Ann. Meeting, Houston, Texas, 1967* p. 20.

Strother, G. K., and Wolken, J. J. (1960). *Nature* **188**, 601.

Strother, G. K., and Wolken, J. J. (1961). *J. Protozool.* **8**, 261.

Sundaresan, B. B., Bovee, E. C., Wilson, D. E., and Lackey, J. B. (1965). *J. Water Pollution Control Federation* **37**, 1536.

Swann, W. F. G., and del Rosario, C. (1931). *J. Franklin Inst.* **211**, 303.

Swann, W. F. G., and del Rosario, C. (1932). *J. Franklin Inst.* **213**, 549.

Szabados, M. (1936). *Acta Biol. Szeged.* **4**, 49.

Szent-Györgyi, A. (1960a). "Introduction to Submolecular Biology," 135 pp. Academic Press, New York.

Szent-Györgyi, A. (1960b). *In* "Structure and Function of Muscle" (G. Bourne, ed.), Vol. III, p. 445. Academic Press, New York.

Taylor, C. V. (1920). *Univ. Calif. (Berkeley) Publ. Zool.* **19**, 403.

Taylor, C. V. (1941). *In* "Protozoa in Biological Research" (G. N. Calkins and F. E. Summers, eds.), p. 191. Columbia Univ. Press, New York.

Taylor, G. (1951). *Proc. Roy. Soc.* **A209**, 447.

Taylor, G. (1952a). *Proc. Roy. Soc.* **A211**, 225.

Taylor, G. (1952b) *Phil. Trans. Roy. Soc. London, Ser. A* **214**, 158.

Tchakhotine, S. (1936a). *Ann. Protistol.* **5**, 1.

Tchakhotine, S. (1936b). *Compt. Rend. Soc. Biol.* **12**, 1162.

Teixeira-Pinto, A. A., Nejelski, L. L., Jr., Cutler, J. L., and Heller, J. H. (1960). *Exptl. Cell Res.* **20**, 548.

Tibbs, J. (1957). *Biochim. Biophys. Acta* **23**, 275.

Ueda, K. (1958). *Cytologia (Tokyo)* **23**, 56.

Uhlehla, V. (1911). *Biol. Zentr.* **31**, 657.

Vavra, J., and Aaronson, S. (1962). *J. Protozool.* **9** (Suppl.), 28.

Verworn, M. (1889). "Psycho-physiologische Protistenstudien: Experimentelle Untersuchungen," 218 pp. Fischer, Jena.

Verworn, M. (1890). *Arch. Ges. Physiol.* **48**, 149.

Verworn, M. (1899). "General Physiology; An Outline of the Science of Life," 615 pp. Macmillan, New York.

Vlk, W. (1938). *Arch. Protistenk.* **90**, 448.

von Dach, H. (1943). *Ohio J. Sci.* **43**, 47.

Wager, H. (1900). *J. Linnean Soc. (London), Zool.* **27**, 463.

Wager, H. (1911). *Phil. Trans. Roy. Soc. London, Ser. B* **201**, 333.

Walker, P. J., and Walker, J. C. (1963). *J. Protozool.* **10** (Suppl.), 32.

Walne, P. L., and Arnott, H. J. (1966). *J. Phycol.* **2**, (Suppl.), 5.

Wenrich, D. H. (1924). *Biol. Bull.* **47**, 149.

Wichterman, R. (1952). *Trans. Am. Microscop. Soc.* **71**, 303.

Wichterman, R. (1955). *Biol. Bull.* **109**, 371.

Wichterman, R., Solomon, H., and Figge, F. H. J. (1958). *Biol. Bull.* **175**, 369.

Wiercinski, F. J. (1955). *Protoplasmatologia* **2** (B/2/c), 1.

Wildeman, E. de (1893-4). *Bull. Soc. Belge Microscop.* **20**, 245.

Willcock, E. G. (1904). *J. Physiol. (London)* **30**, 449.

Wilson, B. W., Buetow, D. E., Jahn, T. L., and Levedahl, B. H. (1959). *Exptl. Cell Res.* **18**, 454.

Wolken, J. J. (1956). *J. Protozool.* **3**, 211.

Wolken, J. J. (1961). "Euglena," 173 pp. Rutgers Univ. Press, New Brunswick, New Jersey.
Wolken, J. J. (1967). "Euglena," 2nd Edition, Appleton, New York.
Wolken, J. J., and Palade, G. E. (1953). *Ann. N. Y. Acad. Sci.* **56**, 873.
Wolken, J. J., and Shin, E. (1958). *J. Protozool.* **5**, 39.
Worley, L. G. (1934). *Proc. Natl. Acad. Sci. U.S.* **19**, 323.
Yalow, R. S. (1959). *Abstr. Natl. Conf. Biophys.*, *Columbus Ohio*, *1957*, Yale Univ. Press,
    New Haven. p. 169.

CHAPTER 4

MORPHOLOGY AND ULTRASTRUCTURE OF *EUGLENA*

Dennis E. Buetow

I. Introduction

The morphology of members of the genus *Euglena* has been a subject of interest since the first description of a *Euglena* by Ehrenberg in 1830. Much excellent work was done by early microscopists, however, many problems of interpretation arose. The electron microscope has helped to resolve many of these problems and to eliminate much of the confusion resulting from numerous conflicting reports published by the early light microscopists. For a review of much of the early literature, the reader is referred to the excellent publications of Chadefaud (1937), Jahn (1946, 1951), Chu (1947), Pringsheim (1948, 1956), Gojdics (1953), Jane (1955), and Fritsch (1965).

It is useful to note here that the early microscopic literature is confusing especially with respect to cytoplasmic inclusions. Staining techniques and resolving power of microscopes did not permit clear distinction among cellular inclusions. For example, mitochondria, first called "chondriosomes" or known collectively as "the chondriome," were apparently sometimes confused with other cytoplasmic particulates and inclusions (Fritsch, 1965). P.A. Dangeard (1919, 1925) recognized three types of cellular inclusions other than mitochondria that he felt were confused under the name chondriosome: "plastidomes" which give rise to plastids, "spheromes" which produce fat and oil globules (Tchang, 1924), and "vacuomes" which develop into vacuoles. From at least the 1920's to the 1950's, however, the *Euglena* literature in most, but apparently not all, cases perpetuated the names chondriome, plastidome, and vacuome to mean, respectively, mitochondria, plastids, and vacuoles (including metachromatic granules or volutin) (Dangeard, 1928; Jahn, 1951; Pringsheim, 1956; Gibbs, 1960). More recently, the electron microscope has helped to show that the cytoplasmic organelles of *Euglena* are similar to those found in most cells. Thus, the older names are seldom used now, e.g., mitochondrion is perferred to chondriome.

The following account will attempt to relate, where possible, the recent electron microscope findings to the earlier light microscope findings. It should be remembered, however, that the light microscopists looked at many *Euglena* species, varieties, and strains, whereas electron microscopy has been used in *detailed* studies of only a few, in particular, *E. gracilis* (several strains), *E. spirogyra*, *E. viridis*, and the closely related *Astasia longa*.

Electron micrographs of a longitudinal section through a green *E. gracilis*, a transverse section through a streptomycin-bleached *E. gracilis*, and a general survey section through the central region of *E. spirogyra* are shown in Figs. 1–3. These figures and Fig. 1 of Chapter 1 should help orient the reader during the following account.

II. Size

The size of a *Euglena* cell ranges from $12 \times 15 \mu$ for *E. minuta* to $530 \times 40 \mu$ for *E. oxyuris* (Gojdics, 1953). Size is not necessarily a constant, however, and there can indeed be variation in size even among members of the *same* species. For example, strains of *E. spirogyra* Ehrenberg show the following lengths: 125μ for strain Dr, 150μ for strain L, and 100μ for strains B and F Leedale *et al.*, 1965a). *Clones* from each strain of *E. spirogyra* do show fairly constant dimensions, however, provided the measurements are always made on stretched, untwisted specimens. The phenomenon of size varieties among species of the class Euglenineae (Euglenophyceae) has been discussed

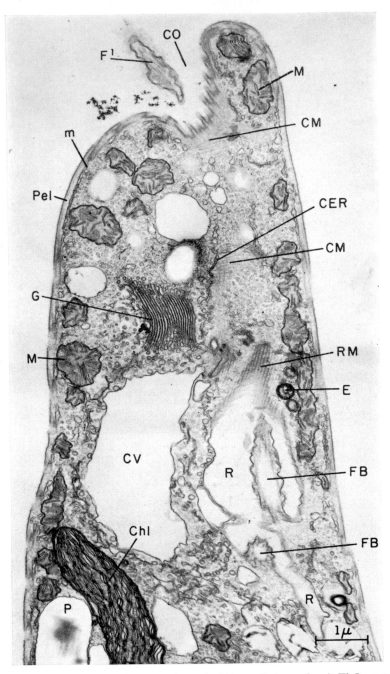

Fig. 1. Longitudinal section of anterior end of *E. gracilis* (green form). F¹, Locomotory flagellum; CO, canal opening (cytostome); CER, portion of canal sheath; CM, canal microtubules; RM, reservoir microtubules; FB, two flagellar bases; R, reservoir; E, eyespot; CV, contractile vacuole; G, Golgi; M, mitochondrion; m, mucus body; Pel, pellicle; Chl, chloroplast; P, paramylon granule. ×14,600. (From Leedale, 1966a.)

by Pringsheim (1948) who concluded that the same basic structural pattern can be shown in several forms apparently differing only in size.

The size of a *Euglena* cell can vary under a variety of conditions. The size of *E. gracilis* changes with the growth cycle (see Chapter 6, Section VII). Furthermore, during the final stages of cytokinesis in *E. gracilis*, violent metaboly occurs and occasionally unequal cleavage produces one large cell and one small cell (Leedale, 1959). Even when daughter cells are of the same size, organelle distribution may be unequal with one cell receiving

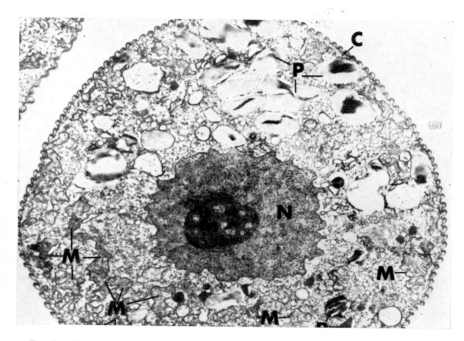

Fig. 2. Transverse section of *E. gracilis* var. *bacillaris* (strain SM-L1, streptomycin-bleached). C, Pellicle; P, paramylon; M, mitochondrion; N, nucleus; n, nucleolus. ×6580. (From Brandes *et al.*, 1964.)

most of the chloroplasts and paramylon. In a rare instance, one daughter cell receives both nuclei (Leedale, 1959). When *E. gracilis* is transferred to fresh medium, the cells show increased volumes (Corbett, 1957).

It is clear from the above discussion that size is only slightly helpful in differentiating species (see also Chapter 1). According to Gojdics (1953), the genus *Euglena*, classified on the basis of *average* size, shows the following groups: (1) those with lower limits under 25 μ, 8%; (2) those measuring 25–49 μ, 25%; (3) 50–74 μ, 25%; (4) 74–99 μ, 19%; (5) 100–124 μ, 10%; (6) 125–149 μ, 2%; (7) 150–174 μ, 6%; (8) over 175 μ, 5%.

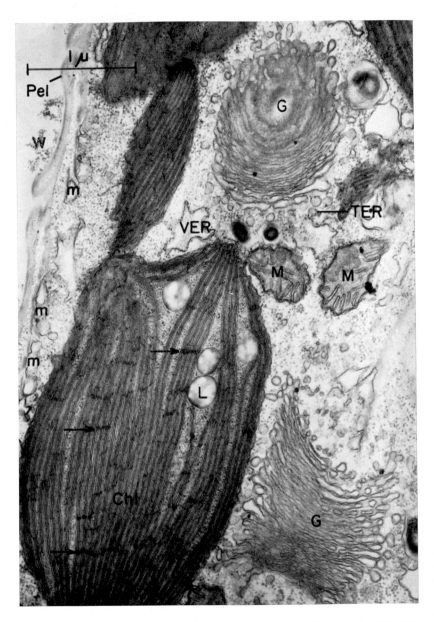

Fig. 3. A general survey cut through the central region of *E. spirogyra*. Chl, Chloroplast; G, Golgi bodies (two sectioned at different angles); L, lipid globule, M, mitochondrion; m, mucus body; Pel, pellicle; TER, tubular endoplasmic reticulum; VER, vesiculate endoplasmic reticulum; W, wart (or protuberance) on pellicle; arrows, "edges" of chloroplast discs. ×29,100. (From Leedale *et al.*, 1965a.)

III. Shape

The shape of *Euglena* species varies from almost spherical, e.g., *E. inflata*, to nearly cylindrical, e.g., *E. acus* and *E. deses* (Jahn, 1946; Gojdics, 1953; Pringsheim, 1948, 1956; see also Chapter 1, Figs. 3 and 4). The largest number are fusiform or spindle-shaped. The posterior end of the cell is drawn out into a cylindrical tail which may be bent to one side as in *E. acus*. Metaboly, or "euglenoid movement," occurs in most if not all species and, therefore, the overall shape of the cell may change at any time (see Chapter 3, Section IV). The size and dry weight of *Euglena* changes with culture age (Buetow and Levedahl, 1962; see also Chapter 6, Section VII) and the body shape of *E. deses* changes with the pH of the medium (see Chapter 1).

There is a spiral tendency in all *Euglena* species and in the class Euglenineae (Euglenophyceae) as a whole. Pellicular striae are spiral or run nearly straight (Gojdics, 1953). Even in those cases in which the spiral tendency is not obvious, the location of the eyespot and canal opening show that the *Euglena* cell never has radial symmetry (Pringsheim, 1956). Furthermore, in many species, the chloroplasts are spirally arranged, and the cells swim by rotating (Pringsheim, 1948). The term dorsal, i.e., the side on which the eyespot lies, and the term ventral have little meaning for *Euglena* (Gojdics, 1953) since the general conformation is a spiral one.

In cross section, most of the species appear nearly circular, although there are variations and some cross sections appear elliptical and some nearly flat. Others appear angular, e.g., *E. tripteris* which has three ridges spirally arranged on its surface and in cross section appears triangular to triradiate except at its anterior and posterior ends (Pringsheim, 1956).

Deformities of various kinds occur and can be induced. In *E. spirogyra*, for example, two-headed monsters are sometimes found in old cultures (Leedale *et al.*, 1965b). This is the result of incomplete division. In addition, structural abnormalities at the posterior end and occasionally at the anterior end of *E. spirogyra* can be induced with an overdose of iron or manganese.

IV. Color

The majority of *Euglena* species are green due to the presence of chlorophyll. In addition, there are many strains that are colorless, the result of "bleaching" by such agents as streptomycin and high temperature (Vol. II, Chapter 12). There are also about a dozen species colored by red granules (0.5 μ or less in diameter) collectively called hematochrome (Jahn, 1946; Gojdics, 1953; see also Chapter 2). When the granules are concentrated at its center the *Euglena* cell appears green because of the peripheral arrangement of the chloroplasts. When the granules are dispersed, the cell appears red. The same or a similar pigment is sometimes found in small quantities in species that do not appear red such as *E. gracilis, E. pisciformis, E. anabaena, E. klebsii, E. stellata*, and the related euglenoid, *Colacium vesiculum* (Hall, 1933; Jahn, 1946).

Reichenow (1910) and Mainx (1927) found that red *Euglena* cells lose hematochrome when placed in a medium rich in nitrogen. The amount of red pigment per cell apparently varies according to the medium; however, in nature, the red species are most common in pools rich in organic matter, especially at temperatures above 30°C (Jahn, 1946; Chu, 1947).

Most of the red species produce a gelatinous sheath (see Section V) around themselves, and some of these species go into a palmella stage (Gojdics, 1953). The sheath firms up the mass of cells which can then be moved on the water surface by the wind. A single-layered palmelloid mass can be quite dense, with one hundred to several hundred cells estimated per square millimeter (Naumann, 1915–1916).

The hematochrome granules disperse and the cells become red at temperatures above 30°C and, in very bright light, even down to 2°C (Johnson and Jahn, 1942). At the low temperature, the blue end of the visible spectrum is most effective in dispersing the granules. The alteration in color from green to red can occur quickly, e.g., within 5 minutes for *E. haematodes* (Gojdics, 1953), or relatively slowly, e.g., 40 minutes at 40°C or 2 hours at 30°C for *E. rubra* (Johnson and Jahn, 1942). The red species become abundant in certain habitats during periods of prolonged hot weather (Hardy, 1911; Kol, 1929; Johnson, 1939; Gojdics, 1953). In contrast, Szabados (1936) reported the absence of red pigment in *E. haematodes* collected in the fall. *Euglena rubra*, found in an Iowa lake in the summer, was observed to change from green to red at sunrise and from red to green at sunset (Johnson, 1939). As a whole, these observations suggest that the pigment protects the cells against heat and excessive light (Jahn, 1946; Gojdics, 1953). More experimentation is needed, however, to determine the nature of this "protection."

V. Cysts

Subpellicular mucus bodies (see Section VIII) eject a substance that forms a sheath around the *Euglena* cell, particularly upon irritation. Extensive gelatinous sheaths were first seen on *E. granulata* by Klebs (1883) and Bütschli (1883–1887) and have been described for at least 37 species of *Euglena* (Gojdics, 1953). Encysted stages have been observed for five other genera of Euglenaceae (Euglenidae) (Jahn, 1951). Hollande (1942) reported that a cyst resulted from the gradual condensation of the mucus sheath.

Subpellicular mucus bodies are reported to be both absent (Pringsheim, 1956; Gibbs, 1960) and present in *E. gracilis* (Leedale, 1966a; see also Fig. 1). In any case, some *E. gracilis* cells form a sheath (or cyst) (Jahn, 1951), especially in a medium containing ethanol at pH 5–7 (Buetow, unpublished). At a more acid pH, a sheath does not form. When the sheath does form, cells divide within it and form palmella (see Section V,A,2).

A. TYPES AND SHAPE OF CYSTS

The cyst wall in *Euglena* is composed of unidentified carbohydrate (Jahn, 1951; Fritsch, 1965). Cysts are usually spherical, but may be flask-shaped *(E. orientalis, E. tuba)* or pentagonal *(Distigma)*. Several types of cysts are formed by *Euglena* species (Gojdics, 1953), and Jahn (1951) has classified them as protective cysts; reproductive or division cysts; temporary, resting, or transitory cysts; and thin-walled cysts.

1. *Protective Cysts*

Protective cysts generally have a heavy, sometimes stratified, wall and contain one cell. Apparently they are protection against unfavorable conditions since they occur in the laboratory in old cultures. Such a cyst is also found in *E. deses* at 0–4°C. Some protective cysts have a thin wall and are joined to a tubular stalk (Gojdics, 1953). In *E. chlamydophora*, the wall is sculptured and has a cap that opens and releases the cell on excystment.

2. *Reproductive Cysts; Palmella Formation*

Reproductive cysts have a thin, elastic, permeable membrane that increases in diameter as the cells divide. Palmella formation occurs among large numbers of cells (Gojdics, 1953). In *E. gracilis* and *E. viridis*, the cyst may contain 32 to 64 cells (Jahn, 1951). The cells are nonflagellated.

3. *Temporary Cysts*

A temporary cyst has a wall that is impervious to water, but which contains a small pore. Jahn (1951) reports that the wall is thick, but Gojdics (1953) and Fritsch (1965) state that it is delicate and elastic. The cell remains flagellated and free to move about within the cavity. The cyst forms in response to strong sunlight and air temperatures of 10–12°C. Günther (1928) proposed that it protects against desication, however, such cysts are also formed on moist substrates by mud-dwelling species such as *E. terricola, E. geniculata,* and *E. sanguinea* (Jahn, 1951; Gojdics, 1953).

9

4. Thin-Walled Cysts

The last category described by Jahn (1951), the thin-walled cysts, have only one cell, e.g., *E. tuba*, which is not known to divide during encystment. Gojdics (1953) does not list thin-walled cysts as a separate category.

B. MULTIPLE ENCYSTMENT DURING THE LIFE CYCLE

Some species of *Euglena* may form more than one of these cysts during the life cycle. For example, *E. gracilis* can form a reproductive cyst or a temporary cyst (Jahn, 1951).

VI. Pellicle Complex

A. GENERAL OBSERVATIONS

The cell exterior of euglenoids was early observed to be differentiated into a pellicle (or periplast). This pellicle could be quite flexible so its shape changed considerably during euglenoid movement ("metaboly") (see Chapter 3), e.g., *E. gracilis*, *E. deses*, and *Distigma proteus;* it could be only slightly flexible so that metaboly was minimized, e.g., *E. trisulcata* and *E. tripteris;* or it could be rigid so that the cell had a fixed shape, e.g., *Phacus*, *Rhabdomonas*, and *Menoidium* (Jahn, 1951; Pringsheim, 1956; Pitelka, 1963; Mignot, 1965). Chu (1947) reported that the periplast of *E. acus*, *E. tripteris*, *E. oxyuris*, and *E. spirogyra* prossessed some degree of rigidity; however, no completely rigid periplast was observed. The role of the pellicle of *Euglena* in swimming and metaboly is covered in Chapter 3, Sections III,A,3,d and IV,A,B,C.

In early studies, it was reported that in some species the pellicle appeared smooth *or* finely striated, e.g., *Astasia torta* and *Distigma sennii;* in others it was striated and also had longitudinally arranged protuberances (also called punctae, granules, papillae, knobs, or warts) which were either simple *(E. spirogyra, Phacus monilata)* or complex in structure *(E. fusca)* (Jahn, 1951). Consequently, the cell exterior structure, particularly the system of protuberances, was widely although incorrectly (Section VI,A,2) used as a specific taxonomic characteristic.

1. Pellicular Striations

Chu (1947) doubted that the pellicle of any species of *Euglena* was really smooth and suggested that the pellicle was always adorned with striations. Electron microscopy supports Chu's suggestion since electron micrographs of the pellicle of those species of *Euglena* so far studied show it to be marked by longitudinal spiral striations—deep "grooves and crests" (Groupé, 1947; Pochmann, 1953; Reger and Beams, 1954; Beams and Anderson, 1961; Grimstone, 1961; Pitelka, 1963). The semiridged rings alternate with soft pliable membranes (Wolken and Palade, 1953; Reger and Beams, 1954; de Haller, 1959; Mignot, 1965).

Characteristic crests (ridges) of pellicular helices have been shown in electron micrographs of *E. spirogyra* (Leedale, 1964), *E. gracilis* (Wolken and Palade, 1953; Reger and Beams, 1954; Roth, 1958; Ueda, 1958; Frey-Wyssling and Mühlethaler, 1960, Gibbs, 1960, Sigesmund *et al.*, 1962; Lefort, 1963; Kirk and Juniper, 1964; Malkoff and Buetow, 1964; Brandes *et al.*, 1964; Sommer, 1965), *A. longa* (Frey-Wyssling and Mühlethaler, 1960; Ringo, 1963; Blum *et al.*, 1965; Sommer and Blum, 1965a), *Entosiphon sulcatum* (Mignot, 1963), *E. viridis* (de Haller, 1959, 1960), *Peranema trichophorum* (Roth, 1959), and *Trachelomonas* sp. (Ueda, 1960).

It appears from electron microscopy that the entire cell, including the pellicular "grooves and crests," is covered by a tripartite membrane called the plasmalemma. This tripartite plasmalemma has been resolved in *P. trichophorum* (Roth, 1959), *E. gracilis* (Frey-Wyssling and Mühlethaler, 1960; Gibbs, 1960; Kirk and Juniper, 1964; Sommer, 1965), *E. viridis* (de Haller, 1960), and *E. sulcatum* (Mignot, 1963).

2. Pellicular Protuberances

Early authors used the system of protuberances, when present, along the striae of the pellicle as a specific characteristic to define certain *Euglena*. An example is *E. spirogyra* (Fig. 4) as reviewed by Leedale *et al.* (1965b). The protuberances in *E. spirogyra* are more or less retangular structures with rounded corners. These protuberances cannot be used as taxonomic characteristics, however, since they often are not present and, in any case, vary according to the culturing conditions. They are well developed only in *E. spirogyra* cultured with soil (biphasic cultures), although not even in all of these, and are regularly absent in axenic monophasic cultures. When the protuberances are present in *E. spirogyra*, they contain ferric compounds (Klebs, 1883; Leedale *et al.*, 1965b) and may contain manganese which could produce the observed brown hue of the pellicle (Pringsheim, 1946). Pringsheim (1956) suggested that the protuberances consist mainly of inorganic gels. Mainx (1926) felt that the protuberances represent exudations through pores in the pellicle. Attempts to produce the protuberances in axenic monophasic cultures by the addition of ferric iron and manganese compounds to the nutrient solution were unsuccessful (Leedale *et al.*, 1965b). On the other hand, cells can be made to produce the protuberances in biphasic culture when ferric iron and manganese are readily available.

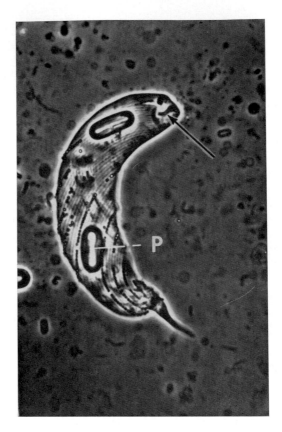

Fig. 4. Euglena spirogyra. Surface striations and a few rows of pellicular protuberances are shown. P, Paramylon granule; arrow, canal at anterior end of cell. Phase
contrast, ×1000. (From Leedale, 1964.)

B. Fine Structure

1. *Euglena spirogyra*

The pellicular structure of *E. spirogyra* Ehrenberg and *E. spirogyra* var.
fusca Klebs as seen under anoptral and phase contrast light microscopy
and under electron microscopy has recently been described (Leedale, 1964,
1966a; Leedale *et al.*, 1965a). The longitudinal helices of the pellicular
striations of *E. spirogyra* are seen in Fig. 5.

Electron microscopy confirms that the pellicle of *E. spirogyra* consists
of flat strips of material that pass along the cell in a helical manner. The
entire surface of the cell (including the main curve of the pellicular strips
and the articulations as described below) is covered by the plasmalemma

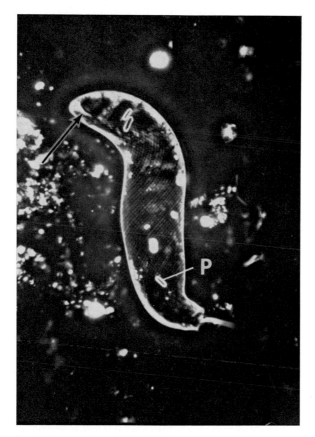

Fig. 5. Euglena spirogyra. Longitudinal helices of the pellicular striations are shown. The canal is lined with pellicular material (arrow). P, paramylon. Anoptral contrast, ×1000. (From Leedale, 1964.)

which is a continuous tripartite membrane, 80–100 Å thick (Fig. 6). Pellicular protuberances lie outside the plasmalemma on a layer of material which is also external to the membrane. The protuberances, when present, always occur on the curve of pellicular strip adjacent to its main peak (Fig. 6).

    a. Pellicle Structure. The pellicle is beneath the plasmalemma and consists of interlocking strips that pass helically along the cell. In cross section, each strip presents an elaborate structure. Along one edge of a strip is a groove facing inward while toward the other edge is a ridge that faces outward and articulates in the groove of the next strip (Figs. 7 and 8). The external surface of the pellicle curves between the groove and ridge. This curve turns back on itself, passes inward and bears two flanges, one

Fig. 6. Euglena spirogyra. Transverse section through pellicular strip covered by tripartite plasmalemma. Pellicular protuberance (wart), partially preserved, is seen outside the plasmalemma on curve of pellicular strip adjacent to its main peak. ×50,000. (From Leedale, 1964.)

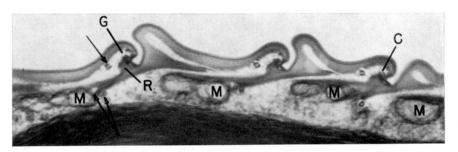

Fig. 7. Euglena spirogyra. Transverse section through pellicular strips. C, Mucus canal; G, groove; M, mucus body; R, ridge; arrows point to fibrils. ×41,500. (From Leedale, 1964.)

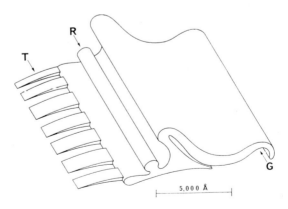

Fig. 8. Isometric projection of deduced shape of a pellicular strip of *E. spirogyra.* G, Grove; R, ridge; T, teeth; ribs are omitted. (From Leedale, 1964.)

of which doubles back beneath the main curve while the other extends beneath the groove of the adjacent strip and bears the ridge which articulates with the adjacent strip. Beyond this ridge the flange extends further beneath the next strip and bears a row of flat teeth on its edge. The teeth vary in width from 500 to 1000 Å and the spaces between them from 300 to 500 Å. In some sections, however, the ends of the teeth do not appear to be free but rather are joined to one another laterally (Leedale, 1964). The inner faces of the main curve and flanges of the pellicular strips carry "ribs" of material parallel to the teeth (Fig. 9). The ribs appear as lines about 80 Å wide and 120 Å apart.

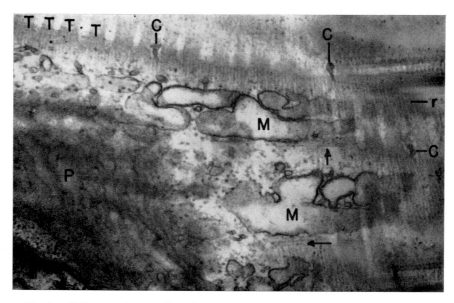

Fig. 9. Oblique surface section of pellicle complex of *E. spirogyra*. C, Mucus canals passing through teeth; M, mucus body; P, plastid; r, ribs; T, teeth; arrows, two pairs of fibrils. ×47,000. (From Leedale, 1964.)

b. Mucus Bodies. Large mucus bodies (see Section VIII) lie, in closely packed rows in the cytoplasm, immediately beyond the end of the ridge-bearing flange and beneath the rows of pellicular teeth (Figs. 7 and 10). Narrow canals from the mucus bodies pass between about every sixth pair of teeth (Figs. 7 and 9) to the groove and then to the exterior. In living cells, the pellicular strips move against one another, presumably as a result of the ridges sliding in the grooves (Leedale, 1964). A possible function of the mucus bodies is to supply lubricant to the articulations. The prominence of the mucus bodies and the amount of mucus substance produced under identical

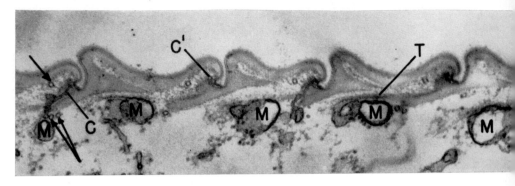

Fig. 10. Euglena spirogyra pellicle complex. C, Mucus body canal passing through a ridge; T, tooth of one pellicular strip cut through; C¹, mucus canal along a ridge; M, mucus body; arrows point to fibrils. ×50,000. (From Leedale, 1964.)

conditions are distinguishing features of different strains of *E. spirogyra* (Leedale *et al.*, 1965a).

    c. Microtubules. Regularly arranged fibrils or microtubules (200–250 Å in diameter) are associated with the pellicle. One is located at one side of the groove (Figs. 7 and 10). Another set of two is found at the end of the ridge-bearing flange, immediately adjacent to the mucus bodies. Oblique and longitudinal sections show that these are continuous fibrils or microtubules of uniform diameter, passing helically around the cell parallel to the pellicular strips.

    d. Variable Number of Pellicular Strips per Cell. The cell has 35 to 45 pellicular strips in the midregion. Therefore, the number of strips is not constant (Leedale, 1964; Leedale *et al.*, 1965a). Sections toward the posterior end of the cell show far fewer strips, and the posterior spine consists of *only* two or three strips. Carbon/platinum surface replicas of *E. pisciformis* seem to resolve the problem in *E. spirogyra* (Figs. 11 and 12). The cell posterior carries striations that start as a whorl and bifurcate several times as they proceed toward the cell anterior so that the cell has 35 to 45 pellicular strips at midregion. Leedale (1964) reports that the progressive bifurcations were seen in all species of euglenoid flagellates he examined. Replicas of the anterior of the cell show that the pellicular strips curve over and continue into the canal (Fig. 13). The striations decrease in number within the canal as they do at the posterior end of the cell. Leedale *et al.* (1965a) conclude that the pellicle is a *single* unit of interlocking strips that fuse into one another at both ends of the cell. This description questions the conclusion of Pochmann (1953) that the euglenoid pellicle consists of independent

Fig. 11. Carbon/platinum replica of posterior end of *E. pisciformis*. The pellicular striations bifurcate several times as they proceed toward the cell anterior. ×15,000. (From Leedale, 1964.)

coaxial closed loops (*Intergyren*) joined together by other loops or gyres (*Gyren*).

   e. Anterior and Posterior Ends of the Cell. The spine at the posterior of the cell consists of solid pellicular material. At the anterior end of the cell, the canal is lined with the same material (Fig. 5). Pellicular material ends at the base of the canal, however, and the reservoir is lined by the cell membrane only (Leedale, 1966a).

2. *Euglena gracilis*

   The pellicle complex of *E. gracilis* var. *bacillaris* (strain SM-L1, streptomycin-bleached) is composed of (1) the cell membrane (plasmalemma),

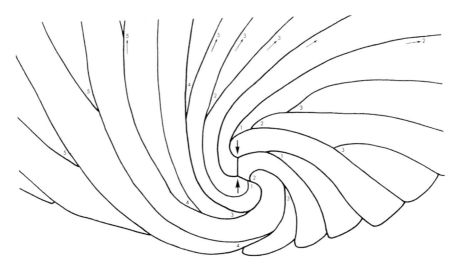

Fig. 12. Interpretation of Fig. 11 with respect to a short bilateral axis (arrows). Four basic striations bifurcate from one to five times. ×40,000. (From Leedale, 1964.)

Fig. 13. Carbon/platinum replica of the anterior end of *E. pisciformis*. Some pellicular striations bifurcate at the point where they curve over to continue into the canal. ×8700. (From Leedale, 1964.)

(2) ridges (each with a notch on one side) and grooves, (3) four fibrils (or microtubules), and (4) a subpellicular tubule of the endoplasmic reticulum (Sommer, 1965).

a. Plasmalemma. The cell membrane (plasmalemma) is a triple-layered structure of unit membrane configuration. The membrane is continuous and covers the ridges and the grooves, the gullet, the reservoir, and the flagella. In many cells, the membrane is covered by a granular material (Fig. 14).

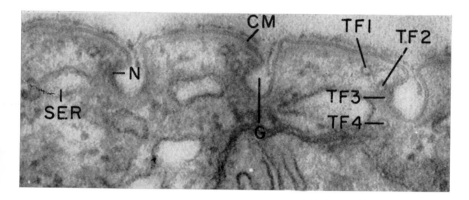

Fig. 14. Pellicle complex of *E. gracilis* var. *bacillaris* (strain SM-L1). G, Groove; CM, cell membrane; N, notch; SER, subpellicular endoplasmic reticulum; TF1,2,3,4, fibrils (or microtubules). ×107,000. (From Sommer, 1965.)

b. Ridges and Grooves; Fibrils. The pellicular ridges show a notch on one side. The "overhang" of the pellicular ridge points toward the posterior of the cell in *E. gracilis* (Kirk and Juniper, 1964) as it does in *A. longa* (Sommer and Blum, 1965a). The number of ridges per cell varies from 32 to 46 according to Kirk and Juniper (1964). They report that, starting at the posterior of the cell, the ridges bifurcate several times as is also the case in *E. spirogyra* (see Section VI,B,1,d). In contrast, Blum *et al.* (1965) report that the number of ridges is invariant (see Section VI,D).

Within the ridge next to the notch are three to five, usually four, fibrils about 250 Å in diameter (Fig. 14). Two of these are always close together with one touching the cell membrane. This account differs from that of Roth (1958) who did not find such fibrils below the ridges of the pellicle of *E. gracilis* and from that of Ringo (1963) who did not find such fibrils in *A. longa*. However, Gibbs (1960) and Pitelka (1963) reported two or three such fibrils in their electron micrograph preparations of *E. gracilis*.

The ridges and grooves of *E. gracilis* strain SM-L1 (Sommer, 1965) are of fairly constant dimensions. The ridges vary in size in the region of the cytostome, however, where alternating tall and small ridges are seen, and in dividing cells in which alternating tall and small ridges cover the entire cell.

In the cytostome itself, some of the small ridges appear to be very shallow, but their presence is emphasized by the four fibrils (microtubules) that are always present. The ridges virtually disappear, however, at the entrance to

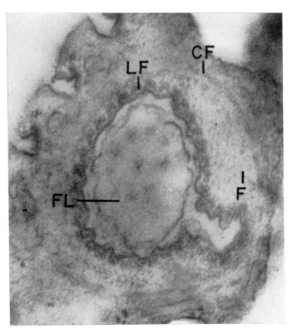

Fig. 15. Transverse section through the canal of *E. gracilis* var. *bacillaris* (strain SM-L1). CF, Circular fibrils; LF, longitudinal fibrils; F, filamentous matrix; FL, flagellum. ×65,000. (From Sommer, 1965.)

the canal (gullet). The fibrils extend further in pairs in single file along the canal just beneath the cell membrane (see longitudinal fibrils, Fig. 15). The canal itself, between the cytostome and the reservoir, is surrounded by a single row (Fig. 15) of approximately 70 to 80 circular (perhaps helical) fibrils (see Section XIII,A).

c. Subpellicular Endoplasmic Reticulum. A characteristic tubule of the endoplasmic reticulum was always present throughout the length of each pellicle complex in *E. gracilis* strain SM-L1 (Sommer, 1965) in agreement with the report of de Haller (1959) on *E. viridis*. The tubules of the subpellicular endoplasmic reticulum were always topographically associated with the notch of the pellicle complex. They also had connections with tubules of adjacent ridges as well as with the endoplasmic reticulum deeper in the cell. In green *E. gracilis* (both strain Z and var. *bacillaris*), the subpellicular endoplasmic reticulum tubules range up to 130 mμ in size (Gibbs, 1960). These tubules appear identical to endoplasmic reticulum tubules of similar size elsewhere in the cell.

d. The Pellicle during External Carbon Deprivation. When cultures of *E. gracilis* are deprived of a carbon source, the cells enter a prolonged stationary, postmitotic phase, and a whole culture will remain viable up to 13 days. During this time, the cells round up, become smaller, and undergo a wide variety of degradative changes (Blum and Buetow, 1963). In spite of these changes, the pellicle retains its characteristic ridged shape for at least 13 days (Malkoff and Buetow, 1964; Brandes *et al.*, 1964).

3. *Pellicle Complex Size Measurements*

Gibbs (1960) has given size measurements for components of the pellicle complex of *E. gracilis* strain Z. Ridges are of variable widths, 0.2–0.4 μ. The pellicle consists of an outer dense layer about 90 Å, a light space about 80 Å wide and an inner dense layer 60–80 Å wide. The outer dense layer (the plasmalemma) of the pellicle can be resolved into two 30-Å membranes separated by a 30-Å space. The pellicle of *E. viridis* is composed of three layers with a combined width of about 300 Å (de Haller, 1959). Reger and Beams (1954) earlier reported that the pellicle of *Euglena* was composed of two membranes: an outer ridged one about 500 Å thick and an underlying relatively smooth one about 300 Å thick.

The ridges of *A. longa*, being 0.3–0.4 μ wide (Ringo, 1963), are similar in size to the ridges of *Euglena*. The "cell wall" of *A. longa* is thinner than that of *Euglena*, being on the order of 100 Å. No separate "cell wall" and "cytoplasmic membrane" were resolved in *Astasia* (Ringo, 1963).

C. DUPLICATION OF PELLICLE COMPLEXES

1. *Euglena spirogyra*

If *E. spirogyra* is burst open along one striation (Fig. 16), the empty pellicle can be flattened out to show that the striations (i.e., edges of the pellicular strips) are arranged in pairs of pairs. On the basis of this patterning, Leedale *et al.* (1965a) suggested that growth of the pellicle is accomplished

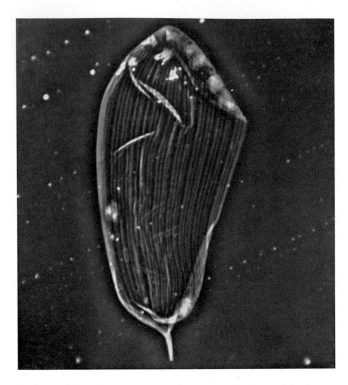

Fig. 16.   Pellicle of *E. spirogyra* opened along one striation and flattened out to show the internal surface. Anoptral contrast, × 1000. (From Leedale *et al.*, 1965a.)

by the laying down of new strips between old ones rather than by the division of existing strips. This is the case in *A. longa* (see below) and apparently also in *E. gracilis* (see Fig. 50 of Chapter 5).

2. *Astasia longa*

Sommer and Blum (1964, 1965a) have described pellicular changes during cell division in synchronized cultures of *A. longa*.

a. The Predivision Cell. In the predivision cell, the surface of the canal (gullet) (see Section XI,A) is convoluted into 18 primordial pellicle complexes (Figs. 17 and 19) which have no noticeable endoplasmic reticulum tubules (the presumptive complexes at the lip of the cytostome do, however). There are nine large ridges alternating with nine smaller ridges. Two tubular fibrils close to the lumen are found in each ridge and two tubular fibrils are found between each ridge; therefore, a total of 72 fibrils (or microtubules) are present (Figs. 17 and 18). An outer fibril (Fig. 19) is associated with the pair of fibrils found between the small and large ridges, and this outer fibril designates the future point of origin of 18 additional pellicle complexes more distal in the canal (Fig. 20). There are 18 large and 18 small pellicle complexes in the region proximal to the cytostome, but only 33 have been preserved in Fig. 21. There are three or four, usually four, tubular fibrils (or microtubules) in each pellicle complex and these are continuous with most, if not all, of the microtubules of the canal (see Section XIII,A). The 36 adult pellicle complexes are found at the anterior end of the lip of the cytostome. These adult pellicle complexes contain a tubule of the endoplasmic reticulum. The pellicle complexes leave the cytostome and proceed posteriorly in the form of a left-handed or β-helix. Near the posterior vortex, the complexes fuse once and then once again, reducing the number about three-fourths, a process similar to that seen in *Euglena* (Groupé, 1947; Pochmann, 1953; Pitelka, 1963; Kirk and Juniper, 1964; Leedale, 1964).

    b. Formation and Development of New Pellicle Complexes. Duplication of the pellicle complex takes place prior to cytokinesis (Sommer and Blum, 1965a) in *A. longa* and occurs at the lip of the cytostome (Fig. 22) where presumptive pellicle complexes arise from each of the adult complexes. Each daughter receives 18 new pellicle complexes alternating with 18 adult ones from the 72 present in the precytokinesis cell (Fig. 23). The 18 small (presumptive) complexes increase in size as the daughter cell grows and become indistinguishable from the original 18 large (adult) ones.

    The presumptive pellicle complexes of the dividing *Astasia* cell grow from the lip of the cytostome and between each adult complex toward the posterior vortex. This finding eliminates the possibility that the pellicle complexes duplicate by some process of longitudinal splitting. Duplication of the tubular fibrils (or microtubules) always associated with the forming, growing, and adult pellicle complexes apparently also takes place at the lip of the cytostome since it has never been observed in the canal or low cytostome. Duplication of the subpellicular endoplasmic reticulum tubule also takes place at the lip of the cytostome.

D. Number of Pellicle Complexes As a Taxonomic Characteristic

    Euglena gracilis is reported to have 20 large pellicle complexes with 20 small ones in the cytostome. This leads to a complement of 40 in the adult cell (Blum *et al.*, 1965). The same result is obtained, as pointed out by Sommer and Blum (1965a), if direct counts are made on the electron micrographs of *E. viridis* published by de Haller (1959). *Astasia longa* has 36 such complexes. Therefore, Sommer and Blum (1965a) suggest that the number of

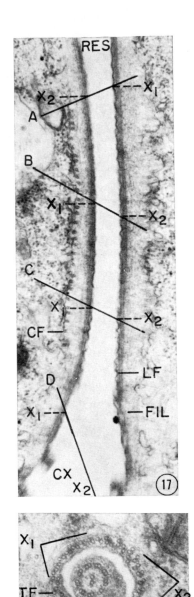

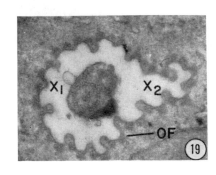

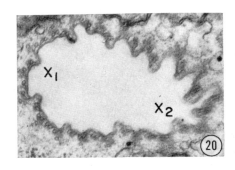

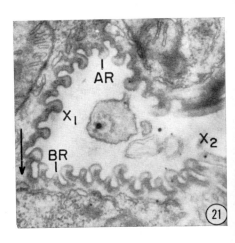

pellicle complexes is a genetically determined taxonomic characteristic of the family Euglenidae (Euglenaceae) and is not numerically related to the taxonomically invariant nine outer fibrils of flagella as suggested earlier by James (1963) for *A. longa*. Leedale (1964) also feels there is no numerical relation as suggested by James (1963). However, Leedale (1964) also finds a *variable* number (35 to 45) of pellicle complexes in *E. spirogyra* cells. In addition, Kirk and Juniper (1964) report a variable number of pellicle complexes in *E. gracilis*. These last-mentioned findings are in opposition to the taxonomic conclusion of Sommer and Blum (1965a).

E. DIRECTION OF THE PELLICULAR HELIX

The direction of the pellicular helix is not random. Gojdics (1953) stated that all species of *Euglena* have a left-handed helix (β-helix) on the pellicle with the exception of *E. oxyuris* (and perhaps two other species) which has a right-handed helix.

The pellicular striations of *E. spirogyra* trace a left-handed helix in a majority of the cells, but a right-handed helix was seen in 5–30% of the cells in any one culture (Leedale, 1964). In cultures of *E. spirogyra* var. *fusca* all the cells examined had a left-handed helix (Leedale, 1964). James

---

Fig. 17. Longitudinal section through canal (gullet) and cytostome region of *A. longa*. The lines labeled A–D are the approximate planes of section corresponding to Figs. 18–21. X_1 and X_2 indicate the approximate points in Figs. 18–21 at which the longitudinal fibrils are cut. CF, Circular tubular fibrils; CX, cytostome; FIL, filamentous matrix; LF, longitudinal tubular fibril; RES, reservoir. The canal is defined by the extent of the circular fibrils (CF). $\times 25{,}730$. (From Sommer and Blum, 1965a.)

Fig. 18. Transverse section of canal close to reservoir, corresponding to plane A of Fig. 17. At X_1, the tubular fibrils (TF) are arranged circumferentially; at X_2, they are skewed. Outer fibrils are absent (compare Fig. 19). $\times 41{,}500$. (From Sommer and Blum, 1965a.)

Fig. 19. Section through canal corresponding to plane B of Fig. 17. Nine large and nine small primordial pellicle complexes are partially elevated. Future pellicle complexes are indicated by the location of an outer fibril (OF). $\times 28{,}640$. (From Sommer and Blum, 1965a.)

Fig. 20. Section through canal corresponding to plane C of Fig. 17. The primordial pellicle complexes at X_2 are now presumptive ones. $\times 42{,}750$. (From Sommer and Blum, 1965a.)

Fig. 21. Canal-cytostome junction. There are now 18 large and 18 small pellicle complexes, although because of the plane of section (corresponds to plane D of Fig. 17), only 33 are seen. Most pellicle complexes have four tubular fibrils as well as tubules of the subpellicular endoplasmic reticulum. Additional tubular fibrils (arrow) are developing probably preparatory to replication of the pellicle prior to cytokinesis. AR, Adult ridge; BR, presumptive ridge. $\times 40{,}260$. (From Sommer and Blum, 1965a.)

(1963) originally reported a right-handed helix for the closely related *A. longa*, but later corrected this (quoted by Leedale, 1964) in agreement with the finding of Leedale (1964) that all cells in cultures of *A. longa* also have a left-handed helix.

No study is available to explain the predominance of the left-handed helix in cells of the genus *Euglena* and the genus *Astasia*. The variability

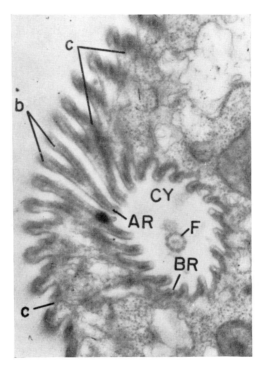

Fig. 22. Distal end of the cytostome of a dividing *A. longa* cell. Eighteen large pellicle complexes alternate with 18 small ones. Some small ones have reached adult proportions (b), thus completing the adult complement of 36. Adjacent to each of these, another presumptive pellicle complex arises (c), resulting in the 72 complexes of the precytokinesis cell (cf. Fig. 23). AR, Adult ridge; BR, presumptive ridge; CY, cytostome; F, flagellum. ×27,000. (From Sommer and Blum, 1965a.)

of this feature in *E. spirogyra* remains puzzling particularly if events at cell division are considered. Since the plane of cleavage follows the helix of the dividing cell, one would expect each daughter cell to have the same direction of helix as the parent cell. Leedale (1964), however, reports clonal culture experiments showing that a mixture of cells with left-handed helices and cells with right-handed helices arise from a single cell.

F. Cell Symmetry

Pochmann (1953) suggested that the euglenoid cell is fundamentally bilateral and that the helical symmetry is secondarily imposed. As Leedale (1964) points out, an alternative view is that the bilateral symmetry is imposed on an underlying helical symmetry. Picken (1960) summarizes the existing evidence that the mass of cytoplasm in euglenoid flagellates may also have a three-dimensional helical order. This suggestion implies a more intimate relation between form and ultrastructure than a mere imposition of the helical form (Leedale, 1964). Cell surface replicas of *Euglena* (Groupé, 1947;

Fig. 23. Transverse section through a dividing *A. longa* cell. There are 36 pellicle complexes (18 adult, 18 presumptive) in the cytostome and 72 outside the cell (36 adult, 36 presumptive). ×10,000. (From Sommer and Blum, 1965a.)

Pitelka, 1963; Kirk and Juniper, 1964; Leedale, 1964) show bifurcations and suggest that pellicular striations originate from a point, but the interpretations are many (Leedale, 1964).

Groupé (1947) and Pochmann (1953), observing *Euglena*, and James (1963), observing *Astasia*, have shown that the anterior ends of the as yet incompletely divided cell rotate around their longitudinal axes. As the parent cell is cleaved, the emerging daughter cells are simultaneously sutured, probably along the groove of one pellicle complex. The mechanism is not known. A further complication is the finding of Leedale (1964) that cells with left-handed pellicular helices and cells with right-handed pellicular helices from a single *E. spirogyra* cell.

The imposition of cellular symmetry in *Euglena* remains an unsolved problem.

G. PELLICULAR ACID PHOSPHATASE ACTIVITY

When *E. gracilis* is deprived of orthophosphate, the cells induce an acid phosphatase enzyme (Price, 1962; Blum, 1965). Cells grown in normal orthophosphate medium have constitutive acid phosphatase activity mainly in the Golgi elements (Brandes *et al.*, 1964; Sommer and Blum, 1965b), but also around the paramylon bodies and in the perireservoir region (Sommer and Blum, 1965b), and in some of the mitochondria (Brandes *et al.*, 1964). Cells with an induced acid phosphatase have constitutive activity as described, but also show strong induced acid phosphatase activity in the pellicle (Sommer and Blum, 1965b).

The induced activity is localized at the notch of each pellicle complex near a group of four microtubules (fibrils) and near the characteristic subpellicular endoplasmic reticulum tubule (Fig. 24). Large and small pellicle complexes alternate in the cytostome region (Section VI,C,2,a) and both show the induced enzyme activity. Localization of the induced enzyme activity near the cell surface suggests that it functions as a nonspecific hydrolase of organic phosphates during the period of external orthophosphate deprivation. Price and Bourke (1966) have, however, recently raised questions about the localization of the induced activity since preparations of *Euglena* in which the pellicles were removed still showed the activity.

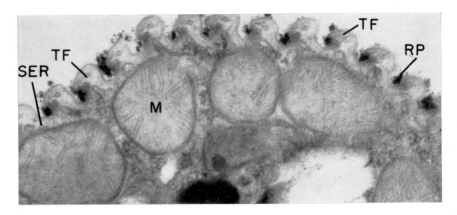

Fig. 24. Euglena gracilis var. *bacillaris* (strain SM-L1) showing induced acid phosphatase activity (RP) at the notch of each pellicle complex. TF, Tubular fibril; SER, subpellicular endoplasmic reticulum; M, mitochondrion. ×49,300. (From Sommer and Blum, 1965b.)

The closely related *A. longa* has low constitutive enzyme activity and does not show induced activity under conditions of orthophosphate deprivation (Blum *et al.*, 1965; Sommer and Blum, 1965b).

VII. Flagella

Through his expert microscopic technique, Fischer in 1894 was the first to describe the flagellum and its thin lateral filaments in a species of *Euglena*, viz., *E. viridis*. Fischer's general light microscope observations on the "Flimmergeisel," although not accepted by other early light microscopists, have been borne out by electron microscope studies (Manton, 1952). Between the time of Fischer's study and the first electron microscope study on the flagellum of *Euglena* by Brown (1945), a considerable amount of literature accumulated. These earlier studies of flagellar structure by light microscopists have been reviewed by Vlk (1938), Brown (1945), Pitelka (1949), Jahn (1951), Manton (1952), and Fritsch (1965). It is clear from this earlier work that considerable detail of the structure of flagella was revealed by careful cytological technique and light microscope observations; however, fine structure has been determined only through the use of the electron microscope.

A. GENERAL STRUCTURE AND LOCATION

The flagellum of four species of *Astasia* is reported to be bifurcated (Lackey, 1934; Hollande, 1942). There are two flagella in *Euglena*, however, not a bifurcated flagellum (Pringsheim, 1956). The larger one is used for locomotion; the importance of the smaller one is uncertain, but it may be a vestigial counterpart of one of the two emergent locomotory flagella found in the related genera, *Eutreptia*, *Eutreptiella*, and *Distigma* (Leedale, 1966a). In Pitelka's terminology (Pitelka, 1963), the larger flagellum is "stichoneme," i.e., bears a single row of fine hairs or mastigonemes (Section VII,B,3). In Round's terminology (1965), the larger flagellum is "pleuronematic," i.e., bears mastigonemes, and the smaller one is "acronematic," i.e., smooth.

Both flagella (see Chapter 1, Fig. 1) arise from conventional kinetosomes (blepharoplasts or basal bodies) located just below the reservoir membrane (Pringsheim, 1956; Beams and Anderson, 1961; Wolken, 1961; Pitelka, 1963). During cell division each kinetosome divides in two, forming the appropriate flagellum (Ratcliffe, 1927; Pringsheim, 1956). The basal 1-μ length of each flagellum, just above the kinetosome, is swollen. The central longitudinal fibrils (see below) arise distil to this swelling (Roth, 1958; Sleigh, 1962; Pitelka, 1963).

Pringsheim (1956) states that the second flagellum is generally thinner and much shorter than the locomotor flagellum, but is attached at its tip to the larger flagellum at about the level of the eyespot. Pitelka (1963) stated that no electron micrograph has yet indicated whether or not the two are attached. However, Sommer and Blum (1965a) more recently showed evidence that the flagella of *A. longa* do "fuse." In any case, near the presumed point of attachment, the locomotor flagellum bears a swelling that is the para-flagellar body (Section X,B).

In *E. spirogyra*, the locomotor flagellum is 40–60 μ long from base to tip, but in some individuals it may reach more than 80 μ (Leedale *et al.*, 1965a,b). The second flagellum is only 7–8 μ long. The two flagella are about the same in thickness.

The locomotor flagellum is of variable length even in the same population of *Euglena* and is easily lost in some *Euglena* under stress or irritation (Pringsheim, 1956; Leedale *et al.*, 1965a,b). Under improved conditions the flagellum regenerates. Flagellatory movement in some *E. gracilis* (strain Z) cells, however, continues even when they are exposed to high hydrostatic pressures, 13,000–15,000 psi, for 10 minutes (Byrne and Marsland, 1965). In contrast, euglenoid movement (metaboly) ceases at lower pressure, about 10,000 psi.

B. Fine Structure of Locomotory Flagellum

Under electron microscopy, the flagellum of *E. gracilis* and *E. spirogyra* consists of an electron-opaque axis (axoneme) surrounded by a lighter region which is thought to be a cytoplasmic sheath, and outside it a mat of lateral filaments on one or both sides of the flagellum (Fig. 25; see also Chapter 3, Fig. 5) (Houwink, 1951; Pitelka and Schooley, 1955; Leedale *et al.*, 1965a). A similar general structure was seen for flagella of other Euglenidae (Euglenaceae), i.e., *Peranema trichophorum* and *Rhabdomonas incurvum* (Pitelka and Schooley, 1955). The structure of the flagellum in relation to locomotion in *Euglena* in discussed in Chapter 3, Section III.

1. *Axoneme*

In unfixed specimens or specimens fixed with conventional fixatives, the axoneme of *E. gracilis* remains intact or nearly so (Pitelka and Schooley, 1955).

Packed longitudinal fibers are frequently evident (Pitelka and Schooley, 1955). The number of fibers present was first reported to be 9 (Pitelka, 1949; Houwink, 1951) and, later, to be 11 (Wolken and Palade, 1953; Pitelka and

Schooley, 1955; Pitelka, 1963). Wolken and Palade (1953) found 11 fibers, arranged as a circle of 9 about a central pair. Once it was found that the two central ones were more susceptible to decomposition than the peripheral nine (Manton, 1953) the problem was resolved. Eleven is the correct number (Fig. 29). This result puts the flagellum of *Euglena* in line with the apparently

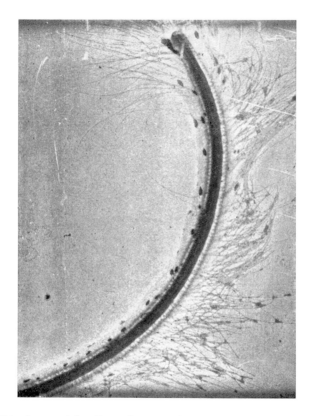

Fig. 25. Distal part of flagellum of *E. gracilis*. Dried without previous fixation; no shadowing. ×7500. (From Pitelka and Schooley, 1955.)

constant 11 major longitudinal fiber arrangement found in flagella and cilia of plants (Manton, 1952) and animals (Fawcett and Porter, 1954).

The width of the peripheral fibers appears to be 30–40 mμ (Pitelka and Schooley, 1955) although earlier estimates ranged from 350 to 600 Å (Pitelka, 1949; Houwink, 1951). Intact flagella show longitudinal fibers following straight parallel courses. No cross striations are found in these fibers. The fibers are surrounded by cytoplasm beyond which they do not protrude, even at the top of the flagellum (Houwink, 1951). Each fiber may consist of at least four subfibers.

The axoneme has an average width of 300–350 mμ (Pitelka and Schooley, 1955) and in several studies was frequently observed to separate into two parallel bundles (Brown, 1945; Pitelka, 1949; Houwink, 1951; Pitelka and Schooley, 1955; Ueda, 1958). Apparently, the correct structure is a single axoneme (Pitelka and Schooley, 1955; Pitelka, 1963).

2. Cytoplasmic Sheath

The sheath extends 100–200 mμ beyond the axoneme and appears as a ribbonlike structure with delicate cross-striations spaced 100–120 mμ apart (Houwink, 1951; Pitelka and Schooley, 1955). Houwink (1951) and Pitelka and Schooley (1955) say that the sheath does not form part of a coiled fiber and does not encircle the axoneme in helical form as originally suggested (Brown, 1945; Chen, 1950; Pringsheim and Hovasse, 1950). There is a fluid component, however, or a material that liquefies readily under certain conditions, within the cytoplasmic sheath region (Brown, 1945; Pitelka, 1949; Houwink, 1951; Pitelka and Schooley, 1955).

Brown (1945) and Houwink (1951) did not find an outer membrane around the sheath, but Gibbs (1960) reported a membrane (Fig. 29). Pitelka and Schooley (1955) report that the sheath appears to have a membranous boundary in about half of their preparations. They further report that the striated sheath material of E. gracilis is very sensitive to chemical changes and may be distorted easily during preparation for electron microscopy. Pitelka and Schooley (1955) then conclude that the sheath in E. gracilis and other euglenoids is "a material with a periodic chemical structure" which "forms a tube enclosing the axoneme." Sommer and Blum (1965a) report a membrane surrounding each flagellum in A. longa.

3. Lateral Filaments

The lateral filaments on the euglenoid flagellum are present in two distinct sizes (Houwink, 1951; Manton, 1952). The mastigonemes are the larger of the two and in E. gracilis have been reported to extend 3 μ (Houwink, 1951) to 4 μ or more (Pitelka and Schooley, 1955) from the surface of the sheath. They are very thin. Two estimates of their width have been made: 15–20 mμ according to Houwink (1951) and 10 mμ or less according to Pitelka and Schooley (1955). The mastigonemes frequently terminate in a short, very fine fibril (Houwink, 1951; Pitelka and Schooley, 1955). Usually, mastigonemes appear to arise in large numbers on only one side of the flagellar axis, generally on the convex side of any curvature (Brown, 1945; Chen, 1950;

Houwink, 1951; Pitelka and Schooley, 1955; Leedale *et al.*, 1965a). Occasionally, they are reported to be present for short distances on both sides of the axis (Pitelka, 1949; Brown, 1945; Pitelka and Schooley, 1955).

The smaller filaments form a mat (100–200 mμ thick in the case of *E. gracilis*) on *both* sides of the flagellum (see Chapter 3, Fig. 5; Pitelka and Schooley, 1955; Pitelka, 1963; Leedale *et al.*, 1965a). The few individual fibers that extend free from this mat may reach 1–2 μ from the surface of the sheath (Houwink, 1951; Pitelka and Schooley, 1955). These smaller filaments are not the mastigonemes. The distinction between the two types of filaments seems clear. There apparently are no intermediate sizes.

The mastigonemes are often observed to be wrapped smoothly and compactly around the flagellum when *E. gracilis* cells are fixed in OsO_4 buffered at a pH the same as or more alkaline than the growth medium (Pitelka and Schooley, 1955; Pitelka, 1963). Chen (1950) interpreted this result as fusion of "mucilagenous" mastigonemes as a result of centrifugation. Pitelka (1963) supports the view of Mainx (1927) and of Pringsheim and Hovasse (1950) that such an arrangement is the normal situation for the mastigonemes and would effectively increase the diameter of the surface of the flagellum. When pointing away from the flagellum, the mastigonemes seemingly could hinder the rapid movement of the flagellum. On the other hand, Leedale *et al.* (1965a) do not support this interpretation. Anoptral contrast microscopy of living *E. spirogyra* shows that the mastigonemes are fully extended (see also Chapter 3, Fig. 5). In any case, the mastigonemes are durable structures and are untouched by trypsin, freezing and thawing, and various chemical manipulations, all of which affect the structure of the axoneme or fragment the cytoplasmic sheath (Pitelka and Schooley, 1955).

C. Duplication of Flagella and Kinetosomes

Cultures of synchronously dividing *A. longa* have been used to determine the sequence of events in replication of kinetosomes (blepharoplasts, basal bodies) and flagella (Sommer and Blum, 1965a). Figures 26 and 27 show longitudinal and transverse sections through kinetosomes with their rootlets. By early prophase, four kinetosomes (Fig. 28) are seen. The two oval-shaped ones touch the reservoir whereas the two smaller newly replicated ones lie at an angle to the original ones. Each pair (presumably one large and one small kinetosome) migrate to opposite ends of the reservoir. The newly formed kinetosome in each pair then synthesizes a new flagellum. In *E. gracilis*, flagella replicate at metaphase (Chapter 5, Section III,A,1).

During mitosis, the nucleus of *Euglena* and *Astasia* moves close to the kinetosomes when they divide in two (Baker, 1926; Leedale, 1958; Leedale *et al.*, 1965a; Sommer and Blum, 1965a). Chromatin filaments often come in contact with the nuclear membrane at its closest point to the kinetosomes (Leedale, 1958; Sommer and Blum, 1965a). Sommer and Blum (1965a) propose that the nucleus triggers the duplication of the kinetosomes. The mechanism is unknown.

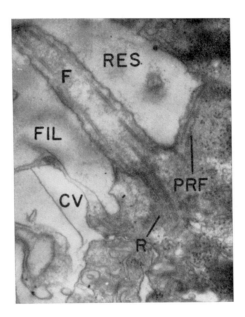

Fig. 26. Longitudinal section through a kinetosome of *A. longa*. R, Kinetosome rootlet; F, flagellum; FIL, filamentous matrix; CV, contractile vacuole; PRF, perireservoir tubular fibrils; RES, reservoir. ×37,000. (From Sommer and Blum, 1965a.)

VIII. Mucus Bodies

Mucus bodies are small spherical or elongate subpellicular bodies which stain vitally with neutral red and cresyl blue and are present in many species of *Euglena*, in *Peranema* and other colorless euglenoids (Jahn, 1951; Gojdics, 1953). They have at various times been mistakenly identified as part of the "vacuome" which develops into vacuoles (Hall, 1931), as probably volutin (Baker, 1933; Patten and Beams, 1936), and as trichite bodies, i.e., homologs of trichocysts in ciliates (Chadefaud, 1934, 1936). Other workers correctly identified these structures as "mucus bodies" in many *Euglena* species (P. Dangeard, 1928; P.A. Dangeard, 1901; de Haller, 1959; Hollande, 1942; Mainx, 1926).

Elongate or fusiform-shaped mucus bodies apparently contribute to the gelatinous membranous sheath (see Section V) formed by several *Euglena* species, particularly upon irritation (Jahn, 1951; Gojdics, 1953; Pringsheim, 1956). Cells within the sheath are nonmotile. Mucin, the gelatinous secretion of the mucus bodies, appears to be ejected upon irritation as shiny and thick threads that fuse perpendicularly in rows along the pellicular striae of the cells and so produce the sheath (Hollande, 1942; Gojdics, 1953; Pringsheim, 1956).

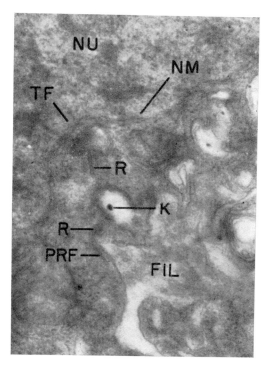

Fig. 27. Transverse section through a kinetosome (K) of *A. longa*. One rootlet (R) is close to the nucleus (NU). A parallel group of tubular fibrils (TF) lies just inside the nuclear membrane (NM) adjacent to the rootlet. Other symbols as in Fig. 26. ×43,000. (From Sommer and Blum, 1965a.)

Mucus bodies apparently have functions in addition to sheath formation since they also are found in species, e.g., *E. viridis*, *E. granulata*, that do not surround themselves with a sheath upon irritation (Gojdics, 1953; Pringsheim, 1956; de Haller, 1959). Leedale (1964) suggests that the mucus bodies of *E. spirogyra* supply a lubricant to the articulations of the pellicular striae (Section VI,B,1,b). An electron micrograph of a mucus body in *E. spirogyra* is shown in Figs. 3 and 31. In *E. gracilis*, the perinuclear space,

tubular endoplasmic reticulum (Section XVI,A,B) and the mucus bodies
form a continuous system (Chapter 5, Section II,C). A further description
of mucus bodies and a discussion of the chemical nature of mucin is given
in Chapter 3, Section V and in Vol. II, Chapter 7, Section II,B.

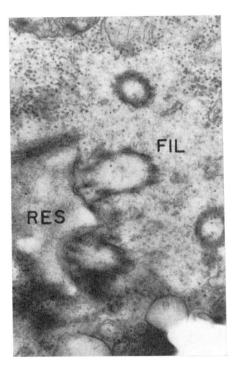

Fig. 28. Kinetosomes of *A. longa* after replication. The two larger kinetosomes are
oval in cross section and attached to the wall of the reservoir. The other two are smaller
and circular. Symbols as in Fig. 26. ×33,000. (From Sommer and Blum, 1965a.)

IX. Paramylon (Paramylum)

Paramylon is a carbohydrate storage compound found in *Euglena* and
in euglenoids in general (Jahn, 1946, 1951; Chu, 1947; Gojdics, 1953;
Pringsheim, 1956; Pitelka, 1963; Fritsch, 1965; Round, 1965). Paramylon
granules in *Euglena* are colorless polysaccharides which can be hydrolyzed
to *d*-glucose (Gottlieb 1850, Habermann 1874). In structure, paramylon
is a 1,3-β-glucan (Kreger and Meeuse, 1952; Clarke and Stone, 1960;
Leedale *et al.*, 1965b). Comparison of a dozen different paramylons from
fourteen Euglenaceae (Euglenidae) indicates a strong chemical uniformity

(Leedale *et al.*, 1965b). The grains can be distributed widely in the cytoplasm, e.g., *E. gracilis* var. *bacillaris;* can form cups over the pyrenoids, e.g., *E. gracilis* Klebs; can be massed together, e.g., *E. virdis;* or can be few, but large, and located in a fairly constant position, e.g., *E. spirogyra* (Gojdics, 1953).

Paramylon is formed as characteristic rods, discs, rings, etc. The kind of paramylon body produced by a *Euglena* is used as a taxonomic criterion, but the quantity produced is not (Gojdics, 1953; Round, 1965; see also Chapter 1). In *E. gracilis*, the granules increase in number in cells grown in the presence of excess carbon (acetate) (Buetow, 1967) and decrease in number in cells deprived of carbon (Padilla and Buetow, 1959; Blum and Buetow, 1963; Malkoff and Buetow, 1964; Brandes *et al.*, 1964). In *E. gracilis*, paramylon grains in the cytoplasm are found in cavities that lack a membrane (Gibbs, 1960; Brandes *et al.*, 1964). Sommer and Blum (1965b), using electron cytochemistry, noted constitutive acid phosphatase activity around paramylon granules.

The shape, size, structure, location, and composition of paramylon granules are treated in detail in Vol. II, Chapter 7 of this treatise.

X. Stigma (Eyespot) and Paraflagellar Body

A stigma (or eyespot) is found in all species of normally green *Euglena*. It is located on the surface of the reservoir next to its outlet into the canal and it faces an enlargement of the flagellum known as the paraflagellar body or photoreceptor (Gojdics, 1953; Wolken and Palade, 1953; Pringsheim, 1956; Gibbs, 1960; and Pitelka, 1963). Many colorless Euglenaceae (Euglenidae) also have an eyespot, but *A. longa* does not (Pringsheim, 1956). The presence of a stigma in *Euglena* has been recognized since the early observations of Ehrenberg (1838) on these cells, although it was thought to be a real eye and the nerve ganglion assumed to accompany it was actually described in a species of *Astasia*. The relation between structure and function of the stigma, paraflagellar body, and flagellum in regard to locomotion in *Euglena* is described in Chapter 3, Section III.

A. STIGMA

1. *Shape*

The stigma of *Euglena* is composed of orange-red droplets in a flat, ill-defined aggregation, generally shield-shaped and more or less curved, with an almost semicircular cross section that fits the rounding of the reservoir (Pringsheim, 1956).

2. *Dispersal of the Stigma*

In some *Euglena*, some of the droplets appear to be separated from the main mass or even scattered throughout the cytoplasm, particularly under unfavorable conditions (Chadefaud, 1937; Hollande, 1942). Hall and Jahn (1929a) found fragmentation of the mass in *Euglena* sp. during prophase. Gojdics (1934) reported that the eyespot broke up during division of *E. deses*, and Hollande (1942) reported a similar occurrence for *E. acus* and *E. mutabilis*. The eyespot may also occur in a dispersed state in cysts of *E. gracilis* (Hall and Jahn, 1929b). In extreme cases, several eyespots seem to exist and the "true" one can not be readily distinguished from the others (Pringsheim, 1956).

3. *A Distinct Cell-Organelle*

The question arose early whether or not the stigma was a distinct well-defined cell-organelle since evidence for a stroma or matrix of a nature different than the rest of the cytoplasm was scanty. In fact, if the droplets, including the carotenoids and the "oily substance," were removed, earlier workers did not observe or at least could not stain a clearcut substratum having the location and shape of an eyespot (Pringsheim, 1956). At least three lines of evidence, however, suggest that the stigma is a functionally sepcialized organelle, i.e., a self-reproducing, defined portion of the cytoplasm: (1) during cell division, the stigma replicates (Grassé, 1926; Baker, 1933; Jahn, 1951, see also Chapter 5, Section III,A,1), (2) the stigma in *Euglena* is completely independent from the chloroplasts and, therefore, differs from a similar structure found in all other flagellates and algae (Pringsheim, 1956; Pitelka, 1963; Leedale *et al.*, 1965a), and (3) in *E. gracilis*, at least, the stigma can be irreversibly lost through treatment of the cells by temperatures of 34–35°C or by streptomycin (Pringsheim and Pringsheim, 1952).

4. *Euglena gracilis*

In *E. gracilis*, the stigma granules appear as dense homogenous globules (Fig. 29) of apparent variable diameter, 150–660 mμ, with denser rims about 100 Å thick (Gibbs, 1960). Wolken and Palade (1953) and Wolken (1958) had previously reported that the stigma was about 2 μ in diameter and 3 μ in length and consisted of 30 to 50 granules which varied in size from 100 to 300 mμ. Ueda (1958) reported that the eyespot consisted of many spherical granules, 200 mμ in diameter, and that each granule was surrounded by a membrane. The granules appear to lie free in the cytoplasm in a loosely packed group (Gibbs, 1960) rather than existing as a closely packed hexagonally shaped group embedded in a matrix as earlier reported by Wolken (1956, 1958) and Wolken and Palade (1953). Between the stigma

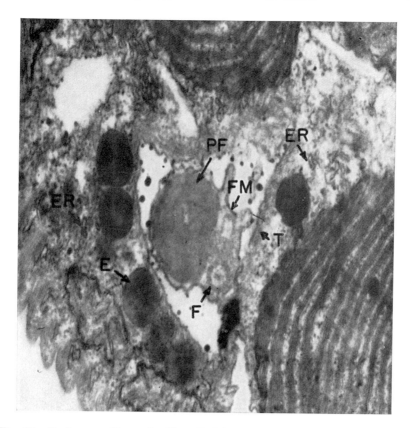

Fig. 29. Euglena gracilis var. *bacillaris*. E, Stigma (eyespot) granule; ER, endoplasmic reticulum; F, fibrils of flagellum; FM, flagellar membrane; PF, paraflagellar body; T, tubules underlying reservoir. ×34,000. (From Gibbs, 1960.)

granules and the limiting membrane of the reservoir (∼100 Å thick) lies a parallel array of tubules, 180–250 Å in diameter (Roth, 1958; de Haller, 1959; Gibbs, 1960; see also Section XIII,B).

Wolken (1956) estimated there were 5×10^5 to 1×10^6 pigment molecules in the eyespot. The eyespot granules of *E. gracilis* strain Z have been isolated by Batra and Tollin (1964). Under electron microscopy (Fig. 30), the sectioned granule appears to be a hollow sphere of varying size. The average outside diameter of the granule is about 400 mμ and the outer shell about 50 mμ thick.

5. *Euglena spirogyra*

The stigma of *E. spirogyra* is a flat plate of red droplets. The droplets may be packed into a compact body, 5–6 μ in diameter, which curves around

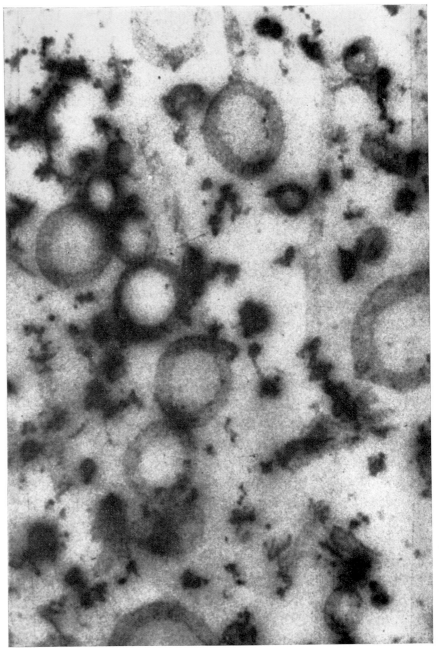

Fig. 30. Sectioned methacrylate pellet containing stigma granules from *E. gracilis* strain Z. ×50,000. (From Batra and Tollin, 1964.)

the neck of the reservoir on the side opposite the contractile vacuole (see Chapter 3, Fig. 6), or they may be less compactly arranged with some of the droplets separated from the main mass (Leedale *et al.*, 1965a). There is no indication of a cytoplasmic matrix and no common membrane uniting the droplets into an integral organelle in *E. spirogyra*.

6. *Loss of the Stigma*

Although heat and streptomycin, particularly under conditions allowing cell division (Mego and Buetow, 1966, 1967), permanently bleach *Euglena* (i.e., permanently affect the chloroplasts), variable results are obtained with the stigma. In some permanently heat-bleached clones of *E. gracilis*, the eyespot persists as long as the cells are illuminated (Pringsheim and Pringsheim, 1952). The action spectrum for preservation of the eyespot corresponds to the light absorbed by the carotenoids of the eyespot (Hutner and Provasoli, 1955). In these same clones, even as late as 6 months after the heat-bleaching treatment, with many subcultures intervening, the stigma was permanently lost once the cultures were deprived of light (Pringsheim and Pringsheim, 1952). Thus, in a majority of strains of heat-bleached *Euglena*, the stigma always disappeared in the dark but was preserved in the light. In a minority of heat-bleached strains, the stigma merely became smaller in the dark. The chloroplasts can be destroyed and the stigma preserved, but the opposite never occurs. Green cells without eyespots are not found (Pringsheim, 1956).

Streptomycin has a similar effect on the eyespot and chloroplasts (Pringsheim and Pringsheim, 1952). Once lost, the stigma, as well as the chloroplasts, does not reappear. However, heterotrophic nutrition and reproductive vigor are not diminished in these heat- and streptomycin-induced colorless strains.

7. *Pigments of Isolated Stigma*

The pigments of isolated eyespots of *E. gracilis* strain Z have been identified on the basis of their chromatographic and spectrophotometric behavior and partition coefficients (Batra and Tollin, 1964; see also Vol.II, Chapter 6, Section II, D). Lutein (or probably antheraxanthin) and cryptoxanthin comprise 83% of the pigments. β-Carotene and an unidentified component comprise the rest. Astaxanthin, an animal carotenoid, was reported present in the red *E. sanguinea* and *E. heliorubescens* (Tischer, 1936, 1941) and in the euglenoid *Trachelomonas volvocina* (Green, 1963), but has not been found in whole cells of *E. gracilis* (Goodwin and Jamikorn, 1954; Krinsky and Goldsmith, 1960) nor in isolated eyespots of *E. gracilis* strain Z (Batra and Tollin, 1964).

B. THE PARAFLAGELLAR BODY

1. Structure and Location

The paraflagellar body (sometimes called "flagellar swelling") of *Euglena* is an oval or slightly plano-corvex body found on the flagellum inside the reservoir at the level of the eyespot (Fig. 29; see also Chapter 3, Fig. 6) (Jahn, 1951; Pringsheim, 1956; Gibbs, 1960; Beams and Anderson, 1961; Pitelka, 1963; Leedale *et al.*, 1965a, Leedale, 1966a). In *E. gracilis* the largest one sectioned for electron microscopy measured 1.5 × 0.95 μ (Gibbs, 1960). Under electron microscopy, it differs from the globules of the stigma (Fig. 29) by its generally lower density, its very low density core, and its lack of a dense limiting rim (Gibbs, 1960). In *E. gracilis*, the paraflagellar body along with the stigma replicates at metaphase (Chapter 5, Section III,A,1). Chadefaud and Provasoli (1939) reported that the eyespot and parabasal body of *E. gracilis* var. *urophora* were connected by a protoplasmic bridge. This condition has not been shown in any other species or strain in the genus *Euglena*.

2. Loss of Paraflagellar Body

When many strains of *Euglena* are treated with heat or streptomycin, the paraflagellar body disappears along with the eyespot and chloroplasts (Pringsheim, 1956). Vavra (1956) found a strain of *E. gracilis*, however, in which the paraflagellar body remained when streptomycin treatment removed the eyespot and the chloroplasts.

3. As Photoreceptor

As early as 1882, Engelmann reported that when a shadow was passed over the body of *Euglena*, no response occurred until the stigma was reached. Mast (1911, 1917) confirmed this observation. Mast (1917) also recognized that the action spectrum for phototaxis was similar to the absorption spectrum of astaxanthin. Thus, astaxanthin was thought to be the photosensitive pigment. This pigment, however, although reported in the red *E. sanguinea*, has not been found in the eyespot of *E. gracilis*.

Wager (1900) discovered that the paraflagellar body was associated with the eyespot. Mast (1911, 1917) thought the paraflagellar body was part of the eyespot (as did Wager, 1900), but considered the paraflagellar body the photoreceptor. Although it has since become clear that the eyespot and the paraflagellar body are separate structures, support for the view that the latter is the true photoreceptor was given by Tchakhotine (1936) who applied UV to *E. gracilis* in localized spots and destroyed the stigma. Such cells swam from light to shade whereas those with intact stigma never did. The consensus of opinion seems to be that the stigma functions as a light-absorbing shield which can shade the paraflagellar body, the true photo-receptor (Hollande, 1942; Jahn, 1951; Pringsheim, 1956; Gibbs, 1960; Leedale *et al.*, 1965a; Leedale, 1966a). As the *Euglena* rotates, the stigma intermittently shades the paraflagellar body.

Wolken (1956, 1958) incorrectly reported that the eyespot and the flagellum of *E. gracilis* act as a *unit* in phototactic response and, thus, incorrectly considered the eyespot the principal light receptor (see Chapter 3, Sections II,G and III,A).

The fine structure of the paraflagellar body has not been resolved (Gibbs, 1960); therefore, it is not yet possible to compare it structurally with other known photoreceptors.

XI. Canal and Reservoir

Euglenoid flagellates have an anterior invagination of the cell from which the locomotory flagellum emerges (Pringsheim, 1956; Fritsch, 1965). This invagination consists of two parts, the canal (also called the "gullet" or the "cytopharynx") and an inner reservoir (see Chapter 1, Fig. 1). The anterior opening of the canal to the exterior is called the cytostome (Gojdics, 1953). An early description of the canal and reservoir was reported by Khawkine (1886) for *Astasia ocellata* and *E. viridis*.

The canal and reservoir are not concerned with ingestion of food particles and terms such as gullet and cytopharynx are probably best avoided (Prings-heim, 1948; Gojdics, 1953). The reservoir itself is permanently connected to the outside via the canal (Jahn, 1951), not sometimes closed as stated in the earlier literature. In the case of *E. spirogyra*, a cell which is 90–100 μ long and 10–20 μ wide, the canal in 10 μ long and 1 μ in diameter and the pyriform-shaped reservoir is 5–6 μ in diameter. (Leedale *et al.*, 1965a).

A. CANAL

In all *Euglena* species, the canal opening is subapical and located on the side of the cell opposite the eyespot (see Chapter 1, Fig. 1). In other euglenoids, e.g., *Eutreptia viridis* and *Astasia curvata*, the canal opening is located medially or nearly so (Pringsheim, 1956). In *E. spirogyra* and other *Euglena*, the anterior end of the cell forms a "lip," while the other side of the canal is shorter (Pringsheim, 1956; Leedale *et al.*, 1965a). In *E. spirogyra*, the canal is curved and almost forms an S-shape in some cases. Pellicular strips form a lining on the anterior part of the canal (Figs. 20, 21 and 31)

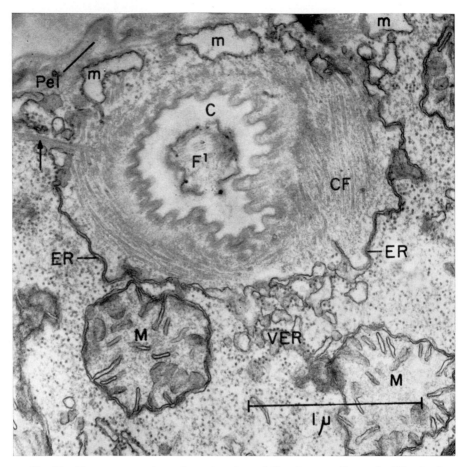

Fig. 31. Transverse section of canal region of *E. spirogyra.* C, canal; CF, circular microtubules cut in oblique longitudinal section; F¹, locomotory flagellum; M, mitochondrion; m, mucus bodies; Pel, pellicle; VER, vesiculate endoplasmic reticulum; ER, cylindrical sheath of endoplasmic reticulum; arrow, occasional groups of microtubules pass through sheath of ER. × 37,600. (From Leedale *et al.,* 1965a.)

of various *Euglena* species and of *A. longa;* however, these thin out and disappear near the reservoir (Pitelka, 1963; Ringo, 1963; Leedale *et al.,* 1965a; Sommer and Blum, 1965a). Only the tripartite plasmalemma (see Section VI,B,2a) covers the posterior region of the canal and the reservoir.

B. RESERVOIR

The reservoir of *Euglena* is a collecting chamber into which contractile vacuoles empty their contents (Pringsheim, 1956). The reservoir is connected to the surrounding medium by means of the narrow canal (see Section XI,A and Chapter 1, Fig. 1). The contractile vacuole (Section XII) is provided fluid by one or more secondary or accessory vacuoles which appear to extract fluid from the cytoplasm (Pringsheim, 1956). The arrangement of this "excretory" system is characteristic of the species and shape of the cell. In narrow forms, e.g., *E. acus*, the reservoir and contractile vacuole lie almost on the longitudinal axis of the cell. In broader forms, they are situated side by side. In *E. gracilis*, the reservoir divides at metaphase (Chapter 5, Section III,A,1).

XII. Contractile Vacuole

In *E. gracilis*, the contractile vacuole is limited by an 80–100 Å-thick membrane and is found close to the reservoir (Gibbs, 1960). A similarly located vacuole is present in *A. longa* (Ringo, 1963; Sommer and Blum, 1965a).

In the older literature the contractile vacuole was called the "secondary vacuole" (Jahn, 1951; Gojdics, 1953). It was thought that the canal and reservoir were not always permanently open to the outside as they are now thought to be. The supposed closed state of the canal and reservoir was called the "primary vacuole" and the contractile vacuole was thus called the "secondary vacuole."

The contractile vacuole of *Euglena* is provided with fluid by one or more accessory vacuoles which appear to extract fluid from the cytoplasm (Pringsheim, 1956). In *E. pseudoviridis* (Chadefaud, 1937), *E. deses* (Hyman, 1938), and *E. acus* (Hollande, 1942), the larger contractile vacuole discharges its contents into the reservoir by "melting" into the base of the reservoir. The small vacuoles move into the place the large one occupied, fuse, and form the next large contractile vacuole. Some earlier workers incorrectly reported that the accessory vacuoles were incorporated into the larger vacuole (Gojdics, 1953).

A recent study on *E. spirogyra* (Leedale *et al.*, 1965a) showed that the single contractile vacuole is delimited by a unit membrane (Fig. 32) and is formed at regular intervals adjacent to the reservoir on the side away from the eyespot. The contractile vacuole is formed by coalescence of continuously arising secondary vacuoles. When full it is 5 μ in diameter and distorts the reservoir. At the time of discharge, the membranes between the vacuole and the reservoir break down and the fluid from the vacuole empties into the reservoir. The reservoir quickly regains its shape, and a new contractile vacuole is formed from the secondary vacuoles. Immediately after the contractile vacuole discharges, the cytoplasmic region where a new vacuole will form contains collapsed secondary vacuoles and numerous spherical vesicles 1000 Å in diameter (Fig. 33). The latter are distinctive in that their walls have a well-defined alveolate pattern. These vesicles are found only in this region of the cytoplasm and are thought to be osmo-regulatory (Leedale *et al.*, 1965a). They may fuse with the secondary vacuoles, with the main contractile vacuole, or with the reservoir.

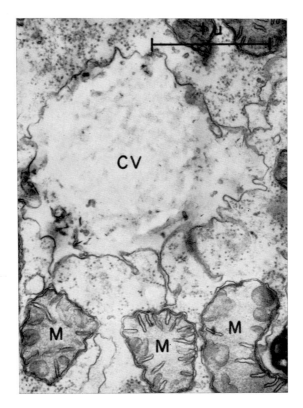

Fig. 32. Contractile vacuole (CV) of *E. spirogyra*, partly collapsed due to processing. M, Mitochondrion. ×30,000. (From Leedale *et al.*, 1965a.)

The contractile vacuole of *Euglena* pulsates rhythmically; however, its rate depends on the temperature (Gojdics, 1953). *Euglena deses* pulsates every 22 seconds at room temperature (Gojdics, 1934) and every 30 seconds at 42°C or 18°C (P. A. Dangeard, 1901); *E. ehrenbergii* pulsates every 22 seconds at 32°C (Klebs, 1883); *E. acus* pulsates every 23 seconds at 14°C (Hollande, 1942); and *E. spirogyra* about every 20 seconds at 20°C (Leedale *et al.*, 1965a).

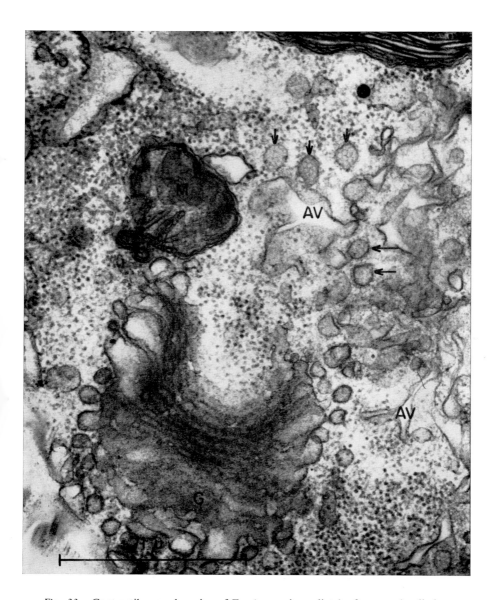

Fig. 33. Contractile vacuole region of *E. spirogyra* immediately after vacuolar discharge. AV, Collapsed accessory vacuoles; G, Golgi; M, mitochondrion. Shown also are numerous alveolate vesicles in surface view (vertical arrows) and section (horizontal arrows). ×47,000. (From Leedale *et al.*, 1965a.)

XIII. Microtubules

A. In Region of the Canal

Several layers of microtubules, 200–250 Å in diameter in *E. spirogyra* (Leedale *et al.*, 1965a), encircle the canal for most of its length. These microtubules (fibrils) are arranged in rows of four to six and form a flat helix around the canal. The microtubules lie in a region of cytoplasm which is separated from the general cytoplasm by a cylindrical sheath of endoplasmic reticulum and appears almost free of ribosomes (Figs. 31 and 43). Occasionally, groups of microtubules pass from the canal collar through the sheath and seem to link with microtubules beneath the pellicle (Fig. 31).

In *E. gracilis*, there also appears to be a single row of 70 to 80 circular or helical fibrils, or microtubules (Fig. 15) around the canal (Sommer, 1965). In *E. gracilis*, however, there also are longitudinal fibrils (Roth, 1958; Sommer, 1965) or microtubules arranged in pairs in single file around the canal just beneath the cell membrane (Fig. 15). These longitudinal fibrils are the same as, or least continuous with, the fibrils or microtubules found in the pellicular ridges (Section VI,B,2b). In some cells, the longitudinal fibrils are separated from the circular (helical) fibrils by a dense filamentous matrix (Fig. 15). In most cases, however, this matrix is especially prominent at the junction of the canal and cytostome.

B. In Region of the Reservoir

The broad collar of circular (or helical) microtubules narrows at the posterior end of the canal in *E. spirogyra*. The reservoir has only a single layer of identical structures underlying the plasmalemma (Leedale *et al.*, 1965a). At the base of the canal, the pitch of the microtubular helix apparently steepens with the result that the microtubules around the reservoir lie almost parallel to the long axis of the cell (Ueda, 1958; Pitelka, 1963; Leedale *et al.*, 1965a; Leedale 1966a). No circular microtubules (fibrils, filaments) are found around the bulbous portion of the reservoir (Roth, 1958). However, the longitudinal tubules (Section XIII,A) are found in the neck of the reservoir and all around its bulb (Pitelka, 1963).

C. In Region of Flagellar Bases (Kinetosomes; Blepharoplasts)

Fibrils (microtubules) in *Euglena* also run from the flagellar bases and connect with the extensive system of microtubules or fibrils surrounding the reservoir and lining the body surface (Roth, 1958; Grimstone, 1961; Sleigh, 1962).

A similar location of microtubules in the canal, reservoir, and flagellar base regions is seen in *A. longa* (Figs. 17–19, 21, 26, 27, and 28) (Sommer and Blum, 1965a).

D. FUNCTIONS OF CYTOPLASMIC MICROTUBULES

The function of the microtubules described above (Section XIII,A,B,C) is not known. Suggestions include the following: (1) contraction with the function of periodically closing the canal (Leedale *et al.*, 1965a), (2) a transporting function in the cell for the whole microtubular system (Leedale *et al.*, 1965a), (3) a function in metaboly (see Chapter 3, Section IV,B and C), and (4) a structural role, i.e., the microtubules form the "cytoskeleton" of the cell (Mignot, 1965).

E. MICROTUBULES IN THE NUCLEUS

An array of microtubules, consistent with the structure of a typical spindle apparatus, has been described within the nucleus of *A. longa* (Blum *et al.*, 1965). These microtubules appear to connect with a basal body. Similar microtubules have been described recently in *E. gracilis*, but they are not attached to basal bodies (Chapter 5, Section III,C,E).

XIV. Mitochondria

A. EARLY STUDIES; GENERAL OBSERVATIONS

The presence of mitochondria in *Euglena* has been long recognized, but many early investigators confused the mitochondria with many other small cell inclusions and lumped them all together under the name chondriosomes (Pringsheim, 1956; Fritsch, 1965). Some of the earlier papers refer to these chondriosomes collectively as the chondriome (Fritsch, 1965). Later, with the advent of improved microscopic and staining techniques, the term chondriome was reserved to denote only mitochondria (Pringsheim, 1956).

In an early report, Causey (1926) claimed an average of only 16 spherical mitochondria (range, 10 to 25) per *E. gracilis* Klebs cell, and further stated that there appeared to be a more or less definite ratio between the number of mitochondria and the volume of the cytoplasm. Early workers also thought they saw division of the mitochondria, but Baker (1933) found no evidence of this in *E. gracilis* Klebs. Although division of mitochondria in *Euglena* has not yet been demonstrated, recent work, including electron microscopy, does show, however, that the mitochondria of *Euglena* can vary in size, shape, structure, and number (Pringsheim and Hovasse, 1948; Wolken and Palade, 1953; Reger and Beams, 1954; Pringsheim, 1956; Gibbs, 1960; Siegemund *et al.*, 1962; Brandes *et al.*, 1964; Malkoff and Buetow, 1964; Leedale *et al.*, 1965a; Leedale, 1966a).

Numerous spherical and rod-shaped mitochondria are found throughout *Euglena* cells, but are most abundant just below the pellicle (Gibbs, 1960).

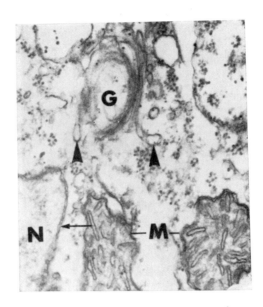

Fig. 34. *Euglena gracilis* var. *bacillaris* (strain SM-L1). Arrowheads, cisternae of Golgi (G) with saccular dilations (vesicles and vacuoles also seen); M, mitochondrion with "undulating" surface, cristae, and typical double outer membrane; N, nucleus with double membrane (arrow). ×48,000. (From Brandes *et al.*, 1964.)

This seemingly favored location for mitochondria is found in dark-adapted (Wolken and Palade, 1953; Leedale *et al.*, 1965a) and streptomycin-bleached *E. gracilis* (Brandes *et al.*, 1964; Malkoff and Buetow, 1964) as well as in light-grown *E. spirogyra* (Leedale *et al.*, 1965a). Electron micrographs reveal an irregular, in some cases "undulating," mitochondrial surface (Figs. 3, 32, and 34) in *E. gracilis* and *E. spirogyra* (Reger and Beams, 1954; Brandes *et al.*, 1964; Malkoff and Buetow, 1964; Leedale *et al.*, 1965a). In fact, the irregular surface appears to be characteristic of euglenoid mitochondria (Leedale *et al.*, 1965a).

B. *Euglena gracilis*; *Euglena spirogyra*

1. *Structure*

In *E. gracilis*, the mitochondrion is limited by two 60-Å membranes enclosing a 60-Å space (Gibbs, 1960). The inner membrane is infolded to form short cristae which extend 33–50% of the way across the mitochondrion (Wolken and Palade, 1953; Gibbs, 1960; Brandes *et al.*, 1964). Ueda (1958) reported cristae to be about 25 mμ thick with the inner membranes 7 mμ thick and the outer membrane about 11 mμ thick. Wolken (1961) reported the outer membrane to be 7–10 mμ thick.

The cristae in general are perpendicular to the long axis of the mitochondrion in *E. gracilis* (Wolken and Palade, 1953) and in *E. spirogyra* (Leedale *et al.*, 1965a). Other alignments are also seen (Wolken and Palade, 1953; Leedale *et al.*, 1965a), but these are thought to be due to the angle of sectioning during preparation for electron microscopy (Leedale *et al.*, 1965a).

The cristae of *E. spirogyra* mitochondria rarely meet or overlap, and the matrix of the mitochondrial lumen is granular and relatively large in volume.

Reger and Beams (1954) reported that *E. gracilis* mitochondria occasionally contain dense granules of unknown significance, and Brandes *et al.* (1964) reported that some mitochondria showed acid phosphatase activity under electron cytochemistry with the sites of the reaction invariably appearing as separate clusters whether in the cristae, the matrix, or both.

2. *Multiple Shapes*

Pringsheim (1956) noted that the mitochondria of *Euglena* as viewed under light microscopy appeared as either relatively long bands which seemed to fuse and branch to form a network or as short rounded shapes. These two types of mitochondria were found even in the same clone, but usually under different growth conditions. A recent example in support of this observation is the case of *E. spirogyra* in which two distinct states are seen for the mitochondria: one typical of cells during the dark period of a day–night culture cycle and one typical of cells during the light period (Leedale *et al.*, 1965a). In cells in the dark, mitochondria are discrete ovoid or elongated bodies, 0.5 μ wide and up to 10 μ long, and are dispersed throughout the cytoplasmic matrix. In cells in the light, the mitochondrial system typically appears as an interconnecting complex, a fine reticulum of threads generally less than 0.5 μ in width and confined to the region between the chloroplasts and the pellicle. At any time, although cells from one culture show all gradations between the two states, one state predominates. In the living cell, in both states, the mitochondria continually move and their configuration continually varies as they slowly fuse, branch, and change their shape.

These mitochondria show an undulating outer membrane and cristae that are constricted at their base.

C. ASTASIA LONGA

Ringo (1963) estimates 250 mitochondria per *A. longa* cell. These mitochondria may be round or rod-shaped, 0.4–0.8 μ in diameter and up to 2 μ in length. There are two mitochondrial membranes, each 60 Å wide, and separated from each other by a 50-Å space. Close inspection suggests an irregular membrane as in *Euglena*. The cristae, each about 150 Å wide, appear, at least in some cases, to run longitudinally rather than transversely.

D. ENVIRONMENTAL EFFECTS

Pringsheim and Hovasse (1948) stated that the chondriome of the naturally colorless *Astasia* as well as that of various colorless euglenas was more "developed" than that of green euglenas. There is a great increase in number of mitochondria under conditions in which chloroplast content is reduced, e.g., dark-adapted *E. gracilis* (Wolken and Palade, 1953), heat-produced colorless *E. spirogyra* (Leedale, 1966a), and a variety of mutant species of *Euglena* (Vol. II, Chapter 10, Section III,D). However, the general organizational pattern of the mitochondria of dark-and-light-adapted *E. gracilis* was the same (Wolken and Palade, 1953).

1. Stationary-Phase Cultures

Old stationary-phase cultures (9–12 days old) of *E. gracilis* strain Z show cells having mitochondria with structurally altered cristae (Siegesmund *et al.*, 1962).

2. Starved Cultures

When streptomycin-bleached *E. gracilis* cells are specifically starved of a carbon source (acetate), many of the mitochondria become encapsulated in membrane-bound cavities (Section XV,D,2) comparable to the cytolysomes (autophagic vacuoles) found in cells of higher organisms. These encapsulated mitochondria (Figs. 35 and 36) are digested away (Brandes *et al.*, 1964). The majority of the remaining mitochondria appear normal, though mainly distributed somewhat linearly along the periphery of the cells (Malkoff and Buetow, 1964; Brandes *et al.*, 1964). In a few cases, a few extremely long mitochondria are seen surrounding the periphery of the cell (Fig. 37) by 7 days of carbon starvation. When the carbon source is resupplied to these cells after 8 days, a surge in oxidative capacity follows, reaching prestarvation

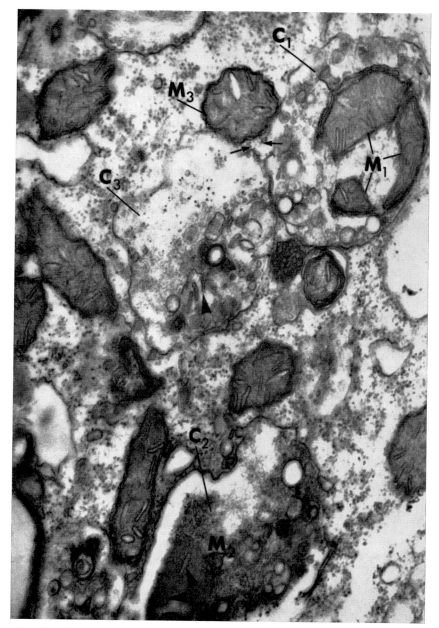

Fig. 35. *E. gracilis* var. *bacillaris* (strain SM-L1) starved of carbon for 2 days. Life cycle of cytolysomes depicted in a single cell. Three cytolysomes (C_1, C_2, C_3) show sequence of digestion of encapsulated mitochondria and other cytoplasmic elements leading to accumulation of low-opacity amorphous material. Mitochondria (M_1) are well preserved in C_1. In C_2, mitochondria (M_2) show marked alterations, but can be recognized by the cristae (arrowheads). In C_3, only a few cristae (arrowhead) remain. C_1 and C_3 are in the process of fusing (arrows) and a mitochondrion (M_3) is about to become trapped in the process. ×48,000. (From Brandes, *et al.* 1964.)

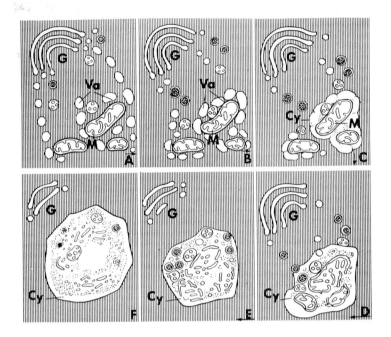

DIAGRAM 1

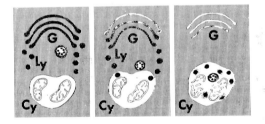

DIAGRAM 2

Fig. 36. DIAGRAM 1. Sequence in cytolysome (CY) formation in carbon-starved *E. gracilis* var. *bacillaris* (strain SM-L1). A: Golgi vacuoles (G, Va) and vesicles converge around mitochondrion (M). B and C: Fusion of Golgi vacuoles leading to segregation of mitochondrion within a cytolysome. During this process, other Golgi components and portions of cytoplasm are trapped within the cytolysome. D–F: Progessive degradation of various structures within the cytolysome.   DIAGRAM 2. Lysosomes (LY) arising from Golgi become incorporated within cytolysome and provide the enzymic component required for the lytic process. (From Brandes *et al.*, 1964.)

levels in 20–25 hours (Blum and Buetow, 1963). As the normal level is approached many hypertrophied mitochoudria appear to be present (Malkoff and Buetow, 1964). By 49 hours mitochondria of normal size and shape are present again.

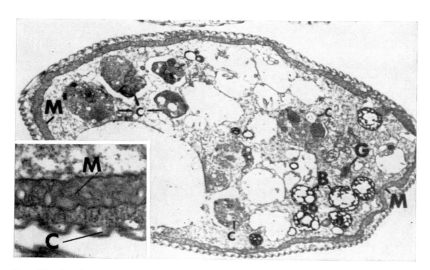

Fig. 37. *Euglena gracilis* var. *bacillaris* (strain SM-L1) starved of carbon for 7 days. Many cytolysomes (c) and cavities with complex walls (B) are present. Numerous "empty" vacuoles appear in the process of fusing and may represent dilated paramylon cavities and cytolysomes in which the visible content has been completely digested. A few extremely large mitochondria (M) are seem at the cell periphery. G, Golgi apparatus. The insert at lower left shows in detail the peripheral mitochondria (M) in relation to the pellicular surface (C) of the cell. ×32,000. (From Brandes *et al.*, 1964.)

E. Summary; Euglenoid Characteristics

In summary, the mitochondria of *Euglena* species show the basic architecture common to all mitochondria (Palade, 1953), but appear to be quite plastic, showing changes in size, shape, and number according to external conditions. In addition, they show at least two apparently characteristic euglenoid features: an irregular outer membrane and a basal constriction of each crista (Leedale *et al.*, 1965a).

XV. Golgi Apparatus (Dictyosome)

The presence of a Golgi apparatus (called dictyosome in some of the literature, as well as Golgi, Golgi body, Golgi complex, Golgi system, or Golgi unit) in *Euglena* was the subject of argument about as long as it was in other cells. In a review of the literature on protozoa, King (1927) concluded that a "normal Golgi apparatus" was present in the Sporozoa, but that the "nature of the Golgi bodies in the other protozoan classes is still under discussion." A little later, Baker (1933) claimed evidence to show that a "normal" Golgi apparatus was present in *E. gracilis* Klebs, but Patten and Beams (1936) were unable to demonstrate a Golgi apparatus in *Euglena* sp. With the development of definitive electron microscope work such as that of Sjöstrand and Hanzon (1954) on exocrine cells of the mouse pancreas, the question could be resolved for *Euglena*. A typical Golgi apparatus is indeed present. The Golgi bodies are well developed and distributed throughout the *Euglena* cell (Beams and Anderson, 1961; Pitelka, 1963; Voeller, 1964; Round, 1965).

A. EUGLENA GRACILIS

In *E. gracilis*, the Golgi bodies (Figs. 34 and 38) are numerous and consist of 8 to 20 stacked (at 70–95 Å), flattened cisternae (also called Golgi lamellae, flattened vesicles, membrane pairs, or paralled plates) that show saccular

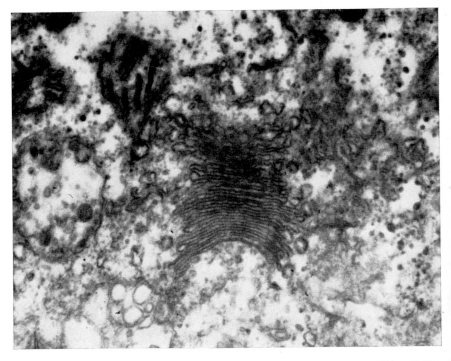

Fig. 38.   Golgi apparatus of *E. gracilis* var. *bacillaris*. ×46,000. (From Gibbs, 1960.)

dilations at one or both ends (Gibbs, 1960; Pitelka, 1963; Brandes *et al.*, 1964). The membranes of the cisternae measure 60 Å and enclose a 60-Å space (Gibbs, 1960). Vesicles ranging in size from 25 to 80 mμ are often concentrated around the Golgi (Gibbs, 1960; Pitelka, 1963; Brandes *et al.*, 1964). A number of vacuoles are also often found nearby (Brandes *et al.*, 1964). The Golgi bodies proliferate large vesicles in cells preparing for division (Chapter 5, Section III,C).

B. *ASTASIA LONGA*

The Golgi apparatus of *A. longa* (Fig. 39) appears to have the same general structure as *E. gracilis* (Ringo, 1963; Sommer and Blum, 1964, 1965a). The Golgi body, 1–2 μ in size, consists of several parallel plates (cisternae) of smooth membranes 60 Å wide which enclose a space of 80 Å. The plates are stacked at about 150 Å apart (Ringo, 1963). Saccular dilations are seen at the ends of the cisternae and numerous vesicles of various sizes surround

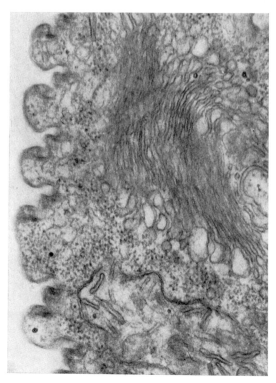

Fig. 39. Golgi apparatus of *A. longa*. A mitochondrion and details of the pellicular surface are also seen. ×40,000. (From Sommer and Blum, 1964.)

the structure. Sommer and Blum (1965a) consistently found a Golgi apparatus located close to the reservoir membrane in *A. longa*.

C. EUGLENA SPIROGYRA

A mature cell of *E. spirogyra* has 10 to 20 Golgi bodies, 1–1.5 μ in size, distributed throughout the cytoplasmic matrix, without any apparent association with other cell organelles (Leedale *et al.*, 1965a; Leedale 1966a). Each Golgi apparatus (Figs. 3 and 33) is a cylindrical stack of 20 to 30 curved and flattened cisternae which appear to interconnect at intervals and are associated with or actively proliferating many small vesicles (800 Å in diameter) from their sacculelike edges. A Golgi apparatus is characteristically seen in the immediate vicinity of the contractile vacuole–reservoir region, as is the case in *A. longa* (Sommer and Blum, 1965a).

D. GOLGI RELATIONSHIP TO OTHER CYTOPLASMIC STRUCTURES

1. *Constitutive Acid Phosphatase Activity*

Under electron cytochemistry, the constitutive acid phosphatase activity found in the Golgi of *E. gracilis* (Brandes *et al.*, 1964; Sommer and Blum, 1965b) has been occasionally observed in the pellicular and perireservoir endoplasmic reticulum, a finding which suggests some functional interrelation between these structures (Sommer and Blum, 1965b). Vesicles of the perireservoir region also show constitutive acid phosphatase activity and resemble Golgi vacuoles more than lysosomes.

2. *Golgi Hypertrophy during Starvation; Cytolysome Formation*

The Golgi bodies of *E. gracilis*, (var. *bacillaris*, strain SM-L1, streptomycin-bleached) are a site of strong acid phosphatase activity (Brandes *et al.*, 1964; Sommer and Blum, 1965b). Under conditions of carbon (acetate) starvation, the Golgi bodies hypertrophy. Saccular distentions at both ends of the cisternae appear to pinch off and give rise to many vesicles and vacuoles that move into the cytoplasm (Fig. 36). Some of the mitochondria and adjacent regions of the cytoplasm become surrounded by Golgi-type vacuoles. Flattening and fusion of these vacuoles results in the isolation of mitochondria and other cytoplasmic components within membrane-bound vacuoles comparable to the cytolysomes (autophagic vacuoles) described in mammalian cells (De Duve and Wattiaux, 1966). Many Golgi components, mainly vesicles and multivesicular bodies, appear embodied in the cytolysome. Encapsulated material is progressively hydrolyzed. The Golgi body appears to supply at least one acid hydrolase, acid phosphatase, to the cytolysome

(Brandes *et al.*, 1964). Along with these degenerative changes, new cytolysomes form in the same cell. This process of autodigestion probably represents a mechanism for providing the cell with breakdown products for utilization in continued maintenance of basal metabolic processes during starvation (Blum and Buetow, 1963; Malkoff and Buetow, 1964; Brandes *et al.*, 1964; Bertini *et al.*, 1965).

Brandes (1965) has suggested that the intracellular site of formation of lysosomes and synthesis of lysosomal enzymes is related to the prevailing type of endoplasmic reticulum (ER) present, i.e., "rough" or "smooth." In cells in which rough-surfaced ER predominates, lysosomal formation and at least some lysosomal enzyme synthesis is mediated through the ER. In cells in which smooth ER predominates, e.g., *Euglena* (see Section XVI), these processes are mediated through the Golgi as described above.

XVI. Endoplasmic Reticulum and Ribosomes

The cytoplasm of *E. gracilis* contains scattered membrane-limited tubules and vesicles, i.e., the endoplasmic reticulum (Ueda, 1958; Gibbs, 1960; Pitelka, 1963). These tubules and vesicles range in diameter from 200 to 600 Å

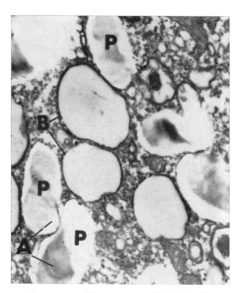

Fig. 40. Euglena gracilis var. *bacillaris* (strain SM-L1). Two types of cavities found in the cytoplasm: type A contains paramylon (P) and lacks a defined membrane; type B does not contain visible material, but shows a membrane. ×16,000. (From Brandes *et al.*, 1964.)

(Gibbs, 1960) with an average of approximately 300 Å (Ueda, 1958). Wolken and Palade (1953) and Wolken (1961) also described tubular and vesicular elements, but their early electron micrographs appear to correspond to Golgi cisternae rather than ER. A discussion of the subpellicular ER appears in Section VI,B,2c.

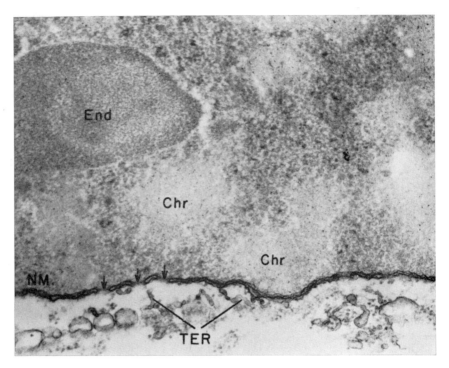

Fig. 41. Euglena spirogyra. NM, Nuclear envelope traversed by pores (arrows); TER, tubular endoplasmic reticulum; Chr, sections of chromosomes; End, endosome (nucleolus). ×40,500. (From Leedale *et al.*, 1965a.)

A. *EUGLENA GRACILIS*; *ASTASIA LONGA*

The ER of *E. gracilis* has been reported to be of the smooth-surfaced type (Gibbs, 1960; Pitelka, 1963; Malkoff and Buetow, 1964, see also Vol. II, Chapter 10, Fig. 7). A similar ER along with "free" ribosomes (120–180 Å in diameter) occurs in *A. longa* (Ringo, 1963); however, *Astasia* appears to have many more ribosomes per unit area of cytoplasm than does *E. gracilis* (Blum *et al.*, 1965). In recent studies with *E. gracilis*, the ER is reported to consist of *both* smooth and rough-surfaced discrete tubular and vesicular elements (Brandes *et al.* 1964, see also Chapter 5, Section II,C). The majority of the ER, however, appears to be of the smooth-surfaced type. Many ribosomes were also scattered throughout the cytoplasm as noted before (Gibbs, 1960; Pitelka, 1963; Malkoff and Buetow, 1964).

Early electron micrographs of carbon-starved *E. gracilis* indicated pronounced vacuolization of the ER caused by depletion of paramylon grains (Malkoff and Buetow, 1964). Subsequent work modified this inter-

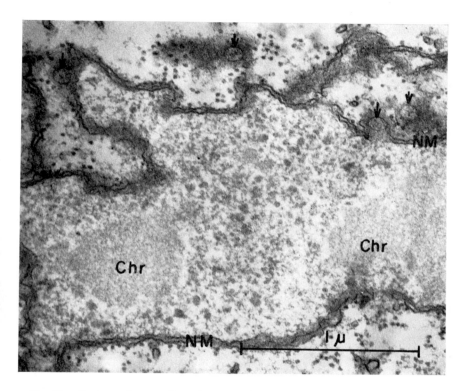

Fig. 42. Euglena spirogyra. Pores (arrows) of nuclear envelope (NM) shown in surface view. Chr, Chromosomes. ×47,500. (From Leedale *et al.*, 1965a.)

pretation. Vacuolization of the cytoplasm does take place, but the paramylon-containing cavities, at least as seen in the nonstarved cell (Fig. 40), appear to be independent of the ER (Brandes *et al.*, 1964).

The large confluent vacuoles (Fig. 37) found in *Euglena* during carbon (acetate) starvation appear to derive from the fusion of depleted paramylon cavities with complex-walled vacuoles and depleted cytolysomes (Brandes *et al.*, 1964), rather than from swelling of the ER as previously suggested (Malkoff and Buetow, 1964).

In *E. gracilis*, the ER serves to interconnect the nucleus, the chloroplasts, and the subpellicular mucus bodies (Chapter 5, Section II,C).

B. *EUGLENA SPIROGYRA*

Tubular extensions of the outer nuclear membrane (Figs. 41 and 42) of *E. spirogyra* extend into the cytoplasm and connect with tubules and vesicles comprising the ER (Leedale *et al.*, 1965a; Leedale, 1966a). The ER contains both smooth and rough surfaces (Figs. 3, 31, 43, and 48). The majority of

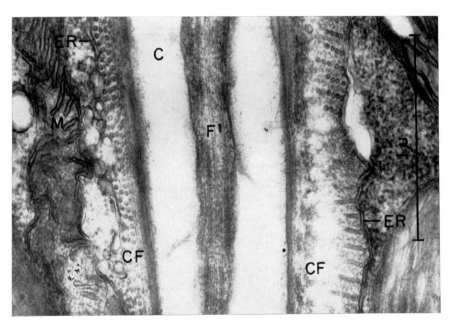

Fig. 43. Oblique longitudinal section of the canal region of *E. spirogyra*. A collar of microtubules (CF) encircles the canal (C) and is delimited from the general cytoplasm by a cylindrical sheath of endoplasmic reticulum (ER). Microtubules are cut transversely and appear in rows of four to six. F¹, Locomotory flagellum; M, mitochondrion. ×57,000. (From Leedale *et al.*, 1965a.)

the tubular ER is found around the nucleus, although occasionally the same elements are found throughout the cytoplasm. The membrane of the tubular ER is smooth. The tubular ER connects with an extensive vesicular ER which is found throughout the cell. The membrane of the vesicular ER is rough. Part of the ER (mostly smooth, but rough in parts) forms a "sheath" that separates the microtubules of the canal (Section XIII,A) from the general cytoplasm (Figs. 31 and 43). This "sheath" is continuous with the vesiculate

ER. The most peripheral elements of the rough vesiculate ER underlie the pellicle and appear to form the mucus bodies that produce mucilage (Section VI,B,1,b and Section VIII). Many ribosomes (100–150 Å in diameter) appear to be free in the cytoplasm (Figs. 3 and 31) in addition to coating the membrane of the vesiculate ER.

XVII. Chloroplasts and Pyrenoids

This topic will be covered only briefly here since it and related topics are covered in detail in Chapter 1 and in Vol. II, Chapters 3, 9, 10, 11, and 12.

A. *EUGLENA GRACILIS*

The chloroplasts of *E. gracilis* arc traversed by 10 to 45 moderately dense bands or lamellae which vary in width from 27 to 210 mμ (Gibbs, 1960). Each band is a stack of from two to five, usually three, closely appressed discs. The thin outer lamellae are 50–60 Å wide, the thicker middle lamellae are 95–120 Å wide, and the intralamellar spaces are usually 70 Å wide, but may occasionally be as much as 200 Å wide (Fig. 44). An irregularly granular material forms the matrix in which the bands of discs lie. The granules range in size from 60 to 220 Å (Fig. 44). Lipid droplets (60–175 mμ) are regularly seen (Fig. 44) and empty vacuoles are occasionally seen (Fig. 45).

The pyrenoid, which has a rim of paramylon, is a region of denser matrix material traversed by widely spaced (40–70 mμ) bands, each consisting of a single or double disc (Fig. 46). The size of the pyrenoid varies greatly and apparently depends in part on the developmental state of the chloroplast.

Usually, a dense membrane 150–200 Å thick appears to delimit the chloroplast. In favorable sections, however, this outer "membrane" is resolvable into two membranes, each 50–60 Å wide and each enclosing a space about 60 Å wide.

The chloroplasts of *E. gracilis* contain a unique species of DNA and evidence has been presented for distinct chloroplast-associated RNA (see Vol. II, Chapter 5). In addition, chloroplast ribosomes are smaller than cytoplasmic ribosomes (Vol. II, Chapter 5, Figs. 13 and 14). The chloroplasts appear to be connected with the nucleus via the endoplasmic reticulum (Chapter 5, Section II,C).

B. *EUGLENA SPIROGYRA*

Euglena spirogyra has numerous chloroplasts (Leedale *et al.*, 1965a) of the simplest euglenoid type (see Chapter 1), lenticular to polygonal with no

Dennis E. Buetow

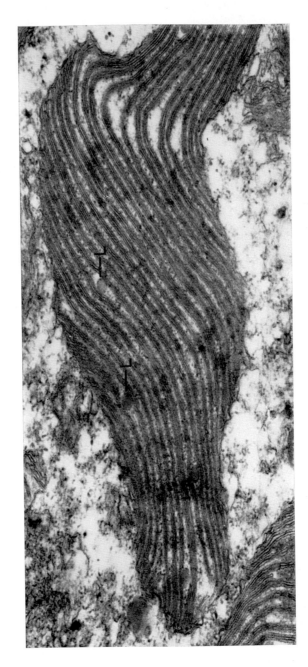

Fig. 44. Chloroplast of *E. gracilis* strain Z. L, Lipid droplet in chloroplast. ×32,200. (Original electron micrograph, courtesy of S. P. Gibbs.)

pyrenoid (Fig. 3). Each chloroplast contains a compound system of outer membranes, chlorophyll-containing lamellae which lie parallel to the flat face of the chloroplast and which usually traverse the full diameter of the chloroplast, a densely granular matrix, and large osmiophilic (presumably lipid) globules lying in the matrix.

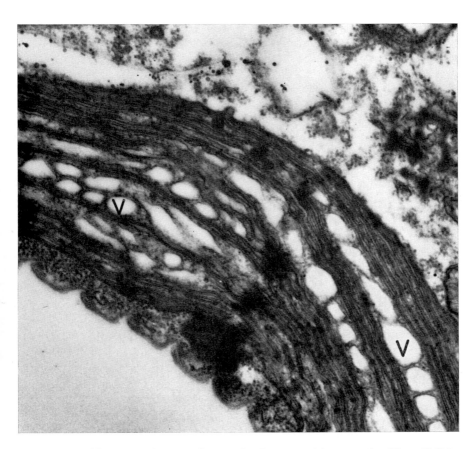

Fig. 45. Chloroplast of *E. gracilis* var. *bacillaris* containing vacuoles (V). ×37,000. (From Gibbs, 1960.)

XVIII. Nucleus

The morphology and ultrastructure of the nucleus of various *Euglena* species are covered in detail in Chapter 5.

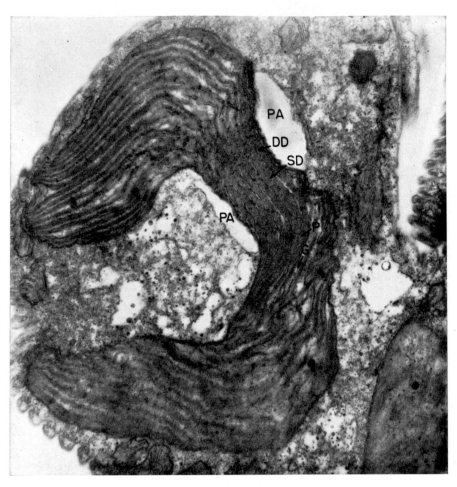

Fig. 46. Pyrenoid of *E. gracilis* var. *bacillaris*. DD, Double disc crossing pyrenoid; G, granular chloroplast matrix; P, dense pyrenoid matrix; PA, paramylon; SD, single disc crossing pyrenoid. ×24,600. (From Gibbs, 1960.)

XIX. Rhizoplast

A rhizoplast, or thin connecting thread, extending from one of the kinetosomes (blepharoplasts or basal bodies) to a granule on the nuclear membrane was described for the family Euglenaceae (Euglenidae) in general (Hall and Jahn, 1929a; Jahn, 1951; see also Chapters 1 and 3). Other investigators did not show a rhizoplast in various species of *Euglena* (Gojdics, 1934, 1939; Krichenbauer, 1938; Hollande, 1942; Pringsheim and Hovasse, 1948;

Roth, 1958; Leedale, 1959, see also Chapter 5, Section II,B). A rhizoplast arising from *each* kinetosome was observed in about 5% of the organisms in a culture of *E. fracta* (Johnson; 1956). Haase (1910) reported two fibers connecting the flagellum to a granule posterior to the nucleus in *E. sanguinea*, but this observation was not confirmed (Gojdics, 1939).

Recent electron micrograph studies on *A. longa* (Sommer and Blum, 1965a) and *E. spirogyra* (Leedale *et al.*, 1965a) do not show a rhizoplast. The existence of rhizoplasts in *Euglena* seems open to question.

XX. Miscellaneous Cellular Inclusions

The following miscellaneous cellular inclusions have been found in *Euglena* and related forms and substantiated by electron microscopy.

A. Vacuoles Containing Dense Conglomerates of Material

This is a category of large vacuoles as defined by Gibbs (1960) for *E. gracilis*. These vacuoles are 0.5–1.7 μ wide and 1.3 μ long, are bounded by an 80-Å wide membrane and contain a matrix of low electron density material (Fig. 47). Large, irregular, vacuolate conglomerates of very dense material are scattered in the matrix. These vacuoles sometimes also contain a variety of membrane-bound structures, vesicles, compound vesicles, and concentric-ringed structures. Gibbs (1960) believes these vacuoles to be the metachromatic granules (or volutin) seen in the "vacuome" of *Euglena* by light microscopists (Baker, 1933; Jahn, 1946, 1951; Fritsch, 1965). These vacuoles also resemble some of the cytolysomes (Section XV,D,2) seen in carbon-starved *E. gracilis* (Brandes *et al.* 1964).

Seemingly related structures are seen in old cultures of *E. spirogyra* and have been called "phospholipid vesicles" (Leedale *et al.*, 1965a). These vesicles contain numerous brownish-orange droplets which are meta-chromatic with toluidine blue and give a positive lipid reaction. Under electron microscopy, they are bounded by a unit membrane and contain fat bodies, myelin figures, and many unit membranes (Fig. 48).

B. Degenerated Chloroplasts

Orange-brown bodies originating from degenerating chloroplasts appear when *Euglena* cells are kept in the dark (Baker, 1933) or are treated with streptomycin under nongrowing conditions (Vavra, 1961). Such bodies also appear in old cultures.

Numerous bodies composed of concentric lamellae appear in streptomycin-bleached cells grown in the light. These bodies are thought to be degenerated or improperly formed chloroplasts (Siegesmund *et al.*, 1962). They do not appear in dark-grown or light-grown normal cells or dark-grown streptomycin-bleached cells.

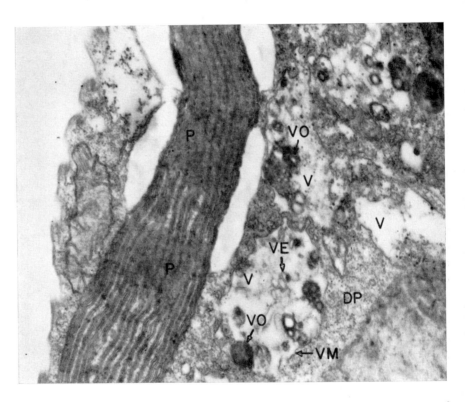

Fig. 47. Euglena gracilis var. *bacillaris* vacuole containing dense conglomerates of material. V, Vacuole; VE, small vesicles within vacuole; VO, volutin; VM, vacuole membrane; P, pyrenoid; DP, dense particles (ribosomes). ×26,910. (From Gibbs, 1960.)

Gibbs (1960) reports the occassional presence in *E. gracilis* of double-membrane–limited bodies that are much smaller than plastids (0.6–1.5 μ) and have only a few unoriented discs and vesicles and some larger empty vacuoles in their granular matrix. Gibbs (1960) could not determine whether these were proplastids or degenerated plastids. It is also possible that they are cytolysomes containing degenerating mitochondria (S. P. Gibbs, personal communication).

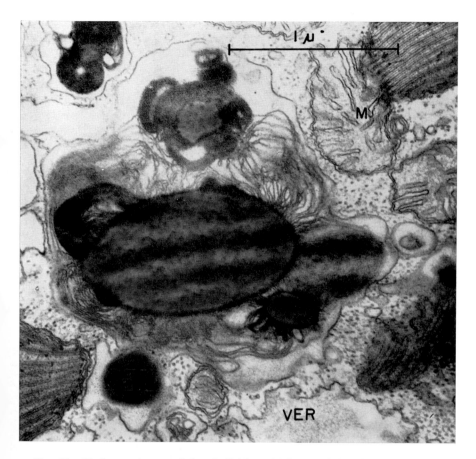

Fig. 48. *Euglena spirogyra* "phospholipid vesicle" containing fat bodies, myelin figures, and proliferations of unit membranes. M, Mitochondrion; VER, vesiculate endoplasmic reticulum. ×45,000. (From Leedale *et al.*, 1965a.)

C. HOMOGENEOUS DENSE BODIES

As described by Gibbs (1960) for *E. gracilis*, these are dense spherical bodies (200–750 mμ in size) scattered in the cytoplasm. They appear to be lipoidal in nature. Some may be dispersed eyespot granules.

D. GRANULAR DENSE BODIES

These spherical or slightly irregularly shaped bodies are frequently observed just beneath the pellicle of *E. gracilis* (Gibbs, 1960). They range from 400 to 750 mμ in diameter. In 1960, Gibbs thought they were not mucus bodies, but better resolution is needed before a conclusion can be drawn (S. P. Gibbs, personal communication).

E. Dark Bodies

These are found in *A. longa* and are about the size of mitochondria (Ringo, 1963). They are about 1.5 μ in diameter, are surrounded by a single 60-Å membrane, are irregular in shape, contain varying numbers of circular membrane profiles and concentric layers of membrane, and appear about as dense as mitochondria under electron microscopy. There are about 20 per cell and seem to surround some mitochondria. They do not appear to be present in actively dividing cells. Some of the dark bodies in *Astasia* resemble cytolysomes (Section XV,D,2) seen in carbon-deprived postmitotic *E. gracilis* (Malkoff and Buetow, 1964; Brandes *et al.*, 1964).

F. Double Membrane Bodies

As described by Ringo (1963), these occur in small number in *A. longa*. They are generally round (about 0.5 μ in diameter) and are bounded by a double membrane. They enclose single- or double-membrane vesicles and ribosomes and have the same density as the surrounding cytoplasm. Siegesmund *et al.* (1962) tentatively identified a similar body in dark-grown *Euglena gracilis* as a proplastid. Since *A. longa* does not form chloroplasts, the function of these bodies remains unknown (Ringo, 1963).

XXI. Parasites

Parasites have been noted in various euglenoids (Jahn, 1946, 1951) since the first such description by P. A. Dangeard (1886). Mitchell (1928) claimed a parasite was present in four species of *Euglena* found in a polluted stream and in roadside pools in Georgia. About 8% of the cells of *E. caudata* and 5% of *E. viridis* were infected. There seemed to be no limit to the number of parasites except the size of the cell itself. No toxic effect associated with the presence of the parasites was noted; however, infected *E. viridis* became round and immobile. The parasite, *Sphaerita dangeardi*, infected some *E. rubra* cells found in an Iowa lake (Johnson, 1939).

Jahn (1951) states that the parasites of euglenoids consist of one species of bacteria and at least five genera of Phycomycetes, all of which are *usually* fatal to the host. The parasites grow in the cytoplasm and undergo division to form zoospores which are released through the wall of the host cell. P. A. Dangeard (1886) and Jahn (1951) say that early workers who claimed to see sexual stages* in *Euglena* were only seeing various stages in the life history of the parasites.

P. A. Dangeard (1933) reported a bacterium that attacks and destroys the euglenoid nucleus. In a recent study, Leedale (1966b) isolated a strain of *E. spirogyra* which contained 20 to 100 rod-shaped bodies within the nucleus. Electron microscopy shows these are distinct from chromosomes and nucleoli and are bacilloid bacteria. They multiply to keep pace with cytokinesis. Infected *Euglena* do not seem harmed or benefited by the bacteria and show the same morphology and growth rate as uninfected cells.

XXII. Final Statements

The morphology and ultrastructure of members of the genus *Euglena* vary somewhat from species to species as might be expected. It is also clear from the above literature survey that the morphology and ultrastructure of a *single Euglena* cell is variable. A single *E. gracilis* cell can be taken as an example here. Under optimal conditions of growth and division, the cell is spindle-shaped (Gojdics, 1953); under conditions of external carbon (acetate) deprivation, the cell becomes smaller and rounds up (Blum and Buetow, 1963). The number of chloroplasts in each cell varies with the largest number seen in cells in the late logarithmic or early stationary phases of growth (Gross and Villaire, 1960). The growth medium is also influential. For example, "excess" acetate in the culture medium appears to repress chlorophyll synthesis (and so delays the final stages of maturation of the chloroplast) but, at the same time, leads to increased paramylon content per cell (App and Jagendorf, 1963; Buetow, 1967). Cells kept in the dark or bleached ("albinized") by various agents no longer contain chloroplasts, but do show increased numbers of mitochondria (Pringsheim and Hovasse, 1948; Wolken and Palade, 1953; Pringsheim, 1956; see also Vol. II, Chapter 10). Cells starved of carbon (acetate) show fewer mitochondria, the majority being digested away in cytolysomes (Malkoff and Buetow, 1964; Brandes *et al.*, 1964). Finally, the volume of the cell increases when it is inoculated into fresh medium (Corbett, 1957).

In order to be of maximum use, therefore, a study of the ultrastructure of a *Euglena* cell should include a *detailed* description of the conditions of growth. A complete description will help eliminate confusion when attempts are made to correlate structure and function of cell particulates.

The genus *Euglena* is composed of many species (Chapter 1) which have adapted to and occupied a variety of ecological niches (Chapter 2). It appears

---

* See Chapter 5, Section V, for a discussion of early claims of meiosis in *Euglena*.

that a comparative ultrastructure study on representative species would be revealing as regards morphological (used in the widest sense here to include ultrastructural) adaptation of cells to differing environmental conditions. As stated above, even cells of a single species of *Euglena* are capable of drastic morphological changes in response to changed environmental conditions. No one species is found in all the ecological niches, however, so several, perhaps many, species should be examined. The effort appears to be worthwhile, particularly if it could be combined with correlated physiological and biochemical measurements.

XXIII. Note Added in Proof

H. J. Arnott and P. L. Walne, [*Protoplasma*, 1967 (in press)] report the presence of two types of bodies associated with the pellicular strips (see Section VI, B) of *E. granulata*. The first type (about 300 mμ in diameter) is distributed along or near the surface of some, but not all, pellicular strips and consists of an aperture and a compartment. The compartment is lined by the plasmalemma (see Section VI, A, 1) and contains a dense, osmiophilic body and a series of microtubules. The second type (about 1–2 μ in diameter) is located below the surface of the pellicle and resembles the "phospholipid vesicles" found in *E. spirogyra* (see Section XX, A and Fig. 48). The functions of these two types of bodies in *E. granulata* are not known.

P. L. Walne and H. J. Arnott [*Planta 75*, 1967 (in press)] recently published some excellent electron micrographs of the stigma (see Section X, A) of *E. granulata*. Fifty to sixty granules (240–1200 mμ in diameter) form the stigma in the cytoplasm near the base of the reservoir and are closely associated with the parabasal body (see Section X, B). The granules exhibit birefringence and, in contrast to other reports (see Section X, A), are enclosed in groups of two or three within a membrane. Microtubules are also found within this membrane. Several hypotheses about the function of the stigma are discussed in terms of its ultrastructure.

ACKNOWLEDGMENTS

I thank the following for permission to reproduce the figures used in this chapter: The British Association for the Advancement of Science for Fig. 1; Academic Press Inc. for Figs. 2, 17–23, 26–29, 34–40, and 45–47; Springer-Verlag for Figs. 3, 16, 31–33, 41–43, and 48; The British Phycological Society for Figs. 4–13; The Rockefeller University Press for Figs. 14, 15, and 24; The University of California Press for Fig. 25; and Elsevier Publishing Company for Fig. 30. Dr. Sarah P. Gibbs kindly supplied the previously unpublished Fig. 44. All electron micrograph figures were reproduced from the original electron micrographs.

References

App, A. A., and Jagendorf, A. T. (1963). *J. Protozool.* **10**, 340.
Baker, C. L. (1933). *Arch. Protistenk.* **80**, 434.
Baker, W. B. (1926). *Biol. Bull.* **51**, 321.
Batra, P. P., and Tollin, G. (1964). *Biochim. Biophys. Acta* **79**, 371.
Beams, H. W., and Anderson, E. (1961). *Ann Rev. Microbiol.* **15**, 47.
Bertini, F., Brandes, D., and Buetow, D. E. (1965). *Biochim. Biophys. Acta* **107**, 171.
Blum, J. J. (1965). *J. Cell Biol.* **24**, 223.
Blum, J. J., and Buetow, D. E. (1963). *Exptl. Cell Res.* **29**, 407.
Blum, J. J., Sommer, J. R., and Kahn, V. (1965). *J. Protozool.* **12**, 202.
Brandes, D. (1965). *J. Ultrastruc. Res.* **12**, 63.
Brandes, D., Buetow, D. E., Bertini, F., and Malkoff, D. B. (1964). *Exptl. Mol. Pathol.* **3**, 583.
Brown, H. P. (1945). *Ohio J. Sci.* **45**, 247.
Buetow, D. E. (1967). *Nature* **213**, 1127.
Buetow, D. E., and Levedahl, B. H. (1962). *J. Gen. Microbiol.* **28**, 579.
Bütschli, O. (1883–1887). *In* "Bronn's Klassen und Ordnungen des Thierreichs," Abt. II, pp. 617–1097. Winter, Leipzig.
Byrne, J., and Marsland, D. (1965). *J. Cellular Comp. Physiol.* **65**, 277.
Causey, D. (1926). *Univ. Calif. (Berkeley) Publ. Zool.* **28**, 217.
Chadefaud, M. (1934). *Bull. Soc. Botan. France* **81**, 106.
Chadefaud, M. (1936). *Ann. Protistol.* **5**, 323.
Chadefaud, M. (1937). *Botaniste* **28**, 85.
Chadefaud, M., and Provasoli, L. (1939). *Arch. Zool. Exptl. Gen., Notes Rev.* **80**, 55.
Chen, Y. T. (1950). *Proc. Conf. Electron Microscopy, Delft* pp. 156–158. Nijhoff, The Hague.
Chu, S. P. (1947). *Sinensia* **17**, 75.
Clarke, A. E., and Stone, B. A. (1960). *Biochim. Biophys. Acta* **44**, 161.
Corbett, J. J. (1957). *J. Protozool.* **4**, 71.
Dangeard, P. (1928). *Ann. Protistol.* **1**, 69.
Dangeard, P. A. (1886). *Bull. Soc. Botan. France* **33**, 240.
Dangeard, P. A. (1901). *Botaniste* **8**, 97.
Dangeard, P. A. (1919). *Compt. Rend. Acad. Sci.* **169**, 1005.
Dangeard, P. A. (1925). *Cellule* **35**, 237.
Dangeard, P. A. (1933). *Botaniste* **25**, 3.
De Duve, C., and Wattiaux, R. (1966). *Ann Rev. Physiol.* **28**, 435.
de Haller, G. (1959). *Arch. Sci. (Geneva)* **12**, 309.
de Haller, G. (1960). *Verhandl. 4th Intern Kongr. Elektronmikroskop., Berlin, 1958* **2**, 517–520. Springer-Verlag, Berlin.
Ehrenberg, C. G. (1830). *Physik. Abhandl. Kgl. Akad. Wiss. Berlin. 1830* p. 1.
Ehrenberg, C. G. (1838). "Die Infusiontierchen als vollkommene Organismen." Voss, Leipzig.
Engelmann, T. W. (1882). *Arch. Ges. Physiol.* **29**, 387.
Fawcett, D. W., and Porter, K. R. (1954). *J. Morphol.* **94**, 221.
Fischer, A. (1894). *Jahrb. Wiss. Botan.* **26**, 187.
Frey-Wyssling, A., and Mühlethaler, K. (1960). *Schweiz. Z. Hydrol.* **22**, 122.
Fritsch, F. E. (1965). "The Structure and Function of the Algae," Vol. I (reprint of 1935 edition). Cambridge Univ. Press, London and New York.
Gibbs, S. P. (1960). *J. Ultrastruct. Res.* **4**, 127.
Gojdics, M. (1934). *Trans Am. Microscop. Soc.* **53**, 299.

Gojdics, M. (1939). *Trans Am. Microscop. Soc.* **58**, 241.

Gojdics, M. (1953). "The Genus *Euglena*." Univ. of Wisconsin Press, Madison, Wisconsin.

Goodwin, T. W., and Jamikorn, M. (1954). *J. Protozool.* **1**, 216.

Gottlieb, J. (1850). *Ann. Chem. Pharm.* **75**, 51.

Grassé, P. P. (1926). *Compt. Rend. Soc. Biol.* **94**, 1012.

Green, J. (1963). *Comp. Biochem. Physiol.* **9**, 313.

Grimstone, A. V. (1961). *Biol. Rev.* **36**, 97.

Gross, J. A., and Villaire, M. (1960). *Trans. Am. Microscop. Soc.* **79**, 144.

Groupé, V. (1947). *Proc. Soc. Exptl. Biol. Med.* **64**, 401.

Günther, F. (1928). *Arch. Protistenk.* **60**, 511.

Haase, G. (1910). *Arch. Protistenk.* **20**, 47.

Habermann, J. (1874). *Ann. Chem. Pharm.* **172**, 11.

Hall, R. P. (1931). *Ann. Protistol.* **3**, 57.

Hall, R. P. (1933). *Trans. Am. Microscop. Soc.* **52**, 220.

Hall, R. P., and Jahn, T. L. (1929a). *Trans. Am. Microscop. Soc.* **48**, 388.

Hall, R. P., and Jahn, T. L. (1929b). *Science* **69**, 522.

Hardy, A. D. (1911). *Victoria Naturalist* **27**, 215.

Hollande, A. (1942). *Arch Zool. Exptl. Gen.* **83**, 1.

Houwink, A. L. (1951). *Koninkl. Ned. Akad. Wetenschap. Proc. Ser. C* **54**, 132.

Hutner, S. H., and Provasoli, L. (1955). *In* "Biochemistry and Physiology of Protozoa" (S. H. Hutner and A. Lwoff, eds.), Vol. II, pp. 17–43. Academic Press, New York.

Hyman, L. H. (1938). *Beih. Botan Zentr. Abt. A* **58**, 379.

Jahn, T. L. (1946). *Quart. Rev. Biol.* **21**, 246.

Jahn, T. L. (1951). *In* "Manual of Phycology" (G. M. Smith, ed.), pp. 69–81. Chronica Botanica, Waltham, Massachusetts.

James, T. W. (1963). *Symp. Intern. Soc. Cell Biol.* **2**, 9.

Jane, F. W. (1955). *New Biol.* **19**, 114.

Johnson, L. P. (1939). *Trans. Am. Microscop. Soc.* **58**, 42.

Johnson, L. P. (1956). *Trans. Am. Microscop. Soc.* **75**, 217.

Johnson, L. P., and Jahn, T. L. (1942). *Physiol. Zool.* **15**, 89.

Khawkine, W. (1886). *Ann. Sci. Nat. Zool.* **1**, 319.

King, S. D. (1927). *J. Roy. Microscop. Soc.* **47**, 342.

Kirk, J. T. O., and Juniper, B. (1964). *J. Roy. Microscop. Soc.* **82**, 205.

Klebs, G. (1883). *Untersuch. Botan. Inst. Tübingen* **1**, 233.

Kol, E. (1929). *Arch. Protistenk.* **66**, 515.

Kreger, D. R., and Meeuse, B. J. D. (1952). *Biochim. Biophys. Acta* **9**, 699.

Krichenbauer, H. (1938). *Arch. Protistenk.* **90**, 88.

Krinsky, N. I., and Goldsmith, T. H. (1960). *Arch. Biochem. Biophys.* **91**, 271.

Lackey, J. B. (1934). *Biol. Bull.* **67**, 145.

Leedale, G. F. (1958). *Arch. Mikrobiol.* **32**, 32.

Leedale, G. F. (1959). *J. Protozool.* **6**, Suppl., 26.

Leedale, G. F. (1964). *Brit. Phycol. Bull.* **2**, 291.

Leedale, G. F. (1966a). *Advan. Sci.* **23**, 22.

Leedale, G. F. (1966b). *J. Protozool.* **13**, Suppl., 22.

Leedale, G. F., Meeuse, B. J. D., and Pringsheim, E. G. (1965a). *Arch. Mikrobiol.* **50**, 68.

Leedale, G. F., Meeuse, B. J. D., and Pringsheim, E. G. (1965b). *Arch. Mikrobiol.* **50**, 133.

Lefort, M. (1963). *Compt. Rend. Acad. Sci.* **256**, 5190.

Mainx, F. (1926). *Arch. Protistenk.* **54**, 150.

Mainx, F. (1927). *Arch. Protistenk.* **60**, 305.

Malkoff, D. B., and Buetow, D. E. (1964). *Exptl. Cell Res.* **35**, 58.

Manton, I. (1952). *Symp. Soc. Exptl. Biol.* **6**, 306.
Manton, I. (1953). *Nature* **171**, 485.
Mast, S. O. (1911). "Light and Behavior of Organisms." Wiley, New York.
Mast, S. O. (1917). *J. Exptl. Zool.* **22**, 471.
Mego, J. L., and Buetow, D. E. (1966). *J. Protozool.* **13**, 20.
Mego, J. L., and Buetow, D. E. (1967). *In* "Le Chloroplaste: Croissance et Vieillissement" (C. Sironval, ed.), pp. 274–290. Masson, Paris.
Mignot, J.-P. (1963). *Compt. Rend. Acad. Sci.* **257**, 2530.
Mignot, J.-P. (1965). *Protistologica I*, Part 1, 5.
Mitchell, J. B. (1928). *Trans. Am. Microscop. Soc.* **47**, 29.
Naumann, E. (1915-1916). *Intern. Rev. Ges. Hydrobiol. Hydrog.* **7**, 214.
Padilla, G. M., and Buetow, D. E. (1959). *J. Protozool.* **6**, Suppl., 29.
Palade, G. (1953). *J. Histochem. Cytochem.* **1**, 188.
Patten, R., and Beams, H. W. (1936). *Quart. J. Microscop. Sci.* **78**, 615.
Picken, L. (1960). "The Organization of Cells and Other Organisms." Oxford Univ. Press (Clarendon), London and New York.
Pitelka, D. R. (1949). *Univ. Calif. (Berkeley) Publ. Zool.* **53**, 377.
Pitelka, D. R. (1963). "Electron Microscopic Structure of Protozoa." Pergamon, Oxford.
Pitelka, D. R., and Schooley, C. N. (1955). *Univ. Calif. (Berkeley) Publ. Zool.* **61**, 79.
Pochmann, A. (1953). *Planta* **42**, 478.
Price, C. A. (1962). *Science* **135**, 46.
Price, C. A., and Bourke, M. E. (1966). *J. Protozool.* **13**, 474.
Pringsheim, E. G. (1946). *J. Ecol.* **33**, 193.
Pringsheim, E. G. (1948). *Biol. Rev.* **23**, 46.
Pringsheim, E. G. (1956). *Nova Acta Leopoldina* **18** (125), 1.
Pringsheim, E. G., and Hovasse, R. (1948). *New Phytologist* **47**, 52.
Pringsheim, E. G., and Hovasse, R. (1950). *Arch. Zool. Exptl. Gen.* **86**, 499.
Pringsheim, E. G., and Pringsheim, O. (1952). *New Phytologist* **51**, 65.
Ratcliffe, H. (1927). *Biol. Bull.* **53**, 109.
Reger, J. F., and Beams, H. W. (1954). *Proc. Iowa Acad. Sci.* **61**, 593.
Reichenow, E. (1910). *Arb. Kaiserl. Gesundh.-Bibliot.*, Berlin **33**, 15.
Ringo, D. L. (1963). *J. Protozool.* **10**, 167.
Roth, L. E. (1958). *J. Ultrastruct. Res.* **1**, 223.
Roth, L. E. (1959). *J. Protozool.* **6**, 107.
Round, F. E. (1965). "The Biology of the Algae." St. Martin's Press, New York.
Siegesmund, K. A., Rosen, W. G., and Gawlik, S. R. (1962). *Am. J. Botany* **49**, 137.
Sjöstrand, F. S., and Hanzon, V. (1954). *Exptl. Cell Res.* **7**, 415.
Sleigh, M. A. (1962). "The Biology of Cilia and Flagella." Pergamon, Oxford.
Sommer, J. R. (1965). *J. Cell Biol.* **24**, 253.
Sommer, J. R., and Blum, J. J. (1964). *Exptl. Cell Res.* **35**, 423.
Sommer, J. R., and Blum, J. J. (1965a). *Exptl. Cell Res.* **39**, 504.
Sommer, J. R., and Blum, J. J. (1965b). *J. Cell Biol.* **24**, 235.
Szabados, M. (1936). *Acta Biol. Szeged* **4**, 49.
Tchakhotine, S. (1936). *Compt. Rend. Soc. Biol.* **121**, 1162.
Tchang, L. K. (1924). *Compt. Rend. Soc. Biol.* **91**, 263.
Tischer, J. (1936). *Z. Physiol. Chem.* **239**, 257.
Tischer, J. (1941). *Z. Physiol. Chem.* **267**, 281.
Ueda, K. (1958). *Cytologia (Tokyo)* **23**, 56.
Ueda, K. (1960). *Cytologia (Tokyo)* **25**, 8.
Vavra, J. (1956). Quoted by Pringsheim (1956).

Vavra, J. (1961). *In* "Progress in Protozoology," Proc. Intern. Congr. Protozool., 1st, Prague, pp. 189–190. Academic Press, New York.

Vlk, W. (1938). *Arch. Protistenk.* **90**, 448.

Voeller, B. R. (1964). *In* "The Cell" (J. Brachet and A. E. Mirsky, eds.), Vol. VI, pp. 81–137. Academic Press, New York.

Wager, H. (1900). *J. Linnean Soc. (London), Zool.* **27**, 463.

Wolken, J. J. (1956). *J. Protozool.* **3**, 211.

Wolken, J. J. (1958). *Ann. N. Y. Acad. Sci.* **74**, 164.

Wolken, J. J. (1961). "Euglena. An Experimental Organism for Biochemical and Biophysical Studies." Rutgers Univ. Press, Rutgers, New Brunswick, New Jersey.

Wolken, J. J., and Palade, G. E. (1953). *Ann. N. Y. Acad. Sci.* **56**, 873.

CHAPTER 5

THE NUCLEUS IN *EUGLENA*

Gordon F. Leedale

I. Historical Review

Studies on the nucleus of *Euglena* date back to the end of the last century. In 1894, Blochmann described nuclear division in *E. deses*, *E. spirogyra*, *E. velata*, and *E. viridis*, and this was immediately followed by a further study of *E. viridis* by Keuten (1895). Keuten records a longitudinal "cleavage" of chromosomes and this observation has since been confirmed by Tschenzoff (1916) for *E. viridis*, Tannreuther (1923) for *E. gracilis*, Baker (1926) for *E. "agilis" (gracilis)*, Ratcliffe (1927) for *E. spirogyra*, Hall (1937) for *E. gracilis*, Krichenbauer (1937) for *E. gracilis* and *E. intermedia*, Hollande (1942) for *E. gracilis*, Leedale (1958a) for *E. acus*, *E. deses*, *E. gracilis*, *E. spirogyra*, and *E. viridis*, Ueda (1958) for *E. gracilis*, and Saitô (1961) for *E. viridis*.

Dangeard (1902, 1910, 1938), Dehorne (1920), and Chadefaud (1939), however, favor the theory that the chromatin in *Euglena* is a continuous spireme which breaks transversely during nuclear division, while Gojdics (1934) reports a transverse cleavage of individualized chromosomes in *E. deses*. Other nonmitotic forms of nuclear division have been described by Haase (1910) for *E. sanguinea*, and by other authors for species of other genera. Some writers, notably Hartmann and von Prowazek (1907), Chatton (1910), Hartmann and Chagas (1910), Alexeieff (1911, 1913), and Drezepolski (1929), have incorporated information on *Euglena* into complex classifications of mitosis.

Most of these studies on nuclear division in *Euglena* (and those on other euglenoid genera) were summarized and critically discussed by the present author in 1958, and the main points to emerge at that time, especially from the work of Keuten (1895), Ratcliffe (1927), Brown (1930), Johnson (1934), Lackey (1934), Hall (1937), Krichenbauer (1937), and Hollande (1942), and from new observations on 7 species of *Euglena* and 17 other euglenoids by Leedale (1958a), may be listed as follows:

(1) Nuclear division in *Euglena* is a mitosis, albeit a peculiar one. The evidence for longitudinal replication of chromosomes is conclusive and theories of transverse cleavage of a spireme or chromosomes must be abandoned. The second criterion of mitosis, that sister chromatids segregate to opposite poles and into different daughter nuclei, is also fulfilled.

(2) During mitosis in *Euglena* a nucleolarlike body (the endosome) persists and divides, retaining at least some of its ribonucleic acid (RNA); the chromosomes become arranged parallel to the division axis at metaphase, after undergoing different patterns of movement and orientation in different species; chromosome replication occurs (or at least first becomes visible) at different stages of the process in different species.

(3) The mitotic mechanism does not appear to involve a spindle or centromeres. Evidence for the absence of the "conventional" spindle of classic mitosis is provided by the lack of inhibition of the mitosis by colchicine, the staggering of anaphase movement (some chromatids reaching the poles while others are still at the equator), the exceptionally long duration of anaphase, and the low chromatid velocity.

(4) The mitosis is intranuclear, the nuclear envelope apparently persisting throughout the entire process.

Knowledge of these and other aspects of mitosis in *Euglena* has been extended subsequently, particularly the time scale of the process (Leedale, 1959a), its periodicity (Leedale, 1959b; Huling, 1960) and, reported here for the first time, its ultrastructure (Section III,C). Naturally, all the authors cited above (and others such as Zumstein, 1900; Bělař, 1926; Günther, 1927; Mainx, 1927; Chu, 1947; Gojdics, 1953; Huber-Pestalozzi, 1955; Pringsheim, 1956; and Grell, 1964) also discuss the interphase nucleus of *Euglena*, studies of which have likewise recently progressed to examination with the electron microscope (Ueda, 1958; de Haller, 1959; Gibbs, 1960; Leedale *et al.*, 1965; O'Donnell, 1965; Leedale, 1967). By contrast, there are relatively few studies on the nuclear chemistry of *Euglena*.

Amitosis has been recorded for three species of *Euglena* (Leedale, 1959c) as a nuclear fragmentation not connected with reproduction. Sexuality has been reported for the genus several times (Köllicker, 1853; Haase, 1910; Biecheler, 1937) but there are no descriptions of meiosis in *Euglena*.

The above-mentioned areas of study have recently been summarized and briefly discussed for an extended range of euglenoid flagellates (Leedale, 1966, 1967). The present account is therefore confined to a more detailed consideration of nuclear structure and chromosome behavior within the genus *Euglena*, especially the ways in which the most recent researches make it necessary to modify ideas on the mechanism of euglenoid mitosis.

II. The Interphase Nucleus in *Euglena*

The interphase nucleus lies in the central or posterior region of the cell in all species of *Euglena* (Figs. 1, 12, and 13), but during euglenoid movement (Chapter 3) it may be moved about the cell and is capable of considerable changes of shape. The nucleus is spherical in most spindle-shaped species (Figs. 1–3, 12, 13, and 35), and ovoid or long and narrow (Figs. 4, 5, and 65) in elongate cells. Chu (1947) has shown a fairly strict correlation between cell size and the size of the nucleus in *Euglena*. In the largest species (such as *E. ehrenbergii*), the nucleus is at least $30 \times 15 \mu$ in dimensions (Figs. 4–6), but in small species the spherical nucleus may be less than 2μ in diameter.

The chromosomes retain their condensed condition throughout interphase and the chromatin threads and endosomes can always be seen in stained and unstained nuclei (Figs. 1–6, 12, 13, 35, 36, and 65–69). Living nuclei can be burst from cells without difficulty, maintaining their normal appearance for some time (Fig. 4). After about one-half hour in water, the chromosomal mass begins to contract away from the nuclear envelope as the nucleus swells (Fig. 5). Manipulation with glass needles demonstrates the plasticity of the nucleus and the strength of the nuclear envelope, despite the fact that the latter has presumably been torn away from the endoplasmic reticular system with which it is normally continuous (see Chapter 4).

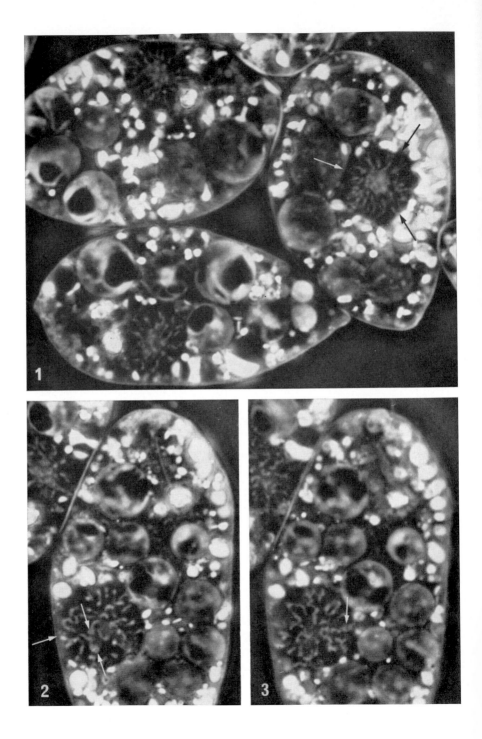

A. Chromatin in the Interphase Nucleus

The interphase chromatin in stained preparations of *Euglena** appears as granular or filamentous threads (Figs. 6, 35, 36, and 65–69). In some species *(E. communis, E. gracilis)* there is a range of nuclear form, from a compact state with thick chromatic threads (Fig. 36) to a diffuse state with chromatin of more granular appearance (Fig. 35). The compact form in *E. communis* is either a stage in reconstitution of the true interphase condition following mitosis, or is the prophase of the *next* mitosis (see page 206). However, this is not the case in *E. gracilis* in which the granular appearance is always established during late anaphase (page 204). In this species, therefore, the occasional compact interphase nucleus is presumably a reflection of a particular physiological state of the cell rather than a normal stage in the mitotic cycle.

Ends of individual chromosomes can be seen in stained nuclei of most species, and relationally coiled chromatids composing the interphase chromosomes are occasionally visible (Figs. 6 and 36) in those species in which this duality is established during telophase of the previous mitosis (page 209). This condition is not uniform throughout the genus. Indeed, in some species *(E. acus, E. spirogyra)* the chromatids become separately visible only during mitotic metaphase, immediately before they segregate into different nuclei (see page 214).

In living nuclei viewed by contrast microscopy, the chromatin may appear more uniformly filamentous (Figs. 1–5, and 13) than in fixed nuclei, suggesting that the granular appearance of the latter is partly due to differential staining. In some nuclei the chromosomes radiate from the central endosome (Fig. 1), while in others (even of the same species) the chromosomes coil haphazardly throughout the nucleoplasm (Figs. 2–5, and 13). The dual nature of the chromosomes can be seen at certain points in appropriate species (Fig. 2),

---

* The most satisfactory stains for euglenoid nuclei are acetocarmine (following a ferric acetate mordant), acetic orcein, the Feulgen reaction, and various hematoxylins (see Leedale, 1958a). The stained cells shown in the present account have been fixed in methanol or Carnoy's acetic alcohol and stained with acetocarmine. In some preparations (e.g., Figs. 36–42) the stain has been removed from the endosome by differentiation with 45% acetic acid, to clarify details of chromosome organization. Fixation and staining methods will be found in Godward (1966). The photographs of untreated living nuclei in the present account have been made using the Reichert system of anoptral contrast light microscopy.

Figs. 1–3. Living cells of *E. deses*. Anoptral contrast, ×2000.

Fig. 1. Three cells to show the shape, size, and position of the interphase nucleus (arrowed in one cell); each nucleus has chromosomes radiating from a central endosome. *Figs. 2 and 3.* Two focal levels of one cell; the dual structure of the condensed chromosomes shows at certain points (arrows).

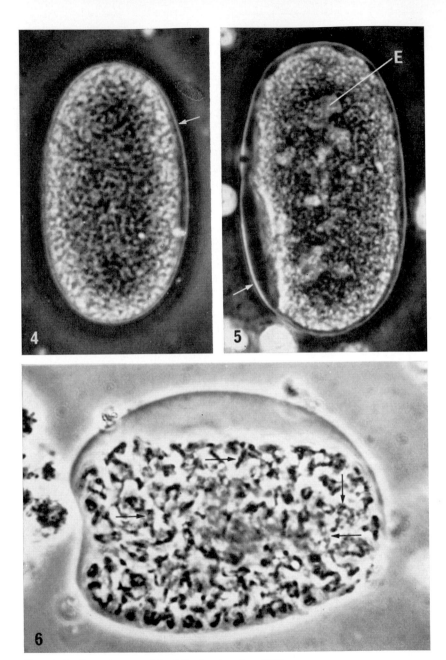

Figs. 4–6. Interphase nuclei of *E. ehrenbergii.* ×2000.

Fig. 4. Freshly isolated living nucleus, focused to show the many interphase chromosomes. Arrow, nuclear envelope. *Fig. 5.* Isolated living nucleus after 30 minutes in water; the chromosomal mass has contracted away from the nuclear envelope (arrow) and this focal level shows the several endosomes (E). *Fig. 6.* Isolated fixed and stained nucleus, phase contrast; some of the interphase chromosomes (e.g., those arrowed) show relationally coiled chromatids.

indicating that the organization of fixed nuclei is not too different from the living condition. Observations with the electron microscope are discussed below (Section II,C), although very little of significance has yet been discovered concerning the ultrastructure of the chromosomes.

B. ENDOSOMES IN THE INTERPHASE NUCLEUS

Species of *Euglena* have one to several endosomes in the interphase nucleus (Figs. 1, 5, 7, 12, 35, 65, 68 and 69). The nature of the euglenoid endosome has been a topic of considerable argument but there now seems little doubt that it is homologous with the nucleolus of other organisms, and has indeed been interpreted as such by numerous authors from Dangeard (1902) onward. It stains with hematoxylins, eosin, acetocarmine, orange G, and light green (although some of these dyes are easily removed again) and is Feulgen-negative (Leedale, 1958a). Staining for light microscopy with pyronin B and methyl green by the Unna Pappenheim technique in combination with nuclease treatment (Brachet, 1940, 1953), indicates the presence of RNA but no deoxyribonucleic acid (DNA). However, O'Donnell (1965) identifies at the fine structure level a component in the endosome which is degraded by DNase. It therefore seems likely that the chemistry of the endosome is similar to that of nucleoli of higher organisms (see, for example, Mirsky and Osawa, 1961; Frey-Wyssling and Mühlethaler, 1965; Kihlman, 1966).

The reason for using the term "endosome" (Minchin, 1912; Calkins, 1926; Leedale, 1958a) instead of "nucleolus" lies in the particular behavior of the body during euglenoid mitosis (its persistence, division, and retention of RNA). At the light microscope level there is so far no direct evidence of association with particular (nucleolar-organizing) chromosomes, but O'Donnell (1965) has described such an association from ultrastructural studies.

The irregular behavior of endosomes at mitosis in some euglenoid species seems to argue against the presence of nucleolar-organizing chromosomes. The several endosomes of certain interphase nuclei can either fuse during prophase and divide as a single body or remain distinct and divide separately (Leedale, 1958a). The number of dividing endosomes in *E. acus* can vary from 2 to 20 in different nuclei. Similarly, the several endosomes of the *E. spirogyra* nucleus fuse prior to mitosis and divide as a single body, but in binucleate cells (Section IV) each "half-nucleus" has its own dividing endosome. Some evidence (page 214) suggests that euglenoids are polyploid (and would therefore have several nucleolar-organizing chromosomes per nucleus); other evidence (page 215) indicates a haploid condition. However, even if polyploidy is the rule, one might expect more regularity of endosome division if it is governed by nucleolar-organizing chromosomes. In higher

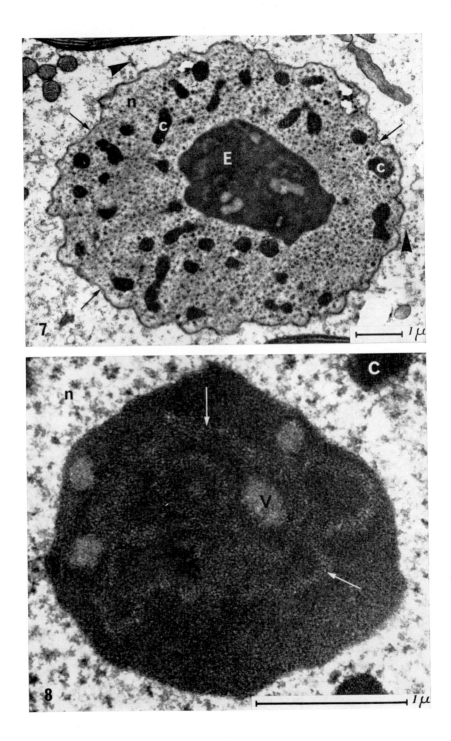

plants there is usually one nucleolar organizer in each haploid set of chromosomes, but the separate nucleoli formed during telophase (two in a diploid cell) tend to fuse so that most interphase diploid cells have only one nucleolus. The ability of euglenoid endosomes to fuse and separate can be claimed as another similarity to nucleoli.

In some species of *Euglena* the endosome appears homogeneous in the light microscope, but vacuoles are often present in the endosomes of stained nuclei (Ratcliffe, 1927) and more densely staining tracks are sometimes seen in the endosome after partial differentiation (Fig. 35). These features presumably relate to differences in density revealed by electron microscopy (Ueda, 1958; de Haller, 1959; Gibbs, 1960; O'Donnell, 1965; Figs. 7 and 8). Leedale (1958a) interpreted the vacuoles in stained nuclei as fixation artifacts but it now seems that they indicate genuine structure, either invaginations of nucleoplasm or possibly nucleolar-organizing regions within the endosome. Fine structural observations on this point are discussed below (Section II,C).

Systems of rhizoplasts from the basal bodies of the flagella to the endosome (or to a related body on the nuclear surface) have been described for *Euglena* by Baker (1926) and Ratcliffe (1927), but Krichenbauer (1937), Hollande (1942), and Leedale (1958a) found no evidence for such structures and believe them to be absent. Recent detailed studies on the ultrastructure of *Euglena* (de Haller, 1959; Gibbs, 1960; Leedale *et al.*, 1965; Leedale, 1967) have also failed to reveal rhizoplasts, while the easy migration of nuclei from one daughter cell to the other during cleavage (Leedale, 1959d) also argues against there being any permanent connection between the nucleus and flagellar bases.

C. Ultrastructure of the Interphase Nucleus

The ultrastructure of the fixed and sectioned interphase nucleus of *E. gracilis* is illustrated in Fig. 7.* Within the confines of a nuclear envelope,

---

* Sectioned material illustrated in the present account has been prepared for electron microscopy by fixation either in 2% osmium tetroxide in 0.1 M phosphate buffer at pH 7, or in variously buffered 2–6% glutaraldehyde with postosmication in 1% osmium tetroxide. All material is Epon-embedded and double-stained with uranyl acetate and lead citrate.

Fig. 7. Interphase nucleus of *E. gracilis*, thin section. C, Chromosomes; E, endosome; n, nucleoplasm; arrows, nuclear envelope; arrowheads, elements of tubular endoplasmic reticulum pushing out into the cytoplasmic matrix from the nuclear envelope. Electron micrograph, × 12,000.

Fig. 8. Endosome of *E. gracilis* in an interphase nucleus, thin section; the substance of the endosome contains "vacuoles" (V) and a tortuous "track" (arrows) of low density, see text. C, Chromosome, n, nucleoplasm. Electron micrograph, × 40,000.

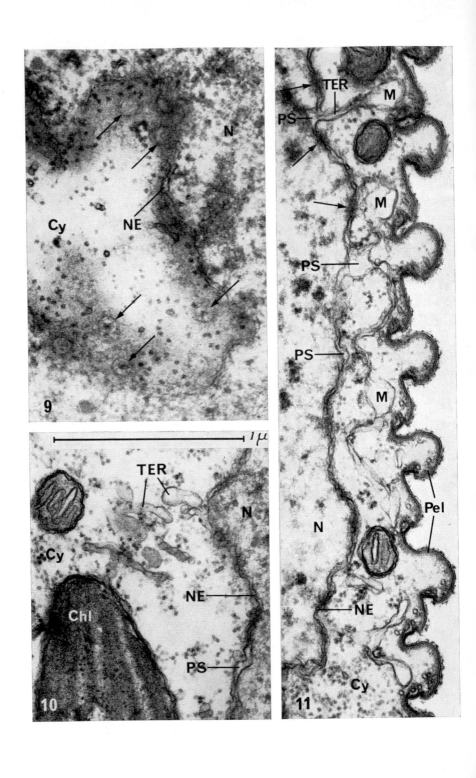

the chromosomes are evenly distributed in the nucleoplasm around the more-or-less central endosome.

As in all eucaryotic cells, the nuclear envelope consists of two unit membranes (Robertson, 1959) separated by the perinuclear space and traversed at intervals by pores. Each membrane of the euglenoid nuclear envelope has a thickness of approximately 80 Å and the rather variable perinuclear space (between the two membranes) averages 200 Å in width (Figs. 10, 11, 57, and 58). The nuclear pores in transverse section (Figs. 11 and 58) are seen to cross both membranes but often appear to be blocked by dense material (Fig. 11). In surface view (Fig. 9) the pores are circular, with a diameter of approximately 800 Å—considerably larger than the pores in the envelopes of most higher animal and plant nuclei (Afzelius, 1955; Whaley *et al.*, 1960). It is generally assumed (see, for example, Frey-Wyssling and Mühlethaler, 1965) that this porosity of the envelope allows for rapid translocation of macromolecules between the nucleoplasm and the cytoplasmic matrix (groundplasm).

The perinuclear space is properly regarded as a special region of the endoplasmic reticulum (ER) (Chapter 4). Tubular extensions of the outer membrane of the nuclear envelope proliferate into the cytoplasmic matrix, connecting the perinuclear space to other elements of the ER (Figs. 7, 56, and 58). For example, branches of this tubular ("smooth") ER link up with the ER sheath associated with each chloroplast (Fig. 10). Such an association between ER and plastids has been demonstrated for Chrysophyta and Phaeophyta (Manton, 1966) but not for Chlorophyta, yet one more piece of evidence that the close taxonomic relationship suggested for *Euglena* and the green algae is a false one (see Leedale, 1967). Another interesting association is that between the perinuclear space and the muciferous bodies (Chapter 4); the latter have been shown (Leedale, 1964) to be the peripheral elements of the vesiculate ("rough") ER in *Euglena*. In regions where the nucleus lies close to the pellicle (Fig. 11), extensions of the perinuclear space can be seen to be immediately transformed into the muciferous bodies which

---

Figs. 9–11. Details of the nuclear envelope in *E. gracilis*. Electron micrographs, × 50,000.

Fig. 9. Glancing section of the nuclear envelope (NE) showing nuclear pores in surface view (arrows). Cy, Cytoplasmic matrix with ribosomes, N, nucleus. *Fig. 10.* Tubular endoplasmic reticulum (TER) connecting the perinuclear space (PS) of the nuclear envelope (NE) to the endoplasmic reticular sheath of a chloroplast (Chl). Cy, Cytoplasmic matrix with ribosomes; N, nucleus. *Fig. 11.* Section of a region where the nucleus (N) lies close to the pellicle (Pel); the perinuclear space (PS), tubular endoplasmic reticulum (TER), and muciferous bodies (M) form a continuous system; see text. Cy. Cytoplasmic matrix with ribosomes; NE, nuclear envelope; arrows, nuclear pores in transverse section.

underlie the pellicular strips in helical array. This intricate ER system of tubes and vesicles must be highly labile, since the nucleus is easily moved about the mobile cell of *E. gracilis.*

The sectioned chromosomes have a densely filamentous construction which O'Donnell (1965) interprets as moderately coiled chromonemata approximately 100 Å in diameter. The nucleoplasm appears as a heavily granular material in a less dense matrix (Fig. 7), with delicate strands resolvable throughout the system. However, the variation of the electron microscopic image is so great with different methods of fixation, different culture conditions, and even within different cells from one and the same fixation, that it seems premature to attempt detailed interpretation of these structures in terms of the molecular organization of the chromosomes. The main point of interest at present is the high degree of condensation shown by the interphase chromosomes of *Euglena* compared with the dispersed condition in most other organisms. Even so, the resolvable filamentous component has approximately the same dimensions as the "elementary chromosome fibril" described for *Tradescantia* by Ris (1961).

In the electron microscope the endosome appears as a body of rather irregular cross section (Figs. 7 and 8), dense in comparison with the surrounding nucleoplasm but slightly less dense than the material of the chromosomes. Within the substance of the endosome are regions of low electron density (Fig. 7) which may be internal cavities or, more probably, invaginations of nucleoplasm. These possibly correspond to the vacuoles seen in the light microscope. In no case are the endosomal cavities membrane-bounded, and neither is the endosome itself ever delimited from the nucleoplasm by a membrane.

Regions of density intermediate between that of the cavities and the main substance of the endosome are also seen (Fig. 7) and, in favorable sections (Fig. 8), these form a continuous, tortuous track which may correspond to the staining "track" seen in certain light micrographs (Fig. 35). Within the dense material of the endosome (Fig. 8) it is possible to resolve coarse granules about the size of ribosomes (150 Å in diameter) and also delicate fibrils. Filaments of varying thickness are described by O'Donnell (1965), but such "nucleonema" originally described for the nucleoli of higher organisms have more recently been ascribed to poor fixation (see Frey-Wyssling and Mühlethaler, 1965) and O'Donnell's highly detailed analysis of filamentous components in the *Euglena* endosome is questionable. The quality of the published micrographs makes it difficult to evaluate her conclusions. In the untreated control cell of *E. gracilis* in O'Donnell's material (O'Donnell, 1965, Fig. 9), the nuclear envelope is not preserved and much of the chromosomal substance has disappeared, hence little reliance can be placed upon the interpretation of enzyme-treated nuclei in which O'Donnell relates the molecular organization of the endosome to that of the chromosomes.

Nevertheless, the overall picture from staining reactions, light microscopy, and electron microscopy (Section II,B and C) does suggest that the euglenoid endosome is a nucleolus with some form of internal organization (which perhaps governs the elongation and distribution of material during mitosis, rather than its dispersion and reformation). The general observation by O'Donnell (1965) that RNase moderately degrades one nucleolar component in *E. gracilis* while DNase and proteinases degrade another component is interpreted by her as evidence that the endosome is an ordinary nucleolus with an organizer of chromosomal nature. At present, it seems eminently reasonable to retain the term "endosome" for this body within the nucleus of *Euglena*, whether it proves to be merely a type of nucleolus which persists throughout the mitotic process, or a structure of more specialized cytological significance.

III. Mitosis in *Euglena*

A. Observations with the Light Microscope

It is the intention to begin this section by combining the descriptions of mitosis given by Keuten (1895), Tschenzoff (1916), Tannreuther (1923), Baker (1926), Ratcliffe (1927), Hall (1937), Krichenbauer (1937), Hollande (1942), Leedale (1958a), and Saitô (1961) into a single paragraph which will serve as a point of departure for more detailed consideration of special variations and controversial aspects of the subject.

Mitosis in *Euglena* begins with a forward migration of the nucleus so that it comes to lie immediately posterior to the reservoir (Chapter 4). In species with several endosomes in the interphase nucleus, these usually fuse to form a single body. The endosome then elongates along the division axis (perpendicular to the long axis of the cell) and the chromosomes orient into the metaphase* position. Three forms of orientation have been distinguished (Leedale, 1958a, 1966, 1967): (1) In *E. gracilis* (Section III,A,1), pairs of chromatids from late prophase orient into a circlet of *single* chromatids, separation and segregation having occurred during orientation (Hall, 1937); (2) In *E. communis* (Section III,A,3) and *E. viridis*, pairs of chromatids from interphase and/or prophase come to lie along the division axis *still as pairs*, parallel to one another and to the elongated endosome; (3) In *E. acus* and

---

* As in previous accounts, the terms *prophase, metaphase, anaphase*, and *telophase* will be retained for the stages of mitosis in *Euglena*, without implying the chromosomal structure and arrangement typical for these stages in classic mitosis.

E. spirogyra (Section III,A,4), *single* chromosomes from prophase line up along the division axis and there undergo duplication into the pairs of chromatids of that mitosis. The degree of variation in the authors' descriptions suggests that these "categories" overlap to a certain extent, species differing mainly in the time at which the double structure of the chromosomes first becomes microscopically visible. In all cases, the endosome continues to elongate (in some polyendosomal nuclei such as that of *E. acus* there may be numerous dividing endosomes at this stage) and the chromatids segregate toward the ends of the endosome into two daughter groups. Separation, segregation, and anaphasic movement of the chromatids is irregular and this, coupled with a very low chromatid velocity, results in an extremely long anaphase (see Section III,D). The end of anaphase is marked by a sudden flowing to the poles of the central region of the elongated endosome, and the persistent nuclear envelope seals around the groups of chromatids and the daughter endosomes to form the telophase nuclei.

This process will now be considered in detail for one example of each of the three orientation types listed above, beginning with previously unpublished observations on living material of *E. gracilis*. The special value of such a study is that the dynamic process can be observed directly, rather than reconstructed from a series of stills. The close agreement between the structures seen in the living nuclei and in stained nuclei considered elsewhere is reassuring to the artifact-conscious cytologist.

1. *The Mitotic Cycle in Living E. gracilis*

The living mitosis illustrated here (Figs. 12–28) is seen in cells which are flattened to allow for sharply focused photography. Most of the cells are flattened beyond the point of recall to continued life, but the mitotic figures themselves are not distorted. It is, in fact, easy to follow the entire mitosis in one and the same cell in a ringed preparation (and this is done when timing the duration of mitotic stages, Section III,D), but such a cell does not permit clear photography of the nucleus. For this reason, the mitotic process is illustrated here in a series of cells rather than in one. The legends to the figures name the stage of mitosis; the following commentary selects salient features of nuclear structure (with occasional reference to other organelles

---

Figs. 12–28. The mitotic cycle in living cells of *E. gracilis* (see text for commentary); the anterior end of the cell is at the top in each figure, the dividing nucleus is seen in side view. Anoptral contrast, ×2000.

Fig. 12. Cell with interphase nucleus (N) showing the endosome (arrow) and "granular" chromatin. R, Reservoir with adjacent contractile vacuolar system. *Fig. 13.* Interphase, the chromosomes in focus. *Fig. 14.* Prophase. R, Reservoir; arrow, endosome. *Fig. 15.* Early metaphase.

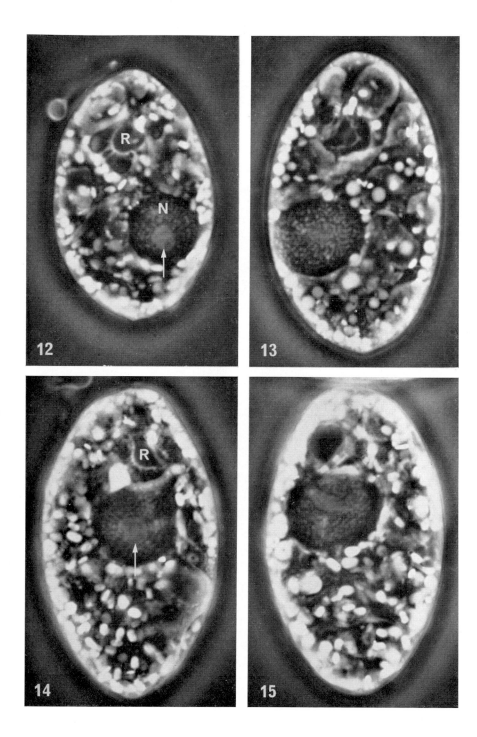

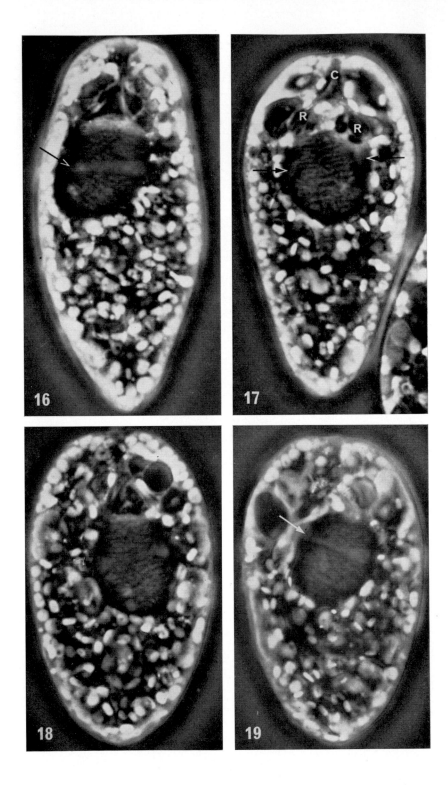

where appropriate). It is hoped that the reader will find it instructive to make a detailed examination of the pictorial sequence for himself.

The interphase nucleus of *E. gracilis* has a single endosome (Fig. 12) surrounded by a mass of chromosomes (Figs. 12 and 13). The nucleus preparing for mitosis migrates from the interphase position toward the anterior of the cell until it lies immediately posterior to the reservoir (Fig. 14). By this time the nucleus may be said to be in prophase, with the filamentous chromosomes still tangled around the endosome but showing some signs of straightening out and replicating into pairs of chromatids. The endosome now begins to elongate along the division axis (Fig. 15) and the chromosomes are moving to the metaphase position. As indicated above, the pairs of chromatids established during prophase apparently orient into a circlet of single chromatids, separation and segregation of the daughter chromosomes occurring during orientation (Hall, 1937; Leedale, 1958a, 1966, 1967). Midway through this process the endosome appears as a cylinder of almost uniform diameter along the division axis (Fig. 16), surrounded by the separating and segregating chromatids. Endosomal material continues to move to the poles and the elongated endosome becomes dumbbell-shaped (Figs. 17–19), with its swollen ends extending beyond the "metaphase–anaphase" circlet of chromosomes. Sharp focusing on the chromosomes (Fig. 17) reveals a preponderance of threads (segregated chromatids) parallel to the endosome and the division axis; other threads which are less regularly parallel are chromatids still in the act of separating (Fig. 18 shows a slightly earlier stage than Fig. 17). It is clear from these division figures (which show only one focal level of the nucleus) that the chromosome number is high, a feature discussed later (Section III,B).

This stage (Fig. 17), with most but not quite all of the daughter chromosomes separated, must be called "metaphase" in *E. gracilis*. It will be seen that during this early-to-late metaphase succession, the locomotor apparatus (flagella, photoreceptor, and eyespot) replicates and the reservoir divides (Fig. sequence 14 → 15 → 16 → 18 → 17). The daughter reservoirs open into the still single canal (Fig. 17), but each now has its own contractile vacuole, eyespot, and two flagella.

The mitotic process continues with elongation of the entire nucleus, further elongation of the endosome, and the start of the anaphase migration of chromatids (Fig. 20). As anaphase progresses, a central gap appears in the chromosomal mass (Fig. 21) and a central constriction of the nucleus

---

Figs. 16–19. Mitosis in living *E. gracilis* (continued). ×2000.

Fig. 16. Early metaphase. Arrow, cylindrical endosome. *Fig. 17.* Metaphase. C, Canal; R, daughter reservoirs; arrows, swollen ends of endosome. *Fig. 18.* Metaphase. *Fig. 19.* Metaphase, focused on the endosome (arrow).

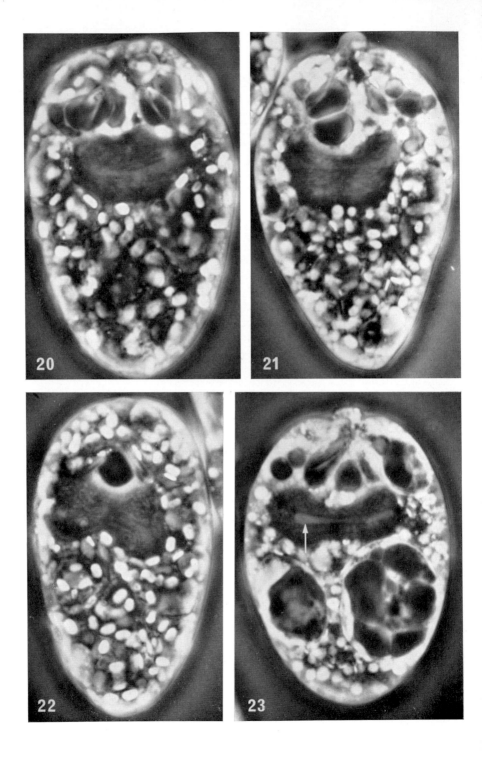

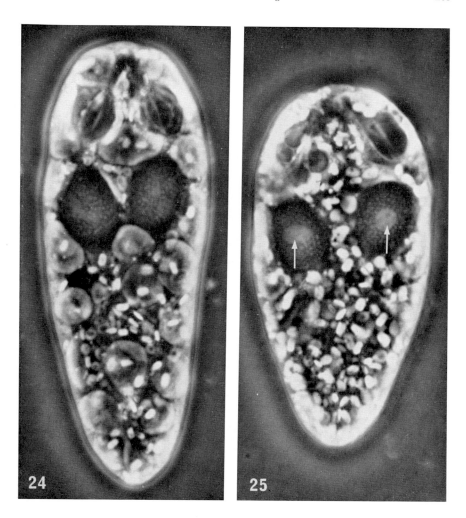

Figs. 24 and 25. Mitosis in living *E. gracilis* (continued). ×2000.

Fig. 24. Very late anaphase.   *Fig. 25.* Telophase, focused on the daughter endosomes (arrows).

Figs. 20–23. Mitosis in living *E. gracilis* (continued). ×2000.

Fig. 20. Early anaphase.   *Fig. 21.* Midanaphase.   *Fig. 22.* Late anaphase. *Fig. 23.* Late anaphase, focused on the endosome (arrow).

becomes apparent. It is also noticeable that the chromatids (daughter chromosomes) are less easily distinguished from the nucleoplasm (Figs. 20 and 21) than at metaphase (Figs. 17 and 18). During late anaphase (Fig. 22) the chromatin returns to the more granular condition typical of interphase nuclei. The elongated endosome is now very thin in the midregion of the division figure (Figs. 21 and 23) and during the final stages of nuclear constriction (Fig. 24) the central endosomal substance moves to the poles to be incorporated into the daughter endosomes. These often show variations in density (Fig. 25) similar to the vacuoles seen in stained preparations (Section II,B).

Once telophase is established (with separate daughter nuclei), one of the two flagella in each daughter reservoir grows to emerge as a locomotory flagellum (Fig. 26). A cleavage line is initiated between the now distinct daughter canals (Fig. 26) and progresses helically backward (Figs. 27 and 28) to separate the daughter cells. The nuclear structure typical for the interphase cell of *E. gracilis* (Figs. 12 and 13) is fully established in telophase (Figs. 25 and 26) before cell cleavage begins.

This descriptive interpretation of the mitotic process in *E. gracilis* incorporates some cytological problems of no mean order. The division of the endosome–nucleolus has already been partially discussed (Section II,B) and is enlarged upon below, but the unexplained nuclear migration, the lack of any reference to centromeres, and the irregularity of chromatid separation and segregation demand some comment. As these problems become even more acute in the species of *Euglena* still to be considered, the reader's indulgence is requested until more general discussion of these points is attempted (Sections III,A,4 and III,E).

2. The Endosome Cycle in E. gracilis

To supplement information on endosome behavior in *Euglena*, brief mention will be made of stained nuclei in *E. gracilis*.

The main point here concerns O'Donnell's (1965) contention that this species has a normal nucleolus which disorganizes at prophase and is reorganized by a particular chromosome in telophase. This is in disagreement with Leedale's (1958a) description of "persistent nucleolus" but O'Donnell

---

Figs. 26–28. Mitosis in living *E. gracilis* (continued). ×2000.

Fig. 26. Telophase, the daughter reservoirs (arrows) now distinct, with two flagella in each, and separate canals (C). *Fig. 27.* Cell cleavage. *Fig. 28.* Cell cleavage. C, Daughter canals; Chl, chloroplasts; e, daughter eyespots; N, daughter nuclei; R, daughter reservoirs (each with two flagella); arrow, flagellar basal bodies on wall of reservoir.

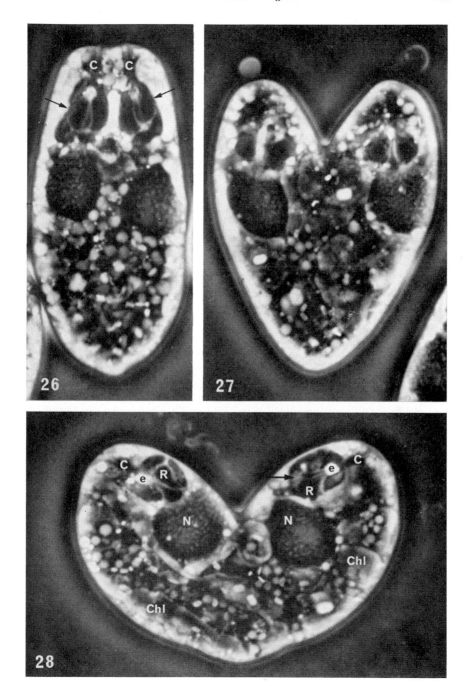

records that "Very early at anaphase, some nucleolar material remained attached to the nucleolar chromosome. This remnant nucleolar material was drawn by each nucleolar chromosome, moving towards the pole."

In the stained prophase nucleus of *E. gracilis*, the spherical endosome is surrounded by the typical prophase chromosomes (Fig. 29). The endosome elongates into a cylinder during early metaphase (Fig. 30) and by full metaphase is dumbbell-shaped with most of the endosomal substance at the poles, *well beyond the circlet of chromosomes* (Fig. 31). Such a division figure squashed flat (Fig. 32) shows the endosome to be one continuous structure (and the chromosome number to be much higher than the four counted by O'Donnell (1965), of which "the nucleolar chromosome was easily identified at all times of mitosis through its characteristic morphology"; see Section III,B). During anaphase (Fig. 33), the chromosomes migrate to the poles and the endosomal material concentrates there also, with a thinning in the central region. Continued elongation of the endosome sometimes leads to a buckling of the center in late anaphase (Fig. 34) and the material finally moves to the poles to form the daughter endosomes. Staining of division figures by the Unna Pappenheim technique in combination with nucleases (page 191) indicates that RNA is retained by the endosome throughout this cycle (Leedale, 1958a).

3. The Chromosome Cycle in E. communis

The interphase nucleus of *E. communis* (Fig. 35) consists of many granular threads encircling a single large endosome. Possible structure within the endosome has already been discussed (Section II,B) and the present section will be confined to a brief account of chromosome organization as seen in stained nuclei, for comparison with the cycle in *E. gracilis*.

During prophase in *E. communis*, the chromosomes become more obviously filamentous and condensed (Fig. 36) and can be seen to consist of paired chromatids. This duality of the chromosomes is more clearly visible as they orient into the metaphase position (Fig. 37). An "equatorial plate" is never formed, but the relationally coiled chromatids line up roughly parallel to the division axis. That all the chromosomes are double at this stage is most easily seen in squashed-out division figures (Fig. 38), with individual chromosomes showing the relational coiling of their chromatids very clearly.

Anaphase movement now begins (Fig. 39) and leads to a more regular separation of chromatids than in some other species of *Euglena* (compare Fig. 47). Even so, the staggered nature of the segregation becomes apparent as anaphase proceeds, midanaphase (Fig. 40) and late anaphase (Fig. 41) stages exhibiting some lagging chromatids at the equator. These finally move into the groups of daughter chromosomes as the two telophase nuclei

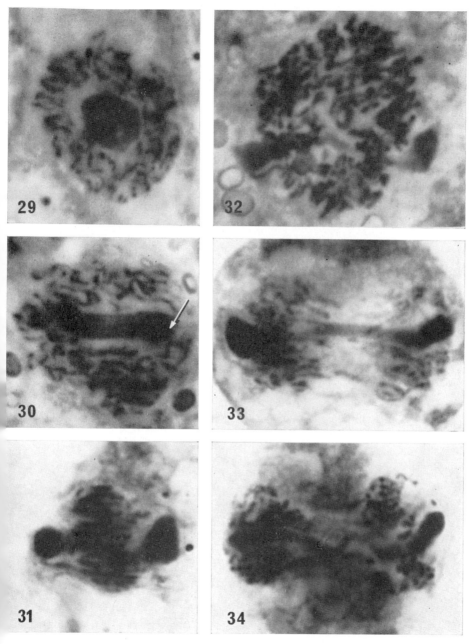

Figs. 29–34. Endosome cycle during mitosis in *E. gracilis* (see text for commentary); the division figure in side view in each case. Acetocarmine stained, ×3000.

Fig. 29. Prophase, endosome "spherical." *Fig. 30.* Early metaphase, endosome cylindrical (arrow). *Fig. 31.* Metaphase, endosome dumbbell-shaped, chromosomes aligned parallel to the division axis. *Fig. 32.* Squashed-out metaphase. *Fig. 33.* Anaphase. *Fig. 34.* Late anaphase, endosome still elongating, chromatin "granular."

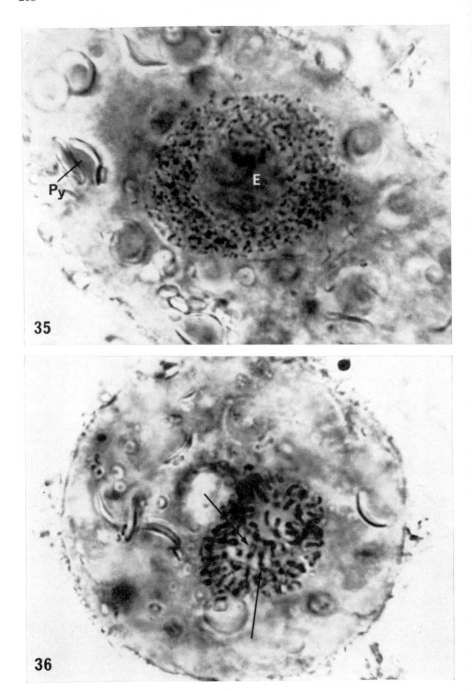

round off (Fig. 42). In some cases it is possible to see the chromatids of the *next* mitosis in individual chromosomes of the telophase nuclei.

This chromosome cycle is similar to that in *E. viridis*, for which species Saitô (1961) has produced elegant micrographs of the relationally coiled chromatids. Saitô claims to see submedian, subterminal, and terminal centromeres in the chromosomes of *E. viridis*, but most authors have failed to observe constrictions or nonstaining regions which they could interpret as centromeres. During anaphase movement, chromatids appear as rods, curves, or U-shapes, the last-mentioned with the curve away from *or toward* the division equator.

4. *The Chromosome Cycle in E. spirogyra*

According to Leedale (1958a), *E. spirogyra* exhibits the third type of prophase-to-metaphase chromosome orientation listed on page 197.

In differentiated stained nuclei, the *single* prophase chromosomes are seen to coil haphazardly around the central endosome (Figs. 43 and 44). The endosome elongates during early metaphase and the still single chromosomes orient into the metaphase position to lie parallel to the division axis. In this position they begin to duplicate longitudinally into chromatids (Fig. 45) and an optical transverse section of full metaphase (Fig. 46) shows each chromosome to consist of two chromatids. Anaphase now occurs by the chromatids separating and segregating into the daughter nuclei. As in other species, separation, segregation, and anaphase movement of the chromatids are staggered, and a typical midanaphase picture is a complete spread of chromatids from equator to poles (Fig. 47). It should be noted that the thickness of each chromatid passing to the poles in anaphase (Fig. 47) is equal to that of one of the two sister chromatids in any metaphase pair (Fig. 46). Continued anaphase migration of chromatids leads to formation of the two daughter nuclei, with a few chromatids lagging at the equator until late anaphase (Fig. 48) in the typical euglenoid manner.

The general problem of mitotic mechanism in *Euglena* is discussed below (Section III,E), but the special problem of chromatid segregation is best considered here. In *E. gracilis* the separation of chromatids as the chromosomes orient into the metaphase position (page 201) also segregates sister

---

Figs. 35–42. Mitosis in *E. communis* (see text for commentary); the division figure in side view in each case. Acetocarmine stained, differentiated, × 3000.

Fig. 35. Interphase nucleus, with endosome (E) showing stained "track"; chromosomes "granular"; pyrenoids (Py) with double paramylon sheaths show clearly in the surrounding cytoplasm.　　*Fig. 36.* Cell with prophase nucleus, the condensed chromosomes showing relationally coiled chromatids (especially at arrowed points).

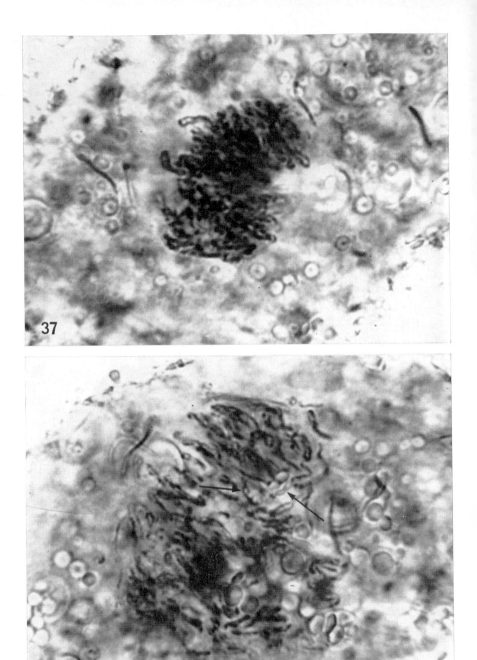

Figs. 37 and 38. Mitosis in *E. communis* (continued). ×3000.

Fig. 37. Metaphase.   *Fig. 38.* Squashed-out metaphase, showing the relationally coiled chromatids of each chromosome (arrows, etc.).

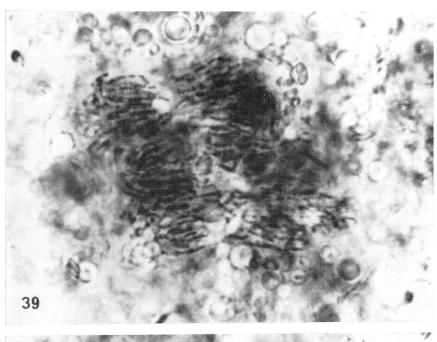

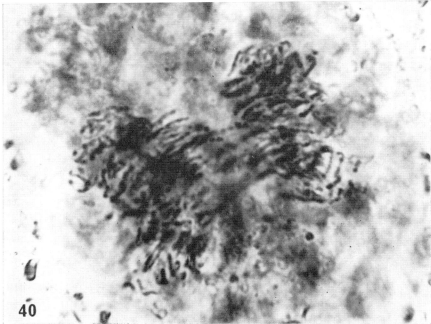

Figs. 39 and 40. Mitosis in *E. communis* (continued). ×3000.

Fig. 39. Early anaphase.   *Fig. 40.* Anaphase.

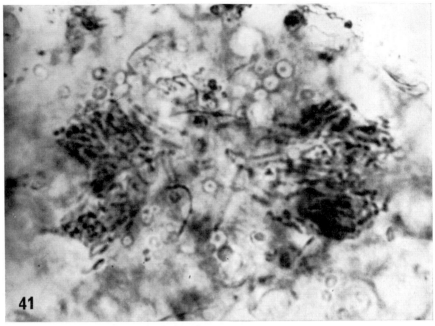

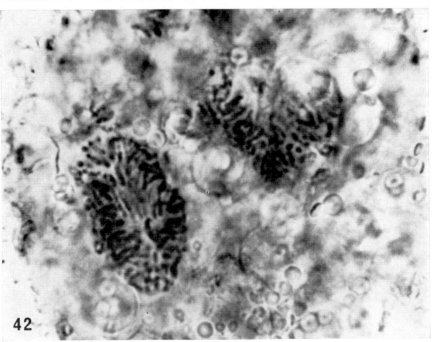

Figs. 41 and 42. Mitosis in *E. communis* (continued). ×3000.

Fig. 41. Late anaphase.　　*Fig. 42.* Telophase.

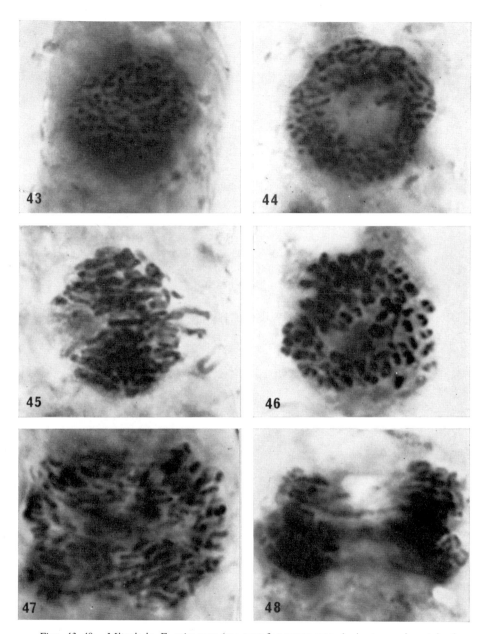

Figs. 43–48. Mitosis in *E. spirogyra* (see text for commentary). Acetocarmine stained, ×3000.

Fig. 43. Prophase, surface in focus.   *Fig. 44.* Prophase, optical section of the same nucleus as in previous figure.   *Fig. 45.* Metaphase, side view.   *Fig. 46.* Metaphase, polar view (optical section), showing each chromosome to consist of a pair of chromatids; this nucleus gives a chromosome count of 86.   *Fig. 47.* Anaphase, side view.   *Fig. 48.* Late anaphase, side view.

chromatids toward opposite poles (Hall, 1937). In *E. communis* and *E. viridis*, although sister chromatids apparently lie side by side parallel to the division axis at metaphase, the chromatids are distinct at prophase and it is quite sensible to suggest that some form of simultaneous separation and segregation occurs, but at a slower rate than in *E. gracilis*. In *E. spirogyra*, however, the sister chromatids are not distinguishable until after the chromosomes have come to lie along the division axis. Why one chromatid of such a parallel pair should move in one direction and the other in the opposite direction (by "parallel displacement") was not explained by either Krichenbauer (1937) or Hollande (1942), the authors who give the clearest description of the phenomenon, and Leedale's (1958a) suggestion that mutual repulsion between sister chromatids deflects them outward and finally into different nuclei has a slight air of desperation. It is worth pointing out that possession of localized centromeres would, if anything, aggravate the problem, since terminal centromeres, for example, would both be toward the *same* pole of the mitotic figure. It must be admitted that on present information the problem is intractable. Interpretation of the dynamic process that occurs in the nucleus of *E. spirogyra* between mitotic metaphase (Figs. 45 and 46) and anaphase (Fig. 47) must await the results of more refined cytological techniques, detailed ultrastructural investigation and, probably most desirable of all, high-speed cinemicrography of the living organism.

B. Chromosome Numbers Species of in *Euglena*

Evidence presented above indicates that high chromosome numbers are the rule for species of *Euglena*. Available counts and estimates are assembled in Table I, but it is only fair to point out that most of the investigators of euglenoid mitosis cited in the present account clearly show similar high numbers of chromosomes in the illustrations for their papers. It will be seen that Dangeard's (1902) counts are of the same order as Leedale's even though he was counting what he identified as "chromospires" of a continuous spireme. The anomalous count by O'Donnell (1965) comes from her electron microscopic study discussed at several points above.

In 1956, Pringsheim wrote of the genus *Euglena*, "The number of chromosomes has not been established in any case, owing no doubt to difficulty caused by their number, length and dense accumulation. This is to be regretted, because polyploidy may be expected in certain groups." The general occurrence of high chromosome numbers is a possible indication of polyploidy (and is a feature of euglenoid flagellates as a whole; Leedale, 1958b). Two species of *Eutreptia* give more positive evidence of polyploidy, the count for *Eutreptia pertyi* ("*lanowii*") being 90 and that for the smaller *Eutreptia viridis* being 44 (Leedale, 1958b). The unchanged morphology

Table I

CHROMOSOME NUMBERS IN THE GENUS *EUGLENA*

Species	Number	Author	Notes
E. geniculata Dujardin	25–30	Dangeard (1902)	"Chromospires"
E. gracilis Klebs	14	Baker (1926)	14 threads in each late anaphase group
E. gracilis Klebs	45	Leedale (1958b)	Metaphase estimate; 30 cells of different strains, including "var. *bacillaris*"
E. gracilis Klebs	4	O'Donnell (1965)	"var. *bacillaris*"
E. pisciformis Klebs	12–15	Dangeard (1902)	"Chromospires"
E. proxima Dangeard	50+	Dangeard (1902)	"Chromospires"
E. spirogyra Ehrenberg	86	Leedale (1958b)	Count; 6 cells; see Fig. 46.
E. splendens Dangeard	35–40	Dangeard (1902)	Rods at anaphase
E. viridis Ehrenberg	42	Leedale (1958b)	Count; 1 cell; late anaphase

of races of *E. spirogyra* with only half, or with one-and-a-half times, their normal chromosome complement (Section IV) also suggests a polyploid condition in this species.

On the other hand, two euglenoids with quite high chromosome numbers, *Colacium mucronatum* (35; Leedale, 1958b) and *Phacus pyrum* (30 to 40; Dangeard, 1902), each have one chromosome which is distinguishable from the others of the complement by its much greater length (Leedale, 1958a, 1966, 1967), suggesting that these two species might be haploid. However, such a cytological situation may equally well be explained as a relic of ancestral polyploidy with some form of chromosomal anomaly (fusion, interchange) preserved within an asexual system.

The apparent absence (or extreme rarity) of sexuality and/or meiosis in *Euglena* (Section V) eliminates the possibility of resorting to meiotic criteria for analysis of the situation, but in the two euglenoids in which a possibly valid meiosis has been recorded, *P. pyrum* (Krichenbauer, 1937) and *Hyalophacus ocellatus* (Leedale, 1962), the complete regularity of chromosome pairing at "diakinesis" suggests that the cell is haploid (even though the chromosome count for *H. ocellatus* at mitosis is approximately 90).

In a taxonomically isolated and caryologically peculiar group such as the euglenoid flagellates, especially where most "species" may be asexual clones, the possible combination of primitively high chromosome numbers, endopolyploidy, ancestral polyploidy with preserved anomalies of varying types, and the sudden loss or gain of half the total chromosome complement, would seem to have a fair chance of defying analysis for some time to come.

C. Ultrastructure of Mitosis in *E. gracilis*

This account consists of preliminary observations from a newly begun study on the ultrastructure of mitosis in *Euglena*. Comments will therefore be confined to points that are relevant to the light microscopy already presented (Section III,A) and to the discussion on mitotic mechanism given below (Section III,E).

At the ultrastructural level, the first sign of approaching division of cell and nucleus in *Euglena* is the appearance of minute daughter pellicular complexes between the mature ones. The nondividing cell shows almost equal-sized pellicular strips in cross section (Fig. 49), associated in a characteristic way with microtubules and muciferous bodies (Leedale, 1964, 1967; see Chapter 4), whereas a cell entering division shows the beginnings of pellicle replication (Fig. 50), small pellicular complexes alternating at certain points with the large ones.

A second, possibly related, indication of a dividing cell is a change in Golgi activity. The relatively quiescent Golgi bodies of the nondividing cell (Fig. 52) show proliferation of small vesicles from the fenestrated cisternal margins. In cells preparing for division, the Golgi bodies are found to be cutting off much larger vesicles with apparently membrane-limited contents (Fig. 53) and, as division proceeds, these vesicles pass through the cytoplasmic matrix to the cell periphery (Fig. 51). The presence of a recognizable time sequence in the sectioned cells (i.e., the later stages of mitosis and cytokinesis) allows for this dynamic interpretation of Golgi activity and prompts the suggestion that the Golgi-derived vesicles might contain materials for the growing pellicle.

Conversely, the importance of two *cytoplasmic* markers for the *beginning* of division cannot be overstressed. Recognizable changes in pellicular organization and Golgi activity allow one to assemble a sequence of sectioned cells from interphase into division and *then* to see what changes occur in the nuclei. Without such external markers, the temptation is to arrange nuclei of varying appearance to fit a preconceived idea of nuclear change from interphase to prophase. O'Donnell (1965) identifies nuclei of *E. gracilis* in "early prophase," "midprophase," and "late prophase" in relation to an expansion, opening out, and disorganization of the nucleolus. There is no indication of how these precise stages of prophase were recognized and, in view of the variability of the electron microscopic image of the nucleus of *Euglena* (page 196), it seems probable that all the stages illustrated in O'Donnell's article are interphase nuclei. This could be the explanation of the discrepancy between her account of endosomal behavior in *Euglena* and those given by other authors.

A nucleus can be confidently identified as being in some stage of mitosis

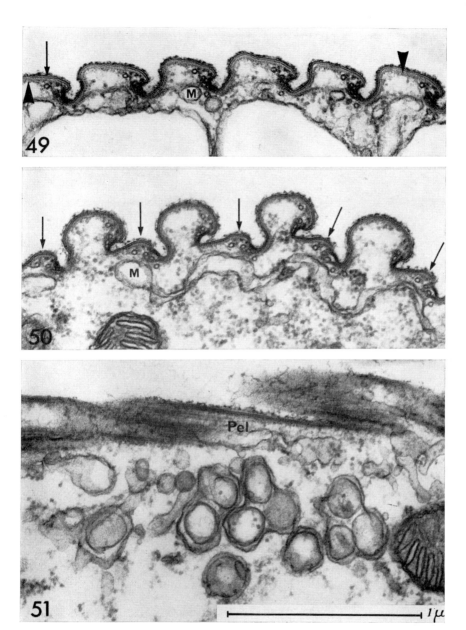

Figs. 49–51. Pellicle structure of nondividing and dividing cells of *E. gracilis.* Electron micrographs, × 50,000.

Fig. 49. Transverse section through several pellicular complexes of a nondividing cell; outside the tripartite plasmalemma (arrow) is a thin layer of mucilage; immediately within the plasmalemma is the material of the pellicular strips (arrowheads), associated microtubules (cut in transverse section) and muciferous bodies (M). *Fig. 50.* Transverse section through several pellicular complexes of a dividing cell; daughter complexes (arrows) are forming between the main ones. M, Muciferous body. *Fig. 51.* Section showing Golgi-derived vesicles near the pellicle (Pel) in a dividing cell (see text).

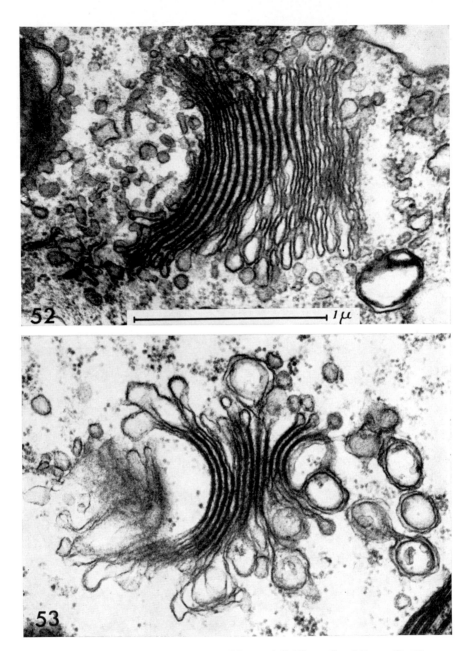

Figs. 52 and 53.   Golgi bodies in nondividing and dividing cells of *E. gracilis*. Electron micrographs, ×50,000.

Fig. 52.   Vertical section of the relatively inactive Golgi body of a nondividing cell.
Fig. 53.   Vertical section of a Golgi body of a dividing cell, showing proliferation of large vesicles from the edges of the Golgi cisternae.

if it lies close to the reservoir in a cell with replicating pellicle and increased Golgi activity. At first sight, such a nucleus in prophase shows little difference from one in interphase. The chromosomes are condensed, but no more so than in some interphase nuclei (see page 196), the endosome is still spherical or ovoid and the nuclear envelope intact. However, close examination of the nucleoplasm reveals a new feature—the presence of microtubules, 250–300 Å (25–30 mμ) in diameter. The origin of the microtubules is obscure. They are absent from interphase nuclei but always present by the time a nucleus can be identified as in mitotic prophase.

From this point on, as the endosome elongates and the chromosomes orient, mitotic stages can be recognized without reference to an external marker (but with close reference to the course of mitosis in living cells, Section III,A,1). By early metaphase, recognized especially by the cylindrical shape of the elongated endosome (Fig. 54; compare Fig. 16), some chromosomes are longitudinal to the division axis and others apparently transverse. This would be the period of chromatid separation and segregation if the interpretation from light microscopy is correct (Hall, 1937; Leedale, 1958a; see above). At full metaphase (Fig. 55) the parallel alignment of chromosomes to the division axis is more apparent, but the occasional branched profile suggests that some segregation is still taking place. As anaphase begins (Fig. 56), sections of the (daughter) chromosomes exhibit similar shapes (rods, curves, and U-shapes) to those seen in the light microscope.

The elongated endosome of early and full metaphase (Figs. 54 and 55) shows density variations similar to those in the endosome in interphase (Figs. 7 and 8), but presence of an internal (or external) nucleolar-organizing chromosome (see Sections II,B and II,C) has not been demonstrated. Microtubules are closely associated with the surface of the elongated endosome, oriented along its long axis (Fig. 59), and this association continues into late anaphase (Fig. 60).

Microtubules are also oriented along the division axis in the nucleoplasm, from early metaphase onward (Figs. 56–60). The arrangement in prophase seems to be more haphazard, but there are difficulties here in determining nuclear orientation. During metaphase and anaphase the microtubules run from one end of the nucleus to the other (Fig. 56), but continuity of any particular tubule across the division figure has not been demonstrated. Equally, no obvious attachment of microtubules to chromosomes has been seen. The microtubules are loosely grouped in bundles of 8 to 12 (Figs. 57 and 58) which certainly pass between chromosomes but not, as yet demonstrated, to points of attachment on them. Regions in Fig. 56 look temptingly like microtubules ending at chromosomes, and yet in the next section of the series it is found that the tubules continue on into the nucleoplasm. At their extremities the microtubules approach closely to the nuclear envelope (Fig. 57). However, there is no evidence that microtubules pass through the envelope, or that there are any related tubules on the cytoplasmic side, or that there is any organizational center at the division pole, or that there are any centrioles. And, although it has been carefully searched for, no relation has so far been found between the microtubules and the flagellar basal bodies (which might be expected to act as centrioles, see Section III,E).

Finally, it is important to note that the nuclear envelope is still intact at metaphase (Figs. 54 and 55) and remains so through anaphase (Fig. 56), late anaphase (Fig. 60), and into telophase (Fig. 61). Thus the persistence of the nuclear envelope throughout mitosis in *Euglena*, originally suggested from light microscopy (Section III,A), is confirmed by studies with the electron microscope. At latest anaphase, as the nuclear envelope seals around the daughter nuclei, microtubules are still present, but they can no longer be found in the daughter nuclei of cells in cleavage. Thus the microtubules of euglenoid mitosis appear within the prophase nucleus, lie along the division axis in metaphase and anaphase, and disappear again during telophase. They are always contained within the nuclear envelope and may therefore be termed "nucleoplasmic" rather than "cytoplasmic." The possible relation of these elements to the spindle of classic mitosis is considered below (Section III,E).

D. THE TIME SCALE OF MITOSIS IN *EUGLENA*

A number of authors have recorded the total time taken for mitosis and cell cleavage in species of *Euglena*. Tannreuther (1923) gives a time of 2 hours for *E. gracilis*, Baker (1926) records 4 hours for the same species, Ratcliffe (1927) 3–4 hours for *E. spirogyra*, Gojdics (1934) 4 hours for *E. deses*, and Leedale (1959a) 4 hours for *E. gracilis* and 3 hours for *E. spirogyra*.

In the most recent of these studies, Leedale (1959a) gives times for the

---

Fig. 54. Approximately transverse section of a cell with the nucleus in early metaphase (compare Fig. 16); the nucleus is sectioned longitudinally to the division axis, with the elongating endosome (E) cut in longitudinal section; the nuclear envelope (arrows) is intact; oblique sections of chromosomes (darkly stained) are seen in the granular nucleoplasm (n); the endosome contains "vacuoles" and other regions of varying density.
Fig. 55. Approximately transverse section of a cell with the nucleus in full metaphase (compare Figs. 17–19); the nucleus is sectioned longitudinally to the division axis (which runs from side to side in the figure); the swollen ends of the endosome (E) are sectioned but the plane of this section does not pass through the very thin central endosomal region; the nuclear envelope (arrows) is intact; numerous chromosomes are cut in approximate longitudinal section along the division axis; the endosomal substance contains "vacuoles" and a low-density "track" (compare Figs. 7 and 8); as in Fig. 54, chloroplasts (Chl), mitochondria (m), and Golgi bodies (G) are seen in the cytoplasm, and also tears in the section caused by paramylon granules (P).

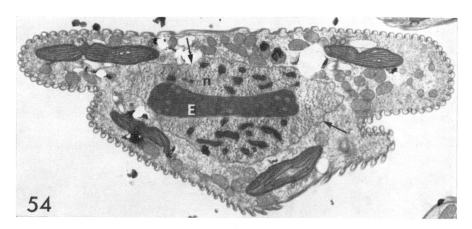

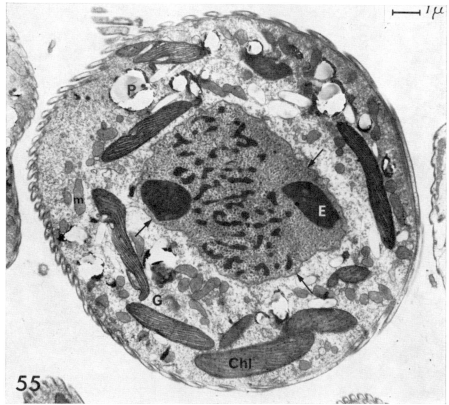

Figs. 54 and 55. Mitosis in *E. gracilis*. Electron micrographs, ×7000.

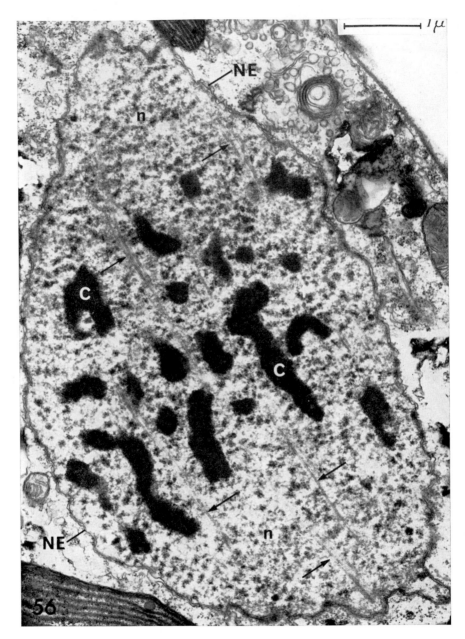

Fig. 56. Mitotic "metaphase–anaphase" in *E. gracilis*, sectioned along the division axis but not through the level of the dividing endosome; microtubules (arrows, etc.) run through the nucleoplasm (n) parallel to the division axis, between (and at some points apparently to) the chromosomes (C); the nuclear envelope (NE) is still intact. Electron micrograph, ×20,000.

individual stages of mitosis in *E. gracilis* and *E. spirogyra* (and also for species of *Astasia, Colacium, Eutreptia, Leponcinclis, Phacus,* and *Trachelomonas*) and compares the results with records for the duration of mitotic stages in living cells of other protista, animal tissues, and plant tissues.

The mean times for 34 timed cells of *E. gracilis* are: prophase (nuclear migration and changes until the start of endosome elongation), 17 minutes; metaphase (chromosome orientation and the first period of endosome elongation), 12 minutes; early anaphase (early chromatid separation until a marked central gap is seen), 71 minutes; late anaphase (continued chromatid separation and endosome elongation until the central region of the endosome breaks), 15 minutes; telophase (2 nuclei, no cell cleavage), 4 minutes; and cell cleavage, 133 minutes. For 12 cells of *E. spirogyra,* the mean times are: prophase, 25 minutes; metaphase, 20 minutes; early anaphase, 44 minutes; late anaphase, 19 minutes; telophase, 2 minutes; and cell cleavage, 69 minutes.

The significant feature of these figures is the long duration of anaphase. Anaphase in *Euglena* takes more than an hour, whereas in nearly 40 other cell types for which records exist (see Leedale, 1959a) anaphase lasts for 1–26 minutes. Anaphase in *Euglena* occupies 57% (*E. spirogyra*) or 71.5% (*E. gracilis*) of the total mitotic period, whereas in all other recorded cells and tissues anaphase occupies 2.5–22% of mitosis.

Two facts account for the long anaphase in *Euglena*. First, the anaphase is staggered (see above, Section III,A): the total time taken for mitotic anaphase is considerably longer than the time taken for the passage of any one chromatid from equator to pole. Second, the anaphase velocity of individual chromatids is lower in *Euglena* than in any other recorded mitosis. Direct measurements of chromatids separating in anaphase give a maximum velocity (in the earlier stages) of 0.1 μ per minute in *E. gracilis* and 0.12 μ per minute in *E. spirogyra*. In all other organisms and tissues for which records exist, the absolute velocity of the chromatids during anaphase separation is between 0.3 and 6.0 μ per minute (Leedale, 1959a).

These differences in anaphase behavior and timing were advanced by Leedale (1959a) as evidence that a nonspindle mitosis operates in *Euglena* (and other euglenoid flagellates). If it is now necessary to modify this view (see below), the data still indicate that euglenoid mitosis justifies treatment as a peculiar form of nuclear division.

E. THE NATURE AND MECHANISM OF MITOSIS IN *EUGLENA*

The studies of Keuten (1895), Tschenzoff (1916), Tannreuther (1923), Baker (1926), Ratcliffe (1927), Hall (1937), Krichenbauer (1937), Hollande (1942), Leedale (1958a), and Saitô (1961) make it quite clear that nuclear division in *Euglena* fulfills the essential criteria of mitosis. Individualized

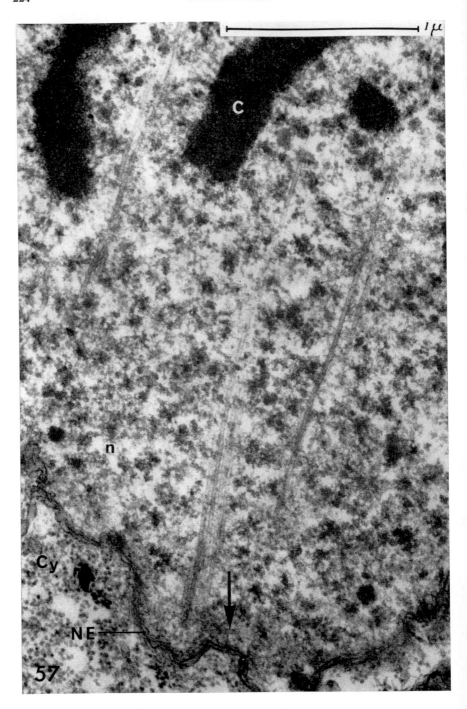

chromosomes replicate longitudinally and sister chromatids segregate to opposite poles and into different daughter nuclei. Accounts of transverse cleavage of a continuous spireme (Dangeard, 1902, 1910, 1938; Dehorne, 1920; Chadefaud, 1939) or of individual chromosomes (Gojdics, 1934) are now discredited.

However, there is also no doubt that mitosis in *Euglena* shows several departures from the "classic" form of this process in eucaryotic cells (Mazia, 1961). In particular, the persistence and division of the endosome (nucleolus), the pattern of metaphase alignment and anaphase separation of chromosomes, and the retention of the nuclear envelope may be selected as the more peculiar attributes of euglenoid mitosis.

In recent years, the main discussion on mitosis in *Euglena* has centered around the mechanism of the process and whether or not there is a mitotic spindle. The available evidence is surveyed in the present account and we can now take stock by assembling points for the two sides of the argument.

Evidence for the absence of a mitotic spindle may be listed as follows:

(1) Chromosomes align along the division axis at "metaphase," with no organization of an equatorial plate.

(2) Anaphase movement of chromatids is staggered.

(3) Numerous investigators have failed to find chromosomal constrictions or nonstaining regions which might be centromeres, or any cytological evidence of "spindle fibers."

(4) Anaphase takes more than 1 hour in *Euglena*, compared with 26 minutes or less in many types of nuclei that are known to have a spindle.

(5) Chromatid velocity in anaphase in *Euglena* is much lower than in nuclei that are known to have a spindle.

(6) Mitosis in *Euglena* is not inhibited by colchicine.

Evidence for the presence of a spindle in euglenoid mitosis consists of the following points:

(1) Saitô (1961) claims to see terminal and subterminal centromeres in the chromosomes of *E. viridis*.

(2) Electron microscopy reveals microtubules in the dividing nucleus. These "arise" in prophase, lie parallel to the division axis during metaphase and anaphase, and disappear in telophase. The microtubules are similar to those identified as "spindle elements" in numerous organisms (see below).

(3) The euglenoid nucleus migrates to the anterior of the cell prior to mitosis, suggesting that the flagellar basal bodies might act as centrioles.

Fig. 57. Microtubules in a mitotic nucleus of *E. gracilis*; higher magnification of the lower right-hand corner of Fig. 56; the microtubules approach closely to the nuclear envelope (NE), with an occasional tubule (arrow) not oriented along the division axis. C, Chromosome; Cy, cytoplasmic matrix with ribosomes; n, nucleoplasm. Electron micrograph, × 50,000.

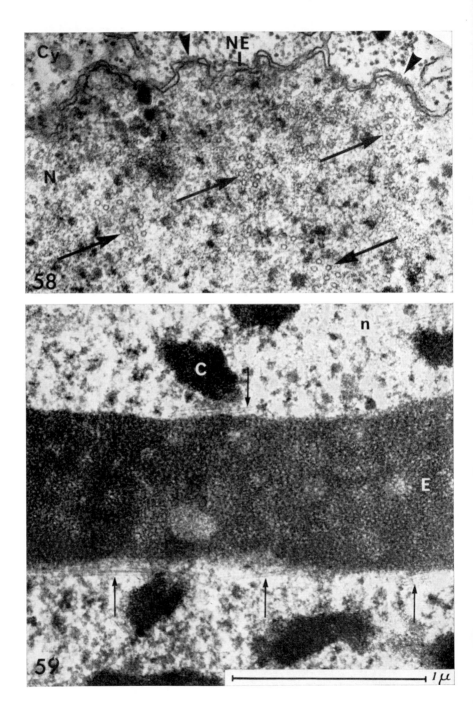

(4) Mitosis in *Euglena* is inhibited by β-mercaptoethanol (Meeker, 1964; see below).

In the apparent absence of centromeres and "conventional" spindle, and with the positive evidence that mitotic movements are very slow and erratic, Leedale (1958a,b) suggested that endosome division and chromosome movements are autonomous in *Euglena*, with mutual repulsion by sister chromatids leading to anaphase. The several forms of chromosome orientation and segregation were explained as merely reflecting differences in the timing of initiation of chromatid repulsion. Retention of the nuclear envelope (now confirmed by electron microscopy) was considered an essential part of the process, resulting in the repelling chromatids becoming marshaled into two groups which move apart to form the cytologically equivalent daughter nuclei.

How far this theory must be modified (or whether it must be jettisoned altogether) is not yet clear. Presence of nucleoplasmic microtubules is not absolute proof that a "normal" spindle is present in *Euglena* during mitosis. However, the microtubules are of the same order of size as those that have been positively identified as spindle elements in a wide range of organisms (see, for example, Harris, 1962; Kane, 1962; Ledbetter and Porter, 1963; Manton, 1964; de-Thé, 1964; Murray *et al.*, 1965). The euglenoid microtubules are aligned along the division axis, that is to say, along the direction of movement of the segregating chromatids. There is no evidence of attachment of microtubules to the chromosomes (which points of attachment would, by homology with spindle structure, be termed "centromeres"), nor is there any apparent relation between the nucleoplasmic microtubules and the flagellar basal bodies or other extranuclear structures. This last point is particularly puzzling. It has already been pointed out (Leedale, 1958a, 1966, 1967) that it is tempting to cite the anterior migration of the nucleus prior to euglenoid mitosis as evidence that the flagellar basal bodies on the wall of the reservoir act as centrioles, determining the division axis and controlling the process. Fine structural confirmation of this seductive hypothesis has been sought but not found.

Fig. 58. Polar region of a mitotic nucleus of *E. gracilis*, sectioned transversely to the division axis and hence to the microtubules; these are seen to be loosely grouped in bundles (arrows). Cy, Cytoplasmic matrix; N, nucleus; NE, nuclear envelope; arrowheads, nuclear pores in transverse section. Electron micrograph, ×50,000.

Fig. 59. Central region of the endosome during mitotic metaphase in *E. gracilis*; higher magnification of the center of Fig. 54; microtubules (arrows) are associated with the dividing endosome (E); other microtubules can be seen in the nucleoplasm (n) between the chromosomes (C). Electron micrograph, ×50,000.

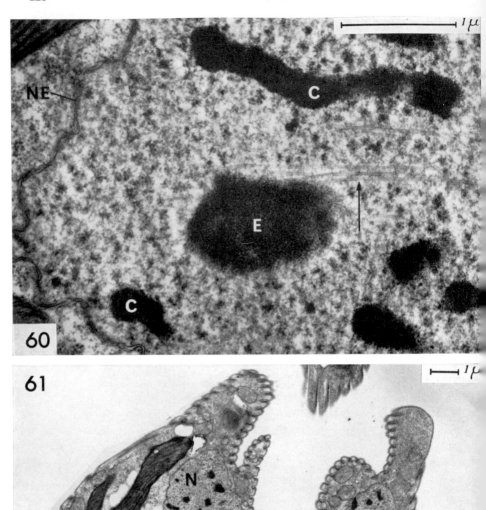

The scanty experimental evidence is also uninformative or conflicting. Leedale's (1958a) attempts to break the chromosomes of *Euglena* with irradiation treatment (to test for localized centromeres by possibly producing acentric fragments which would not move at anaphase) were spectacularly unsuccessful. Neither X-rays nor high-energy electrons (β-rays) from a van de Graff generator (nor γ-rays from a ^{60}Co source; Leedale, unpublished) do any apparent damage to chromosomes or endosomes at nonlethal (300–100,000-rad) or lethal (300,000-rad) doses. Meeker (1964) has demonstrated that cell division in *Astasia longa* is reversibly inhibited by 10^{-3} M β-mercaptoethanol, a compound which is thought to block division in cells of higher animals by disrupting the mitotic spindle. Leedale (1958a) reports that the classic inhibitor of spindle activity, colchicine, has no effect on mitosis in *Euglena* at any concentration from 0.001 to 4.0%. Perhaps the colchicine fails to get into the cell (or into the nucleus), or perhaps the nucleoplasmic tubules of *Euglena* do have some fundamental difference from the microtubules of "normal" spindle.

The intriguing possibility that the euglenoid "spindle" does little other than determine the division axis, perhaps providing a skeletal framework or "guide lines" for autonomously moving chromatids, has been mentioned before (Leedale, 1967), as has the suggestion that a high degree of autonomous chromosome activity may be a more important phenomenon in the mechanism of "classic" mitosis than hitherto supposed. In view of the isolated taxonomic position of the euglenoid flagellates it is dangerous (and therefore fascinating) to speculate as to whether the peculiarities of euglenoid mitosis represent a primitive or reduced condition. The cytological peculiarities are most closely reflected in the nuclear division recorded for some dinoflagellates (Dodge, 1963), which are possibly the euglenoids' closest relatives (Leedale, 1967). Since the function of the spindle in "classic" mitosis is not understood, it seems that further investigation of peculiar mitoses such as those in *Euglena* and various other protista (Grell, 1964) may well throw light on general problems of nuclear division.

Fig. 60. Polar region of a late anaphase nucleus of *E. gracilis*, sectioned parallel to the division axis; microtubules (arrow) are still associated with the endosome (E), of which only the swollen end is sectioned; chromosomes (C) now lie nearer the pole than the end of the endosome (compare Figs. 22 and 23); the nuclear envelope (NE) is still intact. Electron micrograph, ×30,000.

Fig. 61. Section of a cell of *E. gracilis* undergoing cleavage; the daughter nuclei (N) are sectioned, but not through the endosomes; the chromosomes, nucleoplasm, and nuclear envelopes appear much as in interphase (Fig. 7) and during mitosis. Electron micrograph, ×7000.

F. Periodicity of Mitosis in *Euglena*

A nocturnal periodicity of mitotic activity in *Euglena* has been recorded by Dangeard (1902), Baker (1926), Ratcliffe (1927), Gojdics (1934), Chu (1947), Leedale (1959b), and Saitô (1961). In each case, the material was growing phototrophically in a relatively poor medium (split pea infusion, modified Doflein's medium, or soil and water). In biphasic (soil–water) culture (Pringsheim, 1946) at 20°C, under natural light conditions, 2–6% of the cells divide each night (1.9% for *E. acus;* 2.2% for *E. deses;* 4.2–5.7% for *E. gracilis*, according to which physiological strain is used; 3.4% for *E. spirogyra;* and 4.9% for *E. viridis;* mean maximum percentages from Leedale, 1959b).

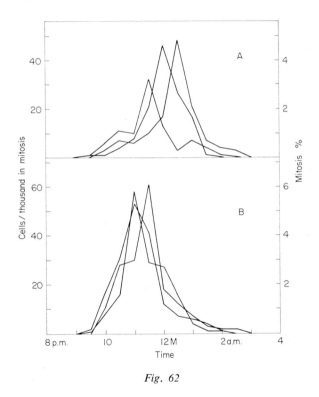

Fig. 62

Figs. 62 and 63. Mitosis in *Euglena*. The number of cells per thousand in mitosis at half-hour intervals during three consecutive nights, plotted as mitosis percentage against time. All results are for biphasic cultures growing at 20°C in the natural day–night cycle, the dark period beginning at 8 P.M.

Fig. 62. A. *Euglena gracilis* strain T. B. *Euglena gracilis* strain Z.   *Fig. 63.* A. *Euglena spirogyra*. B. *Euglena viridis*.

The restriction of mitosis to the dark period has been examined in detail by Leedale (1959b). In different species of *Euglena*, mitosis begins 1–2 hours after the onset of darkness, with the mitotic maximum occurring from $2\frac{1}{2}$ to $4\frac{1}{2}$ hours after dark. Mitosis began at the same time on each of three successive nights in most species (Figs. 62 and 63) but the maxima occurred at different times on successive nights (Figs. 62 and 63A). In *E. viridis*, the mitotic maximum was at the same time three nights in succession (Fig. 63B). The curves for different species of *Euglena* are strikingly similar, and it is interesting to note that there is as much variation between two of the many physiological strains of *E. gracilis* isolated by Pringsheim (Figs. 62A and 62B) as between these and other species (Fig. 63).

In the case of *E. spirogyra*, the number of cells at each stage of mitosis at half-hour intervals during one night has been recorded to give a more detailed picture of the periodicity (Fig. 64). Successive maxima of the mitotic stages occur over a 5-hour period, a wave of prophases being followed by waves of metaphases, anaphases, telophases, and cell cleavage. The size and span of the maxima reflect the relative duration of the stages of mitosis in *Euglena* (see Section III,D).

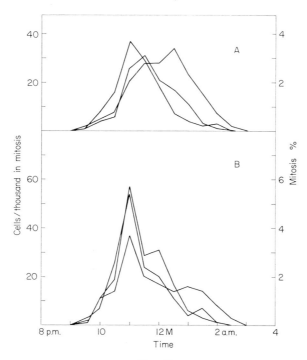

Fig. 63

Unlike mitotic rhythms recorded for many plant and animal cells (see Leedale, 1959b, for references), the rhythm of *Euglena* in biphasic culture is exogenous. It is immediately affected or removed by any change in the light regime or by a change to heterotrophic nutrition. Introduction of an early (artificial) dark period during the natural light period reduces the mitotic index, and mitosis is almost completely inhibited if the day length is

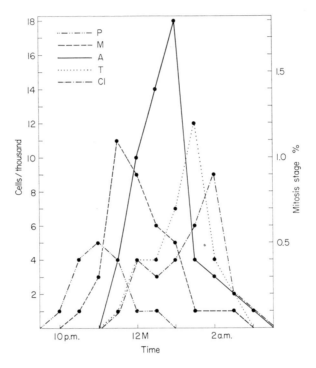

Fig. 64. Mitosis in *E. spirogyra.* The number of cells per thousand in prophase (P), metaphase (M), anaphase (A), telophase (T), and cell cleavage (Cl) at half-hour intervals during one night, plotted as percentage of each mitotic stage against time. The results are for a biphasic culture growing at 20°C in the natural day–night cycle, the dark period beginning at 8 P.M.

cut to much below 12 hours. Similarly, mitosis does not occur in biphasic culture in artificial light (only), continuous light, or continuous darkness. However, once mitosis has begun it will proceed to conclusion even if the dividing cell is then illuminated. There seems to be a threshold period of approximately 1 hour of darkness needed to induce mitosis in *Euglena* following a full-length day. Strains of *E. gracilis* in 0.2% beef extract or a medium containing sodium acetate, beef extract, yeast extract, and tryptone show up to 10% of the cells in mitosis at any time of day or night at 20°C, and the mitotic index can be as high as 30% at 30°C.

Culture of *E. gracilis* var. *bacillaris* in modified Beijerinck's solution has enabled Huling (1960) to study the effects of various photoperiods on cell division. Although divisions can occur during the artificial light periods, most take place in the dark. Short light periods (2 hours in 24) allow almost no cell division to occur; increases in the light period (to 4 or 8 hours in 24) stimulate division. Experiments with split light periods indicate that a dark period per se does not stimulate mitosis; the controlling factor seems to be the total number of light hours received during the 24-hour cycle. Huling correlates his results with those of Leedale (1959b, see above) to suggest that division in *Euglena* is an all-or-none response to a sufficient storage of energy-producing nutrient. However, the inhibition of mitosis in biphasic culture by light (Leedale, 1959b), even after a full-length day which would allow 2–6% mitosis if there was a dark period, suggests a more complex situation than simple dependence on the build-up of carbohydrate reserves. This apparent need for an induction period of darkness in the rhythmic phototrophic system is unexplained.

The mitoses of *E. gracilis* growing heterotrophically in liquid media can be synchronized by means of various 24-hour light or temperature cycles (see Chapter 6), a valuable attribute for physiological studies. Edmunds (1964) has shown that in synchronized cultures of *E. gracilis* strain Z, DNA content per cell doubles stepwise in the few hours prior to mitosis. This obviously relates to the actual period of chromosome replication and is of great interest with respect to the observed cytological events (Section III,A).

IV. Amitosis in *Euglena*

Amitosis, the division of a nucleus without the emergence, orientation, and division of individualized chromosomes, has been described for three species of *Euglena* (Leedale, 1959c). Biphasic cultures (Pringsheim, 1946) of *E. acus* and *E. spirogyra* usually contain a small percentage of cells with two spherical nuclei in the interphase position (Figs. 66 and 67). The two nuclei *together* have the same volume as the normal interphase nucleus (Fig. 65) and the chromatin is often more obviously filamentous. In addition, cells with constricted nuclei are found (Figs. 68 and 69). Nuclei of normal cells of both species contain 1 to 8 (to 20+) endosomes (nucleoli), the commonest numbers being 4, 5, or 6 (Fig. 65). The "half-nuclei" contain 1 to 10 endosomes, the number usually being approximately equal n each nucleus of a binucleate cell or in each part of a constricted nucleus (Fig. 69)

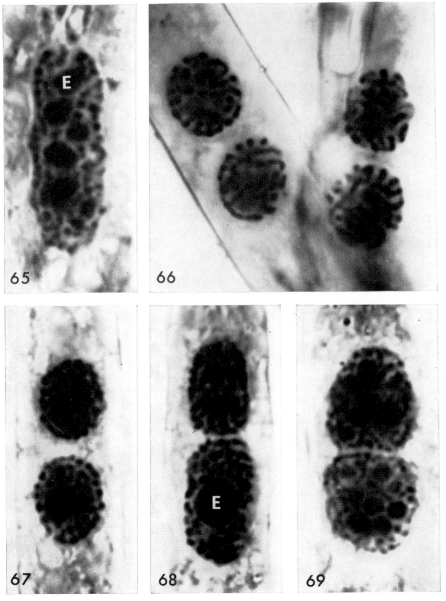

Figs. 65–69. Interphase and amitosis in *E. acus* (see text for commentary). Acetocarmine stained, ×3000.

Fig. 65. Interphase nucleus, showing "granular" chromatin around the central endosomes (E). *Fig. 66.* Central regions of two binucleate cells. *Fig. 67.* "Half-nuclei" in a binucleate cell. *Fig. 68.* Constricted nucleus, with nearly all the endosomal material (E) in one portion. *Fig. 69.* Constricted nucleus, with endosomes equally distributed.

Rarely, one-half of a constricted nucleus contains nearly all the endosomal material (Fig. 68).

In neither species of *Euglena* does cell cleavage occur to separate the two nuclei of a binucleate cell. The amitotic process seems to be a nuclear fragmentation in response to unfavorable conditions (aging of the culture), perhaps with the function of increasing the ratio of nuclear surface to volume (Conklin, 1917; Kater, 1940). It is not certain whether the constricted nuclei are stages in a slowly occurring fragmentation, or whether they are permanently arrested stages of incomplete amitosis.

The amitosis itself is not connected with reproduction, but the binucleate cells of *E. acus* and *E. spirogyra* are nevertheless capable of mitosis and cell division. At the onset of mitosis, the "half-nuclei" migrate to the anterior end of the cell and there divide separately, lying one behind the other, *each* having a single dividing endosome in *E. spirogyra* (Figs. 70–73) and several dividing endosomes in *E. acus*. The chromosome number counted in each of the 4 late anaphase groups in *E. spirogyra* is between 40 and 45 (Figs. 70–73), approximately half the normal chromosome number of 86 in this species (Fig. 46). Subsequent cell cleavage produces two binucleate cells, although occasional miscleavage results in one daughter cell receiving one "half-nucleus" and the other receiving three "half-nuclei" (Fig. 74).

The chromosome counts in *E. spirogyra* make it clear that the two nuclei of the binucleate cell together constitute a single chromosome complement. In the rare case of miscleavage following mitosis in a binucleate cell, individuals are formed that have approximately half or one-and-a-half times the normal genome. Such cells are viable and capable of multiplication, their progeny retaining the size and morphology characteristic of the species. This suggests (Leedale, 1959c) that these species of *Euglena* have a fairly high level of polyploidy, with considerable loss or gain of chromosomes making little difference to the individual cell (see Section III,B).

Amitosis also occurs in *E. viridis* in cultivated material, resulting in nuclear fragments of irregular size and shape. In this species the process seems to be a degenerate one, much as in certain plant and animal tissues (Bucher, 1959), and subsequent mitosis of the part-nuclei has not been observed.

V. The Apparent Absence of Sexuality and Meiosis in *Euglena*

Sexuality and meiosis are generally regarded as lacking in *Euglena*, and in all euglenoid flagellates except for the taxomically questionable *Scytomonas*.

Early accounts of sexuality seem to involve misidentification or misobservation. Köllicker (1853) records "in *Euglena* 4–6 embryos in one indi-

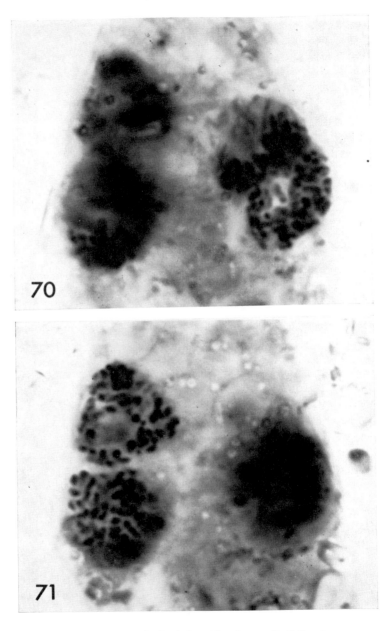

Figs. 70–73.   Late anaphase of mitosis in a binucleate cell of *E. spirogyra.*

Figs. 70 and 71.   Two focal levels of the "double anaphase." Acetocarmine stained, ×3000.

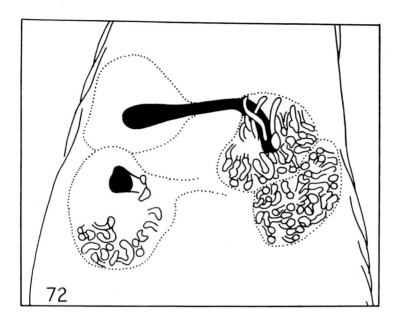

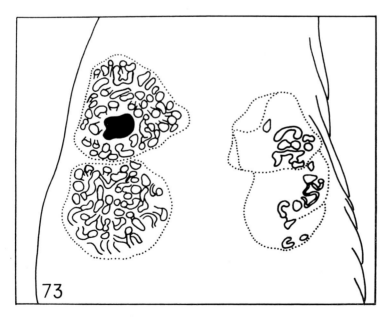

Figs. 72 and 73. Camera lucida drawings of the "double anaphase" at the same two focal levels as in Figs. 70 and 71; endosomal material shown in solid black. ×3000.

vidual, and entirely filling it, which at last, furnished with their red points and cilia, broke through their parent, leaving it an empty case." Was this possibly a *Chlorogonium*, or one of the very elongated species of *Chlamydomonas*? Köllicker's observation has never been repeated for *Euglena*. Carter's (1859) "ovules" in species of *Euglena* and *Phacus* are seen from his figures to be the paramylon granules. Haase (1910) records an isogamous sexual process in *E. sanguinea*, in which small amoeboid

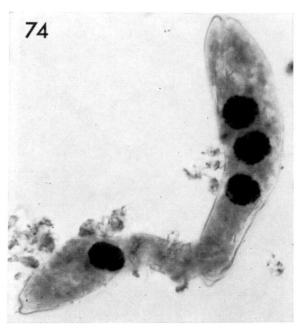

Fig. 74. Miscleavage after mitosis in a binucleate cell of *E. spirogyra*; one daughter cell is receiving one "half-nucleus" and the other is receiving three "half-nuclei." Aceto-carmine stained, × 1250.

gametes, each with two plastids, fused in pairs to form zygotes which immediately divided and released many small vegetative cells. Biecheler (1937) describes a conjugation between two adult cells of an unidentified species of *Euglena*. Cell fusion produced a nonflagellated zygote which encysted within 12 hours. No observations were made beyond this stage and there is no information on the nuclear phenomena accompanying the cell fusion.

Many other workers on *Euglena* have made observations on many generations of cells and have failed to find any evidence of gamete formation or cell fusion. It seems probable that there is no sexuality in the majority of euglenoids, but it remains a possibility that the process does occur as a rare phenomenon; this can never be disproved. It is, however, certain that a sexual process is not a regular feature of the life cycle in *Euglena*.

The main evidence for a meiotic process in euglenoid flagellates is the description by Krichenbauer (1937) of an autogamy and subsequent meiosis in *Phacus pyrum*. The validity of this account has recently been supported by observation (Leedale, 1962) of similar nuclear stages in *Hyalophacus ocellatus*. It may be anticipated that if meiosis is finally discovered for *Euglena* it will be as peculiar in form and as specific to the euglenoids as is the mitosis. If sexuality exists in the genus and can eventually be experimentally induced and controlled, then *Euglena* can add the role of genetic microorganism to its already versatile activities as a tool in cell biology.

References

Afzelius, B. A. (1955). The ultrastructure of the nuclear membrane of the sea urchin oocyte as studied with the electron microscope. *Exptl. Cell Res.* **8**, 147–158.

Alexeieff, A. (1911). Haplomitose chez les Eugléniens et dans d'autres groupes de protozoaires. *Compt. Rend. Soc. Biol.* **71**, 614–617.

Alexeieff, A. (1913). Systématisation de la mitose dite "primitive." *Arch. Protistenk.* **29**, 344–363.

Baker, W. B. (1926). Studies on the life history of *Euglena*. I. *Euglena agilis* Carter. *Biol. Bull.* **51**, 321–362.

Bělař, K. (1926). Der Formwechsel der Protistenkerne. *Ergeb. Fortschr. Zool.* **6**, 235–654.

Biecheler, B. (1937). Sur l'existence d'une copulation chez une Euglène verte et sur les conditions globales qui la déterminent. *Compt. Rend. Soc. Biol.* **124**, 1264–1266.

Blochmann, F. (1894). Ueber die Kernteilung bei *Euglena*. *Biol. Zentr.* **14**, 194–197.

Brachet, J. (1940). La détection histochimique des acides pentosenucleiques. *Compt. Rend. Soc. Biol.* **133**, 88–90.

Brachet, J. (1953). The use of basic dyes and ribonuclease for the cytochemical detection of ribonucleic acid. *Quart. J. Microscop. Sci.* **94**, 1–10.

Brown, V. E. (1930). Cytology and binary fission of *Peranema*. *Quart. J. Microscop. Sci.* **73**, 405–419.

Bucher, O. (1959). Die Amitose der tierischen und menschlichen Zelle. *Protoplasmatologia* **6**(E/1), 1–159.

Calkins, G. N. (1926). "The Biology of the Protozoa." Lea & Febiger, Philadelphia, Pennsylvania.

Carter, H. J. (1859). On fecundation in the two Volvocales and their specific differences; on *Eudorina, Spongilla, Euglena* and *Cryptoglena*. *Ann. Mag. Nat. Hist. III* **3**, 1–20.

Chadefaud, M. (1939). L'infrastructure du protoplasme et du caryoplasme d'après l'étude des Euglènes. *Arch. Exptl. Zellforsch.* **22**, 483–486.

Chatton, E. (1910). Essai sur la structure du noyau et la mitose chez les Amoebiens, faits et théories. *Arch. Zool. Exptl. Gen.* **5**, 267–337.

Chu, S. P. (1947). Contributions to our knowledge of the genus *Euglena*. *Sinensia* **17**, 75–134.

Conklin, E. G. (1917). Mitosis and amitosis. *Biol. Bull.* **33**, 396–436.

Dangeard, P. A. (1902). Recherches sur les Eugléniens. *Botaniste* **8**, 97–357.

Dangeard, P. A. (1910). Études sur le développement et la structure des organismes inférieurs. *Botaniste* **11**, 1–311.

Dangeard, P. A. (1938). Mémoire sur la famille des Péridiniens. *Botaniste* **29**, 3–182.

de Haller, G. (1959). Structure submicroscopique d'*Euglena viridis*. *Arch. Sci. (Geneva)* **12**, 309–340.

Dehorne, A. (1920). Contribution à l'étude comparée de l'appareil nucléaire des infusoires ciliés, des Euglènes et des Cyanophycées. *Arch. Zool. Exptl. Gen.* **60**, 47–176.

de-Thé, G. (1964). Cytoplasmic microtubules in different animal cells. *J. Cell Biol.* **23**, 265–275.

Dodge, J. D. (1963). The nucleus and nuclear division in the Dinophyceae. *Arch. Protistenk.* **106**, 442–452.

Drezepolski, R. (1929). L'évolution du noyau et son rôle chez les Euglènes. *Ann. Protistol.* **2**, 109–118.

Edmunds, L. N., Jr. (1964). Replication of DNA and cell division in synchronously dividing cultures of *Euglena gracilis*. *Science* **145**, 266–268.

Frey-Wyssling, A., and Mühlethaler, K. (1965). "Ultrastructural Plant Cytology." Elsevier, Amsterdam.

Gibbs, S. P. (1960). The fine structure of *Euglena gracilis* with special reference to the chloroplasts and pyrenoids. *J. Ultrastruct. Res.* **4**, 127–148.

Godward, M. B. E. (1966). "The Chromosomes of the Algae." Arnold, London.

Gojdics, M. (1934). The cell morphology and division of *Euglena deses* Ehrbg. *Trans. Am. Microscop. Soc.* **53**, 299–310.

Gojdics, M. (1953). "The Genus *Euglena*." Univ. of Wisconsin Press, Madison, Wisconsin.

Grell, K. G. (1964). The protozoan nucleus. *In* "The Cell" (J. Brachet and A. E. Mirsky, eds.), Vol. VI, pp. 1–79. Academic Press, New York.

Günther, F. (1927). Über den Bau und die Lebensweise der Euglenen, besonders der Arten *E. terricola, geniculata, sanguinea* und *lucens* nov. spec. *Arch. Protistenk.* **60**, 511–590.

Haase, G. (1910). Studien über *Euglena sanguinea*. *Arch. Protistenk.* **20**, 47–59.

Hall, R. P. (1937). A note on behavior of the chromosomes in *Euglena*. *Trans. Am. Microscop. Soc.* **56**, 288–290.

Harris, P. (1962). Some structural and functional aspects of the mitotic apparatus in sea urchin embryos. *J. Cell Biol.* **14**, 475–487.

Hartmann, M., and Chagas, C. (1910). Flagellaten-Studien. *Mem. Inst. Oswaldo Cruz* **2**, 64–125.

Hartmann, M., and von Prowazek, S. (1907). Blepharoplast, Caryosom und Centrosom. *Arch. Protistenk.* **10**, 306–333.

Hollande, A. (1942). Étude cytologique et biologique de quelques flagellés libres. *Arch. Zool. Exptl. Gen.* **83**, 1–268.

Huber-Pestalozzi, G. (1955). *In* "Die Binnengewasser (A. Thienemann, ed.), Vol. 16: Das Phytoplankton des Süsswassers. Part 4. Schweizerbart'sche Verlagsbuchhandlung, Stuttgart.

Huling, R. T. (1960). The effects of various photoperiods on population increases of *Euglena gracilis* var. *bacillaris* Prings. *Trans. Am. Microscop. Soc.* **79**, 384–391.

Johnson, D. F. (1934). Morphology and life-history of *Colacium vesiculosum* Ehrenberg. *Arch. Protistenk.* **83**, 241–263.

Kane, R. E. (1962). The mitotic apparatus. Fine structure of the isolated unit. *J. Cell Biol.* **15**, 279–287.

Kater, J. McA. (1940). Amitosis. *Botan. Rev.* **6**, 164–180.

Keuten, J. (1895). Die Kernteilung von *Euglena viridis*. *Z. Wiss. Zool.* **60**, 215–235.

Kihlman, B. A. (1966). "Actions of Chemicals on Dividing Cells." Prentice-Hall, Englewood Cliffs, New Jersey.

Köllicker, A. (1853). Description of *Actinophrys sol. Quart. J. Microscop. Sci.* 1, 25–34.

Krichenbauer, H. (1937). Beitrag zur Kenntnis der Morphologie und Entwicklungsgeschichte der Gattungen *Euglena* und *Phacus. Arch. Protistenk.* 90, 88–123.

Lackey, J. B. (1934). Studies on the life histories of Euglenida. IV. A comparison of the structure and division of *Distigma proteus* Ehrb. and *Astasia Dangeardi* Lemm. A study in phylogeny. *Biol. Bull.* 67, 145–162.

Ledbetter, M. C., and Porter, K. R. (1963) A "microtubule" in plant cell fine structure. *J. Cell Biol.* 19, 239–250.

Leedale, G. F. (1958a). Nuclear structure and mitosis in the Euglenineae. *Arch. Mikrobiol.* 32, 32–64.

Leedale, G. F. (1958b). Mitosis and chromosome numbers in the Euglenineae (Flagellata). *Nature* 181, 502–503.

Leedale, G. F. (1959a). The time-scale of mitosis in the Euglenineae. *Arch. Mikrobiol.* 32, 352–360.

Leedale, G. F. (1959b). Periodicity of mitosis and cell division in the Euglenineae. *Biol. Bull.* 116, 162–174.

Leedale, G. F. (1959c). Amitosis in three species of *Euglena. Cytologia (Tokyo)* 24, 213–219.

Leedale, G. F. (1959d). Formation of anucleate cells of *Euglena gracilis* by miscleavage. *J. Protozool.* 6 (Suppl.), 26.

Leedale, G. F. (1962). The evidence for a meiotic process in the Euglenineae. *Arch. Mikrobiol.* 42, 237–245.

Leedale, G. F. (1964). Pellicle structure in *Euglena. Brit. Phycol. Bull.* 2, 291–306.

Leedale, G. F. (1966). The Euglenophyceae. *In* "The Chromosomes of the Algae" (M. B. E. Godward, ed.), pp. 78–95. Arnold, London.

Leedale, G. F. (1967). "Euglenoid Flagellates." Prentice-Hall, Englewood Cliffs, New Jersey.

Leedale, G. F., Meeuse, B. J. D., and Pringsheim, E. G. (1965). Structure and physiology of *Euglena spirogyra*. I and II. *Arch. Mikrobiol.* 50, 68–102.

Mainx, F. (1927). Beiträge zur Morphologie und Physiologie der Eugleninen, Ia. II. *Arch. Protistenk.* 60, 305–414.

Manton, I. (1964). Observations with the electron microscope on the division cycle in the flagellate *Prymnesium parvum* Carter. *J. Roy. Microscop. Soc.* 83, 317–325.

Manton, I. (1966). Some possibly significant structural relations between chloroplasts and other cell components. *In* "Biochemistry of Chloroplasts" (T. W. Goodwin, ed.), Vol. I, pp. 23–47. Academic Press, New York.

Mazia, D. (1961). Mitosis and the physiology of cell division. *In* "The Cell" (J. Brachet and A. E. Mirsky, eds.), Vol. III, pp. 77–412. Academic Press, New York.

Meeker, G. L. (1964). The effect of β-mercaptoethanol on synchronously dividing populations of *Astasia longa. Exptl. Cell Res.* 35, 1–8.

Minchin, E. A. (1912). "An Introduction to the Study of Protozoa." Arnold, London.

Mirsky, A. E., and Osawa, S. (1961). The interphase nucleus. *In* "The Cell" (J. Brachet and A. E. Mirsky, eds.), Vol. II, pp. 677–770. Academic Press, New York.

Murray, R. G., Murray, A. S., and Pizzo, A. (1965). The fine structure of mitosis in rat thymic lymphocytes. *J. Cell Biol.* 26, 601–619.

O'Donnell, E. H. J. (1965). Nucleolus and chromosomes in *Euglena gracilis. Cytologia (Tokyo)* 30, 118–154.

Pringsheim, E. G. (1946). The biphasic or soil-water culture method for growing algae and flagellata. *J. Ecol.* **33**, 193–204.

Pringsheim, E. G. (1956). Contributions towards a monograph of the genus *Euglena*. *Nova Acta Leopoldina* **18**, 1–168.

Ratcliffe, H. L. (1927). Mitosis and cell division in *Euglena spirogyra* Ehrenberg. *Biol. Bull.* **53**, 109–122.

Ris, H. (1961). Ultrastructure and molecular organization of genetic systems. *Can. J. Genet. Cytol.* **3**, 95–120.

Robertson, J. D. (1959). The ultrastructure of cell membranes and their derivatives. *Biochem. Soc. Symp. (Cambridge, Engl.)* **16**, 3–43.

Saitô, M. (1961). Studies in the mitosis of *Euglena*. I. On the chromosome cycle of *Euglena viridis* Ehrbg. *J. Protozool.* **8**, 300–307.

Tannreuther, G. W. (1923). Nutrition and reproduction in *Euglena*. *Arch. Entwicklungsmech. Organ.* **52**, 367–383.

Tschenzoff, B. (1916). Die Kernteilung der *Euglena viridis* Ehrenberg. *Arch. Protistenk.* **36**, 137–173.

Ueda, K. (1958). Structure of plant cells with special reference to lower plants. III. A cytological study of *Euglena gracilis*. *Cytologia (Tokyo)* **23**, 56–67.

Whaley, W. G., Mollenhauer, H. H., and Leech, J. H. (1960). Some observations on the nuclear envelope. *J. Biophys. Biochem. Cytol.* **8**, 233–244.

Zumstein, H. (1900). Zur Morphologie und Physiologie der *Euglena gracilis* Klebs. *Jahrb. Wiss. Botan.* **34**, 149–198.

CHAPTER 6

THE CULTIVATION
AND GROWTH OF *EUGLENA*

J. R. Cook

I. Introduction

Fifteen years ago Lwoff (1951) could write that almost nothing was known of the physiology of the euglenoid flagellates, compared to the bacteria. At about that time culture media that permitted luxurious growth of the euglenoids were defined, a development which would not have occurred except for the introduction of chelating agents into such media (Hutner *et al.*, 1950). While it is difficult to assess the real importance of this development on the level of our current knowledge of euglenoid physiology, it is trite to say that the physiology of an organism cannot be studied if it cannot be grown. Certainly it can be said that today our knowledge of euglenoid physiology is considerable. The range of topics covered in this treatise alone is sufficient indication of the interest that has been generated by both the general and unique properties of *Euglena* as a tool in such studies.

The properties of *Euglena* depend heavily on the conditions employed in culture. *Euglena gracilis* can be made large or small, green or white; it can be made to divide rapidly, or very slowly, at the whim of the investigator. The biochemical and physiological flexibility of *Euglena*, coupled with a morphological flexibility that can be easily visualized, doubtless accounts for its growing popularity. This flexibility conforms to patterns described by culture conditions; when culture conditions are suitably controlled, physiological patterns can be predicted, within the limits of our knowledge. It is the purpose of this chapter to review some of the various parameters encountered in the growth of *Euglena*, and where possible to discuss some of the responses of *Euglena* to such parameters.

II. Axenic Culture

One of the great advantages in working with the euglenoid flagellates lies in their simple chemical requirements. The environment can be rigorously controlled with respect to chemical as well as physical factors; the media can be changed chemically in a known direction and to a known extent. To achieve this end, of course, axenic populations must be used.

The importance of using axenic cultures needs no emphasis. In addition to changes in the media caused by contaminants, resulting in subsequent changes in *Euglena* characteristics, physiological properties of the biomass in mixed populations have little meaning in terms of pure euglenoid physiology. The usual contaminants in the laboratory are bacteria and molds. The former most often have growth rates much greater than *Euglena*, and blossom to rapidly deplete the media and inhibit growth of *Euglena*.

Except for *Euglena* newly collected in the wild, methods of obtaining axenic cultures seem almost of historical interest. Such methods have included repeated washing in sterile water, isolation by cloning on agar, and, perhaps more important, the use of antibiotics or drugs that kill or reduce the growth rate of *Xenos*, but permit growth of *Euglena*. Autotrophic growth on salt media, a favorite method of obtaining axenic cultures of photosynthetic bacteria, has apparently never been tried with euglenoids, probably because most euglenoid media usually contain an organic compound in fairly high concentration (either as a growth substrate or chelating agent) and must always contain small concentrations of vitamin B_{12}.

Favorite antibiotics include streptomycin and penicillin. Bowne (1964) has shown that *Euglena* is also more resistant to caffeine than most microorganisms, and indicated its use in obtaining axenic cultures. Streptomycin and penicillin are not without effect on *Euglena;* the bleaching action of the former is well known (Provasoli *et al.*, 1948). Streptomycin also affects the motility of *Euglena* (Goodwin, 1951); concentrations greater than 8.3 mg/liter often result in immobility of more than 70% of the exposed cells. Penicillin has a comparable effect at concentrations greater than 10,000 units/ml (Goodwin, 1951). For an earlier discussion of the effects of streptomycin on the cytology and growth of *Euglena*, see Loefer and Mefferd (1955).

The resistance of *Euglena* to antibiotics is relative, and species differences exist (Goodwin, 1951). They are effective because *Euglena* is more resistant than the common contaminants, but success in their use has generally been coupled with a series of fairly brief exposures alternating with washes in sterile water (Pappas and Hoffman, 1952). Repeated washing in sterile water without antibiotics has also been successful (Leedale *et al.*, 1965).

Axenic stocks of *Euglena* have been available for many years. Many laboratories now working with *Euglena* make it an easy matter to obtain axenic stocks without resorting to the time-consuming techniques mentioned here. Table I lists some of the varieties available from the principal culture collections.

III. Techniques of Cultivation

A. Choice of the Method

Euglena is a hardy cell and extremely flexible in terms of culture approaches. The choice of a culture technique will of course depend on the end desired. With care, single-cell studies can be handled with conditions controlled as rigorously as in studies using 20-liter batch cultures.

In addition to the different media that have been developed for *Euglena*, mentioned below, a variety of culture techniques have been devised or

Table I

SPECIES AND STRAINS OF *Euglena* WITH SOURCE OF AVAILABILITY AND CULTURE CONDITION[a]

Species	Strain[b]	Condition[c]	Isolated by	Source[d]
E. acus	—	Ag	Pringsheim	2
E. agilis (E. pisciformis)	—	Ag	Mainx	7
E. anabaena var. *minor*	—	Ax	Mainx	1, 3
E. deses	—	Ag	Pringsheim	1, 2, 3
		Ax	Pringsheim	9
E. deses var. *carterae*	—	Ag	Pringsheim	3
E. deses aff. *minor*	—	Ag	Lefèvre	4
E. deses var. *vermiformis*	—	Ag	Pringsheim	2
E. geniculata	—	Ax	Vischer	1, 2, 3
E. gracilis	1	Ax	Mainx	1, 2, 3
	2	Ax	Elmore-Saver	1, 2, 3
	3	Ax	Vischer	1, 3
	4	Ax	Pringsheim	1, 3
	17 strains (5, 6, 8, 10–13, 17–21, 23, 24, 26)	Ax	Pringsheim	1, 3
	7	Ax	Lackey	1, 3
	9(T)	Ax	Pringsheim	1, 3
	16	Ax	Provasoli	1, 3
	22	Ax	Hartshorne	1, 3
	25(Z)	Ax	Pringsheim	1, 3
	Dusi I, II	Ag	Dusi	8
	Dusi III	Ag	Dusi	8
		Ag	Damon	7
E. gracilis var. *bacillaris*	15	Ax	Cori	2, 3
	8 apochlorotic strains	Ax	Gross, Loefer, Provasoli	6
E. gracilis var. *saccharophila*	—	Ax	Pringsheim	1, 3
E. gracilis var. *urophora*	14	Ax	Provasoli	2, 3
E. granulata	—	Ax	Provasoli	1, 2, 3
E. klebsii (E. mutabilis var. *mainxi)*	—	Ax	Mainx	9
E. mesnili (E. deses?)	—	Ag	Dusi	1, 2, 3
E. mutabilis	—	Ax	Pringsheim	1, 2, 3
		Ax	Mainx	1, 3
		Ax	Lewin	3
E. mutabilis var *lefevrei*	2 strains	Ag	Lefevre	4
E. pisciformis (E. agilis)	—	Ax	Mainx	9
E. proxima	1224/11c	Ag	Pringsheim	3
	1224/11d	Ag	Pringsheim	3
		Ag	Grell	5
E. stellata	—	Ax	Mainx	1, 2, 3

Table 1 (continued)

Species	Strain[a]	Condition[b]	Isolated by	Source[c]
E. thinophila	1224/15b	Ax	Pringsheim	1, 2, 3
	1224/15c	Ax	Pringsheim	1, 3
E. viridis	—	Ax	Pringsheim	1, 2, 3
E. viridis aff. minor	—	Ag	Lefèvre	4
E. viridis var. halophila	—	Ag	Pringsheim	3
E. viridis var. maritima	—	Ag	Pringsheim	3
E. viridis var. paludosa	—	Ax	Pringsheim	1, 3
E. viridis var. terricola	—	Ax	Lewin	1, 2, 3

[a] Condensed from Anonymous (1958). *J. Protozool.* **5**, 1.

[b] Strain designations for those *Euglena* indicated as number/number are those employed by the Culture Collection of Algae and Protozoa, The Botany School, Cambridge University, Cambridge, England.

[c] Culture condition: Ax = axenic; Ag = agnotobiotic (culture in the presence of an unknown microbial flora and/or fauna).

[d] Source: *1*, Algensammlung, Pflanzenphysiologisches Institute, Göttingen, West Germany; *2*, Curator of the Culture Collection of Algae, Botany Department, Indiana University, Bloomington, Indiana; *3*, The Curator, The Culture Collection of Algae and Protozoa, The Botany chool, Downing Street, Cambridge, England; *4*, P. Bourelly, Muséum National d'Histoire Naturelle, Laboratoire de Cryptogamie, 12 Rue de Buffon, Paris, France; *5*, K. G. Grell, Max Planck Institute für Biologie, Hausserstrasse 43, Tübingen, West Germany; *6*, J. A. Gross, Life Sciences Research Institute, Illinois Institute of Technology Research Institute, Chicago, Illinois; *7*, R. P. Hall, Biology Department, New York University, University Heights, New York, New York; *8*, O. Jirovec, Parasitologie, Charles University, Vinicna 7, Prague, Czechoslovakia; *9*, L. Provasoli, Haskins Laboratories, 305 East 43rd Street, New York, New York.

borrowed. Estimates of population density are also an integral part of most growth studies, and these procedures have increased in sophistication along with culture techniques.

B. LIQUID AND SOLID MEDIA

Euglena is generally grown in liquid media, but most strains will also grow in semisolid or solid media, and in many cases the latter is the method of choice. In bleaching studies, for example, white or green colonies descending from single cells are easily scored in pour plates (Lyman *et al.*, 1961; Cook, 1963b). Such plates are made by flooding a small aliquot of cell suspension with media made to about 0.8% with agar. This concentration of agar remains liquid after autoclaving if held at 40°–45°C, and rapidly solidifies at room temperature. The brief exposure of *Euglena* to this temperature apparently has no permanent effect.

Euglena will also form colonies on filter pads (Millipore filters) placed on nutrient agar, in much the same way as bacteria. Maintenance of sterility is more of a problem than with pour plates, and this technique has apparently received little use, but in some cases—e.g., isolation of mutants—it could prove very satisfactory.

C. STOCK CULTURES

Stocks are maintained either on agar or, perhaps more commonly, in liquid media. Prime conditions for this necessary chore are those that keep *Euglena* in a healthy state with a minimum amount of work for the investigator. Tubes of 1–2% proteose peptone (Difco) or the equivalent are easily prepared, and support growth of most strains of *Euglena*. If these are kept at 14°–15°C transfers need be made only infrequently; in the case of *E. gracilis* Z and *E. gracilis* var. *bacillaris*, monthly transfers are quite adequate. Apparently, light is not essential except for the obligate phototrophs; stocks of the two strains just mentioned have been kept in the dark for nearly 10 years with no loss of greening ability.

If the media to be used in experimental studies differs from the maintenance media, then care should be exercised to prevent carry-over in transfer when this is important to the experiment. If time permits, the use of serial transfers is the most convenient method of doing this. Alternatively, washing by suspension in sterile water or media can be used; screw-cap centrifuge tubes make this a fairly simple procedure.

D. COUNTING AND SAMPLING

Estimates of population density are made by direct counting methods or by photometric devices. The former are to be preferred in general, since photosensing is influenced not only by the population density, but also by the size of the cells and by pigment concentrations. Since cell size and biochemical profile change continuously with culture age (see Section VII), a photometric readout may not always be proportional to cell number. Such methods are convenient, however, and with proper calibration can be used confidently.

1. *Photometric Methods*

The colorimeter is commonly used to measure light absorption by tube cultures of *Euglena*. A wavelength at which pigment absorption is minimum is best. Elliott (1949) has evaluated the use of the Klett–Summerson instrument in estimating growth of *Euglena* and other protozoans.

The nephelometer is a photoelectric device that functions essentially as a turbidimeter by measuring light scattering; the photoelectric cell is placed at right angles to the incident light path. The latter method may provide better estimates of cell number than the colorimeter. Apparently, the first nephelometer was constructed by Richards and Jahn (1933).

Satisfactory use of photometric devices requires that the media be as clear as possible, and not become discolored in growth. Some media permit such dense growth that the optical density in ordinary test-tube cultures exceeds 2.0 well before the maximum population density is reached. Appropriate dilutions can be made in these cases. The error in replicate estimates of population densities can be less than 2% (Elliott, 1949).

2. Direct Counting Methods

Aliquots may be plated and counted after colonies are large enough to be visible. This method is not often used to follow population growth of *Euglena*, but has been used to advantage with *Chlamydomonas* (Bernstein, 1960). At least a week is required for colony development, but since repeated counts can be made on the same plate, the only important source of error is in pipetting.

Fixed cells, well mixed, can be counted with the microscope (Hall *et al.*, 1935). A Sedgwick–Rafter counting chamber of 1-mm depth is filled with the cell suspension and covered with a slide. After the cells have settled, the chamber is scanned with a microscope fitted with a calibrated Whipple disc. The method is tedious and time-consuming, and the error in replicate counts approaches 5%.

The Coulter electronic cell counter is the most convenient and accurate means for direct estimation of population numbers. Although initially developed for clinical counting of red blood cells, the Coulter counter has proved quite satisfactory for *Euglena* and other protozoa. The count is rapid, and error in replicate counts is less than 2%. Other commercial instruments read out as a rate, and calibration permits conversion to numbers, but these are less satisfactory in most cases.

James and Anderson (1963) have described the construction of a device that, when coupled to the Coulter counter, gives a continuous recording of cell number in large populations. Such instruments are of particular advantage in long-term studies, or when population expansion is not strictly exponential (as in synchronized populations). Padilla (1960) used a pipetting apparatus that sampled at preselected intervals determined by a timer, the samples being delivered to a fraction collector for later counting. Petropolous (1964) has also described such an apparatus.

E. CULTURE VESSELS

1. *Single-Cell Studies*

We have explored the use of capillary tubes employed so successfully by Prescott in following the growth of single *Tetrahymena* (1959), but they are not well suited for the smaller and less active *Euglena*. Refraction by the walls and meniscus frequently make it difficult to find the cell. Hanging-drop cultures, however, work quite well. A small drop containing the cell is pipetted onto a cover slip which is then inverted over a depression slide and sealed with stopcock grease (Fig. 1a). The whole preparation may be submerged

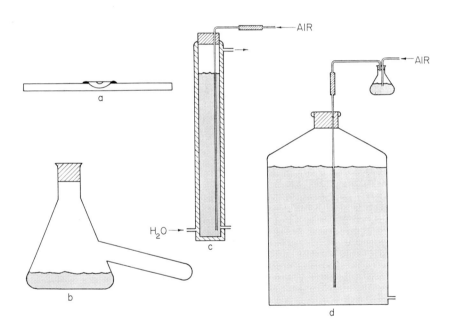

Fig. 1. Culture devices used for growth of *Euglena*. a, Hanging-drop preparation for single-cell studies; b, Erlenmeyer flask fitted with a test tube to permit optical readout; c, water-jacketed Pyrex cylinder that provides constant temperature and even illumination (by vertical fluorescent tubes); d, Pyrex carboy used in mass culture.

in a water bath. It should be mentioned that some greases are toxic even when not in direct contact with the medium. If attention is paid to sterility and environmental shocks are avoided, the average growth rate of such cultures is the same as that of large populations, other conditions being equal (Cook and Cook, 1962).

2. Small and Intermediate Volume Cultures

A surface area that provides for adequate gas exchange is necessary for optimum growth. In our laboratory, unshaken tubes of *E. gracilis* Z have a generation time of 23 hours at 30°C, compared to the optimum of 10–11 hours. Erlenmeyer flasks fitted with a test tube (Fig. 1b) permit optical readout coupled with adequate surface area in cultures large enough to be used for physiological studies.

In illuminated cultures, temperature control is often difficult, particularly in light-synchronized populations. We have found that a flask of the type shown in Fig. 1c works very satisfactorily. The flask is made cylindrical to provide for even illumination by vertical fluorescent tubes. Water is pumped from a thermostatically controlled bath through the outer jacket. The suspension is stirred with a magnetic bar, and any gas mixture desired can be bubbled through the culture. Cells are removed either by a siphon tube or, if desired, a drain tube can be built through the jacket as shown in Fig. 1c. The tip of the draining device can be kept in a beaker of dilute acid to prevent contamination through this route. These vessels, which are constructed at a modest cost by any of the commercial glass-blowing firms, can be made to any desired volume. The smallest we have used has a capacity of 100 ml, and the largest 4 liters. They are equally convenient for heterotrophic growth of *Euglena*.

Rapid temperature changes are not possible with the flask shown in Fig. 1c. In studies of temperature synchronization, this is a real disadvantage. Padilla and James (1964) have described a culture vessel containing internal steel coils that does permit rapid temperature changes. The vessel has been used to synchronize cell division in large populations of the colorless flagellate *Astasia longa* (Blum and Padilla, 1962) as well as *E. gracilis* (G. M. Padilla, personal communication).

3. Cultures of Large Volume

Bach (1960) has described a method for growing mass cultures of *Euglena*. It consists essentially of growing the cells in a 10- or 20-liter carboy provided with adequate gas exchange. Difficulties inherent to this procedure include temperature control and nonrandom distribution of cells. *Euglena* tends to settle to the bottom in cultures that are not well agitated; generally, however, mass cultures are to be used in studies in which knowledge of the absolute growth rate is not a prime requisite.

We have used vessels of the type shown in Fig. 1d, which can be obtained commercially at modest cost. The tubulature makes the flask slightly more expensive, but is convenient in harvest, and can be used with little risk of contamination for periodic cell counts. Visual estimates of population density are quite unreliable in cultures of these volumes.

Many incubators will accommodate these carboys for heterotrophic growth. A constant-temperature room and often a water cell are necessary for temperature control in autotrophic growth, although incubators with glass windows will suffice. The carboys may also be partially submerged in a constant-temperature water bath; aeration provides agitation sufficient to hold the culture temperature within a degree or so of the bath temperature, even with illumination.

Bach (1960) has also described a gas humidification vessel for use with such cultures to prevent loss of water by evaporation. We have found that gases are sufficiently humidified by bubbling through water contained in an Erlenmeyer flask (Fig. 1d). Loss of water from the culture is negligible.

The yield in these mass cultures depends of course on the conditions of growth and the cell density at harvest. At 30°C, 20 liters of *E. gracilis* Z grown heterotrophically on the medium of Cramer and Myers (1952) with glucose as the carbon source, harvested at 500,000 cells/ml, yielded 30 ml of packed cells (ca. 10 gm dry weight).

F. QUASICONTINUOUS CULTURE

Some experiments call for repeated sampling of relatively large volumes. Common examples include most analyses of synchronized populations, and studies of many induction phenomena. The effort expended in establishing cultures of several liters can be profitably reduced by constructing an in-line reservoir of sterile media. An exhausted culture can then be diluted to any desired population density for initiation of a new growth cycle. In addition to convenience, studies of an extended nature can be conducted in full confidence that growth of *Euglena* is started in media of invariant composition. Periodic dilution of this sort can hardly qualify as "continuous culture" in the context normally used (see the next Section), and is referred to here as quasicontinuous. An extended discussion of this technique as used in euglenoid studies has been presented by Padilla and James (1964), and Lorenzen (1964) has emphasized its use in populations of synchronized *Chlorella*.

Figure 2 is an illustration of the culture arrangements we have used for quasicontinuous growth studies. Autoclaves of large capacity permit several large carboys, connected by T-tubes, to be made in-line with the culture flask. Appropriate insertion of T-tubes with ground-glass ball-and-socket joints permit additional autoclaved reservoirs to be aseptically added to the culture. Autoclaves of small capacity which will accept the culture vessel will also usually accept a two-tubulature filter flask fitted with a sterilizing filter, either of the vacuum or pressure type, which with more work will serve the same end as large autoclaved reservoirs.

In theory, quasicontinuous cultures could be continued indefinitely. In practice, it is found that adherence of cells to the inner surface of the culture vessel is visibly obvious after a few weeks, and becomes exponentially worse; accurate cell counts are no longer possible. Elaborate "windshield wipers" have been developed to alleviate this problem in true continuous cultures of bacteria, but in quasicontinuous cultures of *Euglena* we have found it more convenient to start new cultures.

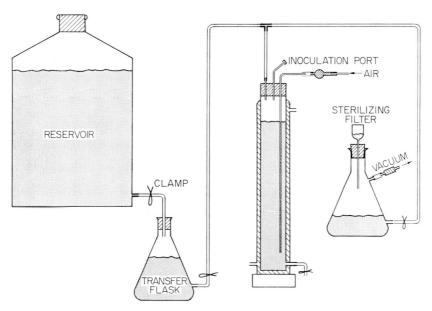

Fig. 2. Culture apparatus used for periodic dilution, or "quasicontinuous" culture of *Euglena*. The culture vessel stands on a magnetic stirrer. The culture can be diluted as desired from the sterile in-line reservoir of fresh media; frequently an intermediate flask of smaller size simplifies the transfer of media. Alternatively, the culture vessel may be autoclaved with a filter flask fitted with a sterilizing filter, shown here for convenience as being in-line with the reservoir.

G. Continuous Culture

In spite of the many advantages offered by continuous culture, this important technique—so successfully exploited in microbial genetics—has received remarkably little attention in studies of protozoan physiology. The chemostat keeps cell populations in balanced exponential growth, a condition never obtained in ordinary means of culture (see Section VII), and furthermore this condition of balanced growth can be maintained for extended periods of time. Examples of studies in which continuous culture is a prerequisite for truly controlled experiments are too numerous for mention; any response of *Euglena* to experimental variables requiring more than a few generations for completion, and indeed any physiological analysis ideally based on the premise of balanced growth, calls for this culture approach.

In principle, the chemostat is simply a culture of constant volume which is diluted with fresh media at a constant rate. One of the nutrients is in limited concentration, and the cells find and maintain a growth rate determined by the availability of that nutrient. The growth rate is thus never as great as in the presence of an excess of the limiting nutrilite, and in effect the cells are in some early stage of stationary phase, although a constant rate of exponential expansion is maintained.

Many devices have been described that yield continuous culture, ranging from the simple type with external control similar to that of Novick and Szilard (1950) for bacterial cultures, to the more elaborate optically controlled apparatus described by Myers and Clark (1945) for growth of *Chlorella*, which apparently was the first attempt at continuous culture. These approaches have been reviewed by James (1961).

Bacterial populations are easily studied in continuous culture. Because of their fast growth rates, relatively long-term experiments—those requiring many generations—are completed in a day or two. For this reason, less than precise control of dilution rates can be tolerated. The more extended generation times of the euglenoids offer special problems. We have found that some of the devices used with bacteria are unsuited for *Euglena* because the dilution rate drifts at a rate that could be serious after a few weeks. In addition, very large cultures must be used if the overflow is to be harvested in amounts adequate for biochemical analysis. This in turn raises the practical problem of sterilizing an appropriately large reservoir of medium.

Figure 3 is a schematic drawing of one approach we have used with some success in extended continuous culture of *E. gracilis* Z (Cook, 1966e). In effect, a constant-rate infusion-withdrawal pump (Harvard Apparatus Co., Dover, Massachusetts) pumps fresh media into a water-jacketed culture vessel which may be as large as 4-liter capacity (Ace Glass, Inc., Vineland, New Jersey), and at the same time removes the cell suspension at the same rate. The culture volume is thus kept constant. The growth rate (i.e., the dilution rate) can be adjusted either by changing the pumping rate or the volume of the culture.

The pump is equipped with motor-driven syringes. Operation of a single chemostat requires two channels—one for input and one for output. We have used a four-channel pump for the simultaneous operation of two chemostats. Since both the input and output syringes operate in unison, a by-pass (a in Fig. 3) has been inserted into the output line; a clamp at the appropriate point on this line thus permits the media to be pumped into the vessel while only air enters the output syringes. This by-pass is necessary in the initial filling of the culture vessel or whenever the culture volume in the vessel must be increased.

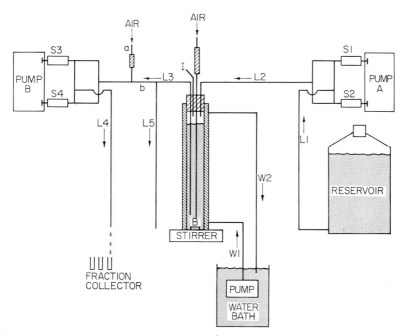

Fig. 3. Continuous culture apparatus, or "chemostat," used for growth of *Euglena*. Pump A transfers media from the reservoir to the culture vessel via the syringes S1 and S2, while pump B removes cell suspension at the same rate via the syringes S3 and S4. In practice, all four syringes are incorporated into a single infusion-withdrawal pump, assuring equal input and output rates. The line L5 is used to reduce the volume of the culture. In operation, line a is clamped off and L3 is open at b; clamping b and opening a permits filling of the culture vessel, since the output line pulls only air. W1 and W2 are lines used to circulate water through the jacket of the culture vessel. The culture is kept homogeneous by bubbling air and by the magnetic stirring bar B.

A draining siphon (L5 in Fig. 3) is also built into this chemostat, the tip of which is kept in a beaker of dilute acid to avoid contamination. The siphon is necessary to reduce the volume of the culture when necessary.

The pump used here has a wide range of rates—nominally 0.0008–38 ml/minute—but these rates are only approximate, and the overflow or output volume must be measured to obtain an accurate estimate of the dilution rate.

A fraction collector may be used for this, but if regular checks are made a graduated cylinder suffices. If all lines are of small volume and made air-free, drift in the dilution rate is negligible or nonexistent.

Euglenoid cells tend to settle out in the lines and in the syringes. Unless the culture volume is large and the dilution rate relatively great, the output suspension may not be representative of the culture suspension itself. In this situation we have used the siphon (b in Fig. 3) to sample, refilling the culture to its initial volume by letting the output syringes pull air instead of cell suspension. This procedure makes the apparatus something less than a chemostat, but if the volume removed is relatively small and the refilling proceeds at the same rate previously used in controlled dilution, it can be shown that the growth rate does not change significantly in refilling.

We have found the chemostat described here to be quite satisfactory for continuous culture of *Euglena*. Modifications or other approaches will no doubt work equally well. Continuous culture of *Euglena* is eminently desirable; there seems to be no reason why studies of *Euglena* requiring this culture method cannot be approached with the same sophistication routinely found in the study of microbial systems.

H. Synchronous Culture

1. *Synchronization by Light*

Euglena grown on a diurnal light–dark cycles tend to divide at night if nutrition is autotrophic (Leedale, 1959). This phenomenon permits synchronization in the laboratory by culture on light–dark cycles (Cook and James, 1960; Huling, 1960; Edmunds, 1964). The temperature may be held constant.

If the lengths of the light and dark periods are appropriately chosen with respect to the incubation temperature, about 100% of the cells will divide in each light–dark cycle; cell fissions are confined for the most part to the dark periods. We have reviewed an approach that permits approximation of the correct light–dark regime (Padilla and Cook, 1964). This varies among strains of *Euglena*; with the first strain synchronized (erroneously referred to at the time as *E. gracilis* var. *bacillaris* and which we now call *E. gracilis* L), the best program found was a temperature of 20°C and light : dark periods of 16:8 hours (Cook and James, 1960). With *E. gracilis* Z, on the other hand, a temperature of 21.5°C and a light : dark cycle of 14:10 hours is best (Cook and Hess, 1964; Cook, 1966a). Hoogenhout (1963) has reviewed the literature concerning synchronous cultures of green cells, including *Euglena*. Some of the physiological aspects of synchronous growth in *Euglena* are discussed below (Section VIII,D).

2. Synchronization by Temperature

a. Heat Shocks. Pogo and Arce (1964) have recently reported that *E. gracilis* Z, grown heterotrophically on a complex organic medium, will divide synchronously after being subjected to supraoptimal temperature shocks. Their regime consists of 11.0 hours at 24°C and 13 hours at the higher temperatures, this program being offered repetitively. Temperatures of 36°–38°C block cell division, with an approximate doubling of cell number occurring at each 24°C sojourn. Heat shocks of 35°C or less are not effective in synchronizing cell division. These authors found that the synchronizing heat shocks do not impair cell growth as measured by volume increase.

b. Cold Shocks. *Euglena gracilis* Z can also be synchronized by a shift from the optimal temperature (29°–30°C) to a colder temperature, which blocks cell division (G. M. Padilla, personal communication). The approach is similar to that used so successfully in synchronization of the colorless flagellate *A. longa* (Padilla and James, 1960). The temperature program is offered repetitively and a doubling of cell number occurs in each warm period.

The heat-shock approach relies on temperatures that bleach *Euglena;* Pogo and Arce (1964) report that seven heat shocks of 37°C result in 14% bleached cells. The usefulness of this approach will be limited because of this effect. Synchronization by cold shocks, however, has some roots in the ecological history of *Euglena* and presents a welcome comparative approach to studies of the physiology of *Euglena* synchronized by light–dark shifts.

IV. Gross Biochemical Composition

The biochemical profile of *Euglena* is greatly influenced by the conditions of culture and the stage of growth. Some of the effects of these variables are discussed in Sections VII and VIII. Table II is a summary of the composition of an average *E. gracilis;* the values are pooled from a large number of sources in which *Euglena* were grown under a wide variety of conditions, and the table is meant to be no more than a rough guide to ranges of values expected.

Kempner and Miller (1965a) have made an elemental analysis of *E. gracilis* Klebs. The cells were grown on a light–dark cycle in a defined medium containing 3.0 gm *l*-glutamic acid per liter. The cells were harvested in log growth, washed, and lyophilized for analysis. Table III is a summary of their results. As these authors point out, the values are similar to those reported for other microorganisms.

Table II

GROSS BIOCHEMICAL COMPOSITION OF *E. gracilis*[a]

Fraction	mg/10[6] Cells
Total dry mass	0.7–1.0
Protein	0.2–0.4
Paramylum	0.04–0.4
Lipid	0.11–0.3
Chlorophyll	0–0.035
RNA	0.015–0.045
DNA	0.0027[b]–0.0043[c]

[a] These ranges are estimated from a wide variety of sources. The cells were generally grown under optimal conditions, and do not represent extreme values that can be found under other growth conditions (see *e.g.*, Section VIII).

[b] *Euglena gracilis* Z.

[c] *Euglena gracilis* var. *bacillaris* SM-L1.

The DNA content of *Euglena* makes up about 2–4% of the total mass, depending on how the cells are grown. Chloroplast DNA comprises about 3% of all the DNA (Edelman *et al.*, 1964) and is about 75% adenine plus thymine (Brawerman and Eisenstadt, 1964; Edelman *et al.*, 1964; Ray and

Table III

ELEMENTAL ANALYSIS OF LYOPHILIZED *E. gracilis*

Element	mg/gm dry cells
Carbon	466[a]
Oxygen	331[a]
Hydrogen	72.9[a]
Nitrogen	80.2[a]
Phosphorous	11.5[a]
Sulfur	6.7[a]
Potassium	1.98[a]
Chlorine	0.99[a]
Iron	0.33[b]
Magnesium	0.83[b]
Zinc	0.17[b]
Manganese	0.01[b]
Copper	0.02[b]
Calcium	0.04[b]

[a] From Kempner and Miller (1965a).

[b] From Price and Vallee (1962).

Hanawalt, 1964), in contrast to the nuclear DNA, which is 45–50% adenine plus thymine. The use of base ratios as an aid in taxonomy and evolution among the protozoa and algae has been attempted (Schildkraut *et al.*, 1962; Serenkov, 1962).

Several studies have been made relative to the fatty acids (e.g. Korn, 1964), amino acids (e.g. Kempner and Miller, 1965a; Kott and Wachs, 1964), pigments (Goodwin and Gross, 1958; Krinsky and Goldsmith, 1960), etc. in *Euglena*, but most of these are described in detail in other parts of this treatise and need not be considered further here.

V. Requirements for Growth

A. Common Media

A comprehensive discussion of culture media for *Euglena* has recently appeared (Hutner *et al.*, 1966). Effective media contain a chelating agent like citrate or ethylenediamine tetraacetate (EDTA). Such agents serve to reduce the background level of trace contaminants, and at the same time permit the addition of ions in concentrations that yield heavy growth without being initially toxic (Hutner *et al.*, 1950).

Fogg (1965) has recently made an interesting comparison of the composition of a freshwater lake with that of several culture media for algae. We have made a similar comparison, shown in Table IV, with two media commonly used for *Euglena*. In some of the components striking differences occur. A direct comparison can be deceptive, however; as pointed out by Fogg, natural waters contain organic matter (ca. 5 mg/liter) that may have biological functions of which we are ignorant, including the possibility of lowering the requirement for some of the components—a sort of sparing action. The example is quoted in reference to the observations of Rodhe (1948) who found concentrations of 0.04 mg/liter of phosphorous necessary for maximum growth of *Asterionella formosa* in artificial media, while 0.002 mg/liter sufficed in natural lake water.

B. Inorganic Requirements

A few of the inorganic factors required for the growth of *Euglena* have received increasing attention in recent years, although studies of the utilization of organic compounds is still more important from the standpoint of expended effort. The metabolism of the inorganics, and their role in the biochemical growth of *Euglena*, are much better understood today than a decade ago.

Table IV

COMPOSITIONS OF SYNTHETIC MEDIA COMMONLY USED FOR GROWTH OF *Euglena* COMPARED
TO THAT OF A TYPICAL FRESHWATER LAKE[a]

Component	Kettle Mere, Shropshire, England[b]	Cramer–Myers medium[c]	Hutner's medium[d]
Na	7.6	120	3.6
K	8.6	287	115
Ca	23.2	7.3	200
Mg	2.9	40.0	4.9
HCO_3	34.8	—	300
Cl	13.9	12.8	0.335
SO_4	26.8	81.7	100
$N (NO_3)$	0.05	0.06	—
$N(NH_3)$	—	212	42.4
$P(PO_4)$	0.004	480	424
SiO_2	1.0	—	—
Fe	—	0.06	0.05
B	—	—	0.10
Mn	—	0.50	19.2
Mo	—	0.12	7.9
Co	—	0.26	0.40
Cu	—	0.005	0.206
Zn	—	0.09	40
I	—	—	0.02
dl-Malic acid	—	—	2000
L-Glutamic acid	—	—	5000
Citrate[e]	—	482	—
EDTA	—	—	500
Vitamin B_{12}	—	0.0005	0.0002
Vitamin B_1	—	0.01	1.0

[a] Amounts in milligrams per liter.
[b] Gorham (1957).
[c] Cramer and Myers (1952).
[d] Greenblatt and Schiff (1959).
[e] As $Na_3C_6H_5O_7 \cdot 11 \ H_2O$.

In many cases, minute amounts of a required inorganic will suffice for optimal growth. Thus, Hutner *et al.* (1950) calculated that 4900 molecules of vitamin B_{12} are needed to produce a single cell of *Euglena*. Under their culture conditions, maximum growth of *Euglena* (to 7×10^6 cells per milliliter) required 0.15 ng B_{12} per milliliter; this is equivalent to 0.6 pg cobalt per milliliter of culture medium. The same growth response required a (highly purified) iron source that contained over 30 times as much cobalt as that found in the B_{12} requirement. Low-level requirements coupled with difficulties in purification are inherent problems in studies of the metabolism of inorganic components of *Euglena* media. The more carefully conducted work reviewed in this section has recognized this difficulty, and appropriate interpretation of results has been made by the investigators.

1. *Water*

Although *Euglena* will grow in sewage (see e.g., Ludwig *et al.*, 1951; Kott and Wachs, 1964; Chapter 2) controlled growth clearly calls for clean water. Reproducible growth can be obtained when the water is distilled and then deionized; addition of EDTA to the boiling flask is a further aid, and simplifies cleaning of the vessel.

2. *Sulfur*

Sulfate present as a trace contaminant in other salts permits considerable growth of a streptomycin-bleached strain of *E. gracilis* var. *bacillaris* (Buetow, 1965). Buetow found that 0.01 mg sulfate sulfur per milliliter yielded maximum cell densities as great as higher concentrations. Concentrations of 0.001 mg/ml yielded about half this number of cells. The generation times of cells grown in less than optimal amounts of sulfate did not differ from that of cells grown with adequate amounts of sulfate.

Most animal cells require sulfur in a more reduced form, but *Euglena*, in common with higher plants and many microorganisms, utilizes sulfate both in the light and in the dark. Methionine serves as well as sulfate in terms of final cell density (not true of *Chlorella*, however; see Schiff, 1959); cysteine and homocystine support growth but at reduced final levels; while taurine and homocysteic acid are inadequate as sulfur sources for *Euglena* (Buetow, 1965).

Kempner and Miller (1965b) and Buetow (1965) have studied the distributions of radiosulfur in *Euglena*, the former authors using cells totally labeled after long-term growth and the latter author using cells grown for one generation in the presence of radiosulfur. Slightly different extraction procedures were used, but both sets of data are comparable. Protein contained 40–50% of the total sulfur, while lipid solvents removed 5–15% of the total label. The nucleic acid fraction (i.e., hot trichloroacetic acid extraction) contained 10–25% of the sulfur. Buetow was able to show that this fraction in *Euglena*, unlike most other microorganisms, contains an appreciable amount of protein.

Transfer of ^{35}S-labeled *Euglena* to sulfur-deficient media (Buetow, 1965) or to a nonradioactive medium (Kempner and Miller, 1965b) results in a rapid decrease of alcohol- and acid-soluble label. In sulfur-deficient media, the protein-^{35}S remains constant in level for about half a generation, after which a slow decline sets in; this suggests a mobilization of soluble pools into residual protein. After about 1.5 days of sulfur starvation, *Euglena* cells start dying and after 15 days only 25% remain viable. Buetow (1965) feels that the surviving cells are those that are able to regulate the extensive redistribution of sulfur reserves.

Goodman and Schiff (1964) have examined the products formed from radiosulfate by *E. gracilis* var. *bacillaris* grown in the light on the pH 3.5 medium of Hutner. They found cysteine, cystine, and glutathione, but no methionine, homocysteine, or adenosylmethionine in the soluble pools; the last-mentioned two compounds are present in soluble pools of *Chlorella*. The authors suggest that steady-state concentrations of methionine may be too low to detect in *Euglena* because of the demands of protein synthesis. Methionine is a constituent of *Euglena* proteins.

Sulfur-containing nucleotides or nucleotide–peptides have been found in *Euglena*. Abraham and Bachhawat (1965) and Davies *et al.* (1966) have demonstrated the presence of 3′-phosphoadenosine 5′-phosphosulfate ("active sulfate," PAPS). Recent work in our laboratory has shown that at least two soluble nucleotides containing sulfur are present in *Euglena* (Whitney, 1966). These have not yet been completely characterized, but in both cases the radiosulfur is easily removed by alkaline hydrolysis, after which it can be quantitatively precipitated as barium sulfate. Ion-exchange chromatography of digested RNA also separated a sulfur-containing nucleotide that has many characteristics of a derivative of uracil. Purified preparations of DNA from *Euglena* grown in the presence of radiosulfur were found to have no detectable radioactivity. On the other hand, hot PCA extracts prepared in the usual way showed significant incorporation of radiosulfur into this fraction (Cook and Hess, 1964). The activity found in hot PCA extracts may be due to peptide–nucleotide complexes. This view is substantiated by Buetow's finding (1965) that the hot-acid fraction from *Euglena* contains significant amounts of protein. Sulfur-containing peptide–nucleotide complexes were first found in *Chlorella* (Hase *et al.*, 1960) and have been studied in some detail in that alga (Hase *et al.*, 1961; Schmidt, 1966). Shibuya *et al.* (1963) described a nucleoside diphosphate sulfoquinovose from *Chlorella*. In synchronized *Euglena*, the sulfur content of the hot PCA fraction is proportional to the DNA content of the cells (Cook and Hess, 1964), implying that the sulfur-containing component may be important in the structural organization of the DNA molecule.

Lipids contain an important amount of the total sulfur in *Euglena* (Buetow, 1965; Kempner and Miller, 1965b). The major sulfolipid in photosynthetic cells is 6-sulfo-6-deoxy-α-D-glucopyranosyl-1,1′-diglyceride (6-sulfoquinovosyl diglyceride) (Benson *et al.*, 1959; Daniel *et al.*, 1961). The major fatty acid components are palmitic acid and linolenic acid (O'Brien and Shibuya, 1964). Benson *et al.* (1962) showed that levels of this sulfolipid are low in etiolated *Euglena*, increasing with greening after exposure to light. It was suggested that the sulfolipid is a requirement of active photosynthetic structures, and is intimately related to carbohydrate metabolism of the chloroplast. Davies *et al.* (1966) have shown that the sulfur from cysteic acid can serve as a source of the C-S unit for sulfolipid biosynthesis, although cysteine sulfur is not incorporated into the sulfolipid of *Euglena*. These authors also found that molybdate inhibited the incorporation of sulfate into sulfolipid. Since molybdate inhibits the formation of adenosine 5'-sulfatophosphate (Wilson and Bandurski, 1958) and therefore PAPS, Davies *et al.* (1966) suggest that PAPS is on the route from sulfate to sulfolipid in *Euglena*. Davies *et al.* (1965) also found that the amount of sulfolipid in light-grown *Euglena* is almost five times that of dark-grown cells, although the total uptake of radiosulfur was about the same in both green and etiolated *Euglena*.

Wedding and Black (1960) showed that sulfate uptake by *Chlorella* is enhanced by light. Abraham and Bachhawat (1965) found that the rate of sulfate incorporation by *Euglena* is higher in photosynthesizing cells than in etiolated cells, and still higher than is found in streptomycin-bleached *Euglena*. Under their conditions of culture, however, the nonphotosynthesizing cells had a lower growth rate, and the sulfate uptake on a dry weight basis was the same regardless of culture conditions. Stimulation of sulfate uptake in *Chlorella* by light, and its inhibition by dinitrophenol (Wedding and Black, 1960) implies that sulfate incorporation is an energy-requiring process, a finding confirmed in *Euglena;* incorporation is inhibited by dinitrophenol and azide (Abraham and Bachhawat, 1965).

Reduction of sulfate in *Euglena*, as in other organisms (Robbins and Lipmann, 1956), involves the initial formation of a high-energy intermediate, PAPS. Abraham and Bachhawat (1963, 1965) have described some of the properties of the enzyme forming PAPS in *Euglena*. The nucleotide requirement is specific for ATP, and the rate of the reaction is enhanced by sulfhydryl compounds. The specific activity of the PAPS-forming enzyme changes with the stage of growth, but photosynthesizing as well as etiolated and bleached cultures of *Euglena* had essentially the same specific activities when measured at the same stage of growth. These authors also described an enzyme from *Euglena* that degrades PAPS into sulfate and 3'-phosphoadenosine 5'-phosphate (PAP). ADP is a potent inhibitor of this enzyme. Adenosine 5'-phosphosulfate (APS) and inorganic phosphate, degradation products of PAPS found in rat liver extracts (Lewis and Spencer, 1962), could not be detected in *Euglena* by Abraham and Bachhawat (1965).

Later stages of sulfate reduction have not been studied much in *Euglena*, and indeed detailed knowledge of the enzyic pathway of the reduction of the PAPS-sulfate is limited to a few organisms. The fact that sulfur-amino acids will satisfy the sulfur requirements of *Euglena* (Buetow, 1965) suggests the interesting possibility that the whole assemblage of sulfate-reducing enzymes may be inducible. The demonstrated importance of an energy source in sulfate reduction has not been examined from the standpoint of sulfate utilization in growth of *Euglena*, although Kylin (1964a,b) has initiated such studies with *Scenedesmus*.

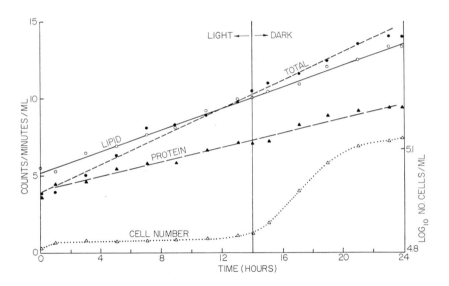

Fig. 4. Incorporation of radiosulfate by light-synchronized *E. gracilis* Z. Cells were synchronized by repetitive culture on a light–dark cycle (1000 ft-c, 14:10 hours) at 21.5°C. $^{35}SO_4^{2-}$ (10 Ci/liter) was added at the beginning of the light period preceding the one shown here. Incorporation is expressed as counts per minute per milliliter of culture to facilitate comparison of the interphase and division phases of the cell cycle. In order to make this plot, observed values for total incorporation were multiplied by 10^{-3}, those for lipid by 10^{-2}, and those for protein by 0.5×10^{-2}.

In light-synchronized *E. gracilis* Z (Sections III,H and VIII,D), total radiosulfate incorporation occurs at about the same rate in both the 14-hour light period and the 10-hour dark period (Fig. 4). In this experiment the cells were grown in the presence of $^{35}SO_4^{2-}$ (10 mCi/liter) for one full light–dark cycle before harvests were initiated, so that the cells were already heavily labeled at the beginning of the cycle shown in Fig. 4. Aliquots of equal volume were removed periodically from the culture over this cycle and analyzed by the usual fractionation procedures for the intracellular distribution of label. The components containing the major portion of radiosulfur—protein and sulfolipid—both showed continuous incorporation over the light and dark periods (Fig. 4). Autoradiograms of the ethanol-soluble fraction revealed a number of radioactive spots. It may be noted that the most prominent sulfolipid in autotrophic *Euglena*—sulfoquinovosyl diglyceride (Benson *et al.*, 1959)—is apparently synthesized in both light and dark in synchronized cells, although synthesis of the photosynthetic pigments is strictly confined to the light periods (Cook, 1966a; Section VIII,D). Sulfoquinovose is largely confined to the chloroplasts (Davies *et al.*, 1965), and has been assigned a functional role in photosynthesis (Benson *et al.*,

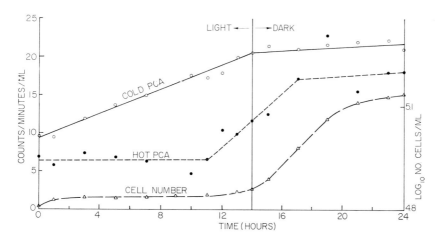

Fig. 5. Incorporation of radiosulfate into the cold acid– and hot acid–soluble fraction of synchronized *Euglena*. Culture conditions were the same as those described under Fig. 4. The cold 5% perchloric acid fraction increases continuously in the light, with little further increase in the dark. The hot perchloric acid fraction increases discontinuously, correlated but not synchronous with the increase in nuclear DNA (cf. Fig. 18). Observed counts were multiplied by 10^{-2} (Cook and Hess, 1964).

1962). It is of interest to note that the capacity for photosynthesis in synchronized *Euglena* is better correlated with protein (and sulfoquinovose?) concentrations than with chlorophyll concentration (Cook, 1966a).

Not all sulfur-containing fractions of synchronized *Euglena* behave with such regularity as protein and the sulfolipids. The cold perchloric acid fraction increases steadily in the light period, but little or no increase is seen in the 10-hour dark period (Fig. 5). Ion-exchange chromatography of this fraction showed the presence of two sulfur-containing components with spectral characteristics of nucleotides (Cook and Hess, 1964). This fraction also contained free sulfate. The lack of net accumulation in the dark period with a continued increase in protein and sulfolipid sulfur in the dark, implies that different mechanisms of incorporation may operate in the light and in the dark in synchronized *Euglena*.

The hot perchloric acid extracts from synchronized *Euglena* also contained two sulfur-containing compounds resembling nucleotides. The activity in this fraction increased at discontinuous rates proportional but not synchronous with the time course of DNA synthesis (Cook and Hess, 1964; Fig. 5). It was proposed that these compounds are comparable to the peptide–nucleotides found in *Chlorella* by Hase *et al.* (1960; see also Schmidt, 1966). The increased incorporation of sulfur into these nucleotides, starting soon before onset of the dark period, suggested that the compounds may be of importance in mitosis or cell division.

Sulfurated compounds are unquestionably required in processes that must be closely involved in mitosis of the flagellates. Thus, temperature synchronization of *A. longa* on a chemically defined medium is not possible unless a sulfhydryl compound is added (Padilla and James, 1960). The role of the sulfhydryl is unknown. Autoradiographs of *Astasia* incubated with thioglycolate-^{35}S show the label to be concentrated in the region of nuclear division (James, 1960). Contraction of the protein isolated from sea urchin eggs by Sakai (1962) can be controlled by varying the concentration of sulfhydryl or disulfide groups. It is possible that sulfhydryls may act in a similar manner in mitosis or cytokinesis of the flagellates.

3. Iron

Price and Carell (1964) found that even purified chemicals still contained enough iron as a trace contaminant to support some growth of *Euglena*. They found that $5 \times 10^{-8} M$ iron was limiting, while a concentration of $5 \times 10^{-7} M$ supported the same growth rate and yielded the same final density as $3.6 \times 10^{-5} M$ concentrations. Intracellular iron concentrations of 0.5 μg or less per milligram of protein were found in cells showing reduced growth rates in iron-limited media (Karali and Price, 1963).

The rate of chlorophyll formation in *Euglena* is linearly correlated with the intracellular concentration of iron. The absolute rate depends in part on the concentration of iron in the medium; under the conditions used by Price and Carell (1964), the rate of chlorophyll synthesis is twice as great at $10^{-3} M$ as it is at either $10^{-9} M$ or $10^{-5} M$.

These workers have used the iron requirement by *Euglena* to study the hypothesis that iron acts in the conversion of coproporphyrinogen to protoporphyrin (Carell and Price, 1965). They found that in *Euglena*, as in other microorganisms, iron appears to be required for this conversion. However, they feel that this iron requirement is satisfied at concentrations that still do not permit optimal rates of chlorophyll synthesis; accumulation of coproporphyrinogen and decreased rates of protoporphyrin formation occurred only in iron deficiency so extreme that growth was limited. The implication is that larger amounts of iron are also necessary in some later, or perhaps ancillary, steps in chlorophyll biosynthesis.

It is perhaps appropriate to mention here some of the definitive work done with iron-containing proteins in *Euglena*. Wolken and Gross (1963) described some of the properties and a method for the isolation of cytochrome c-552, first discovered by Nishimura (1959). Gross and Wolken (1960) also discovered another cytochrome (c-556) in *Euglena*. Smillie (1963) found a b-type cytochrome (b_6) in *Euglena* chloroplasts. Perini *et al.* (1964b) showed that cytochromes c-552 and b_6 are functionally localized in the chloroplasts, while c-556 and an a-type cytochrome (a-605, *Euglena*) appear to be associated with the respiratory system. That the chloroplast cytochromes are intimately involved in photosynthesis is indicated by the fact that the number of chlorophyll molecules was always proportional to the number of cytochrome molecules (the ratio was 350:1) when chlorophyll concentrations were varied by growth conditions.

A detailed chemical and physical analysis of some of these cytochromes has been described (Perini *et al.*, 1964a). Perini (1963) has speculated on the possible role of the cytochromes in respiration and photosynthesis.

4. Zinc

The final cell density attained by populations of *E. gracilis* Z is a linear function of the initial zinc added, over the range 0–15 μg/liter (Price and Vallee, 1962). The growth rate was independent of zinc concentration between concentrations of 10^{-6} *M* and 3×10^{-5} *M*. Cells grown in zinc-sufficient media contained almost 7 μg zinc per milligram of protein nitrogen in exponential growth, and about 3.8 μg in the stationary phase. On the other hand, cells grown in zinc-deficient media (up to 26 μg zinc added per liter) contained only about 0.4 μg zinc per milligram of protein nitrogen, measured in the stationary phase.

Since the pyridine nucleotide-dependent dehydrogenases are zinc metalloenzymes, Millar and Price (1960; Price and Millar, 1962) examined the possibility that low zinc levels limit growth by reducing respiration. They found, however, that oxygen consumption was still great enough to support more growth than was found, and concluded that the observed rate of ethanol oxidation in zinc deficiency does not account for the cessation of growth. Presumably the pyridine nucleotide-dependent dehydrogenases are present in relatively large concentrations even in zinc-deficient *Euglena*. Price (1961), however, has shown that the activity of a D-lactate dehydrogenase is decreased in zinc-deficient growth of *E. gracilis* Z, and the oxidation of lactic acid by intact cells is virtually abolished. Rutner and Price (1964) have described some of the properties of a purified preparation of this enzyme.

As might be expected, zinc deficiency results in profound alterations in the biochemical profile of *Euglena* (Wacker, 1962a,b). Protein and RNA decrease, while an increase is seen in free amino acids, all of which indicates an impairment of protein synthesis in zinc deficiency. It might be mentioned that comparable results could probably be expected when growth is limited by any number of required factors. Wacker (1962b) also found that zinc deficiency causes a massive increase in acid-insoluble polyphosphate, as well as mitotic arrest as indicated by a doubling of the DNA content of the cells.

5. *Nitrogen*

Nitrogen metabolism in *Euglena* has received relatively little attention, compared to that given other photosynthetic cells. Some early reports indicated that ammonium salts are better nitrogen sources than nitrate salts (Ternetz, 1912; Mainx, 1928; Dusi, 1933), while Pringsheim (1914) felt that nitrate served as well as ammonia as a nitrogen source. Cramer and Myers (1952) studied this problem under more controlled conditions with *E. gracilis* var. *bacillaris*, and found negligible growth with nitrate either in light or in the dark, butyrate serving as carbon source in the latter case.

Huzisige and Satoh (1960) reported that nitrite will not support cell multiplication of *E. gracilis* var. *bacillaris;* on the other hand, cells grown with ammonia as nitrogen source showed very high activities of nitrite reductase. No hydroxylamine or ammonia accumulated during these reactions, in which intact cells were used, from which the authors concluded that nitrite reduction by *Euglena* is of the assimilatory type. The rate of reduction was accelerated by light or by the addition of hydrogen donors, of which malate was the most effective of those tested. Aeration enhanced the rate of the reaction in the absence of an exogenous hydrogen donor; however, aeration in the presence of exogenous malate resulted in a rate of reduction less than that in unaerated controls containing malate. These very interesting studies have apparently not been pursued. Nitrate or nitrite reduction is intimately involved with the respiratory machinery (Nason, 1962) and the photochemical apparatus (Evans and Nason, 1953) in many cell systems. The report of Huzisige and Satoh (1960) suggests that possible nitrite or nitrate utilization by *Euglena* might well be re-examined from standpoints other than the support of cell multiplication.

Amino acids are frequently added to culture media as a carbon source. Buetow (1966) has shown that some amino acids—but not all—will also serve as sole nitrogen source for *E. gracilis* var. *bacillaris* SM-L1. The neutral aliphatic amino acids and the dicarboxylic amino acids (as well as their amides) yield final densities comparable or in excess of those found with $(NH_4)_2HPO_4$, although the generation time is sometimes extended. Glutamic acid is the poorest source of nitrogen in this group. Not used were the aromatic and basic amino acids, and those containing secondary amino groups or sulfur. As pointed out by Buetow (1966), some of the amino acids not serving as nitrogen source are known to penetrate *Euglena*. The related flagellate *A. longa* is much more restricted, utilizing only asparagine and glutamine. Buetow's findings are summarized in Table V. His paper also reviews the earlier literature; there are several scattered reports of amino acids serving as nitrogen sources for *Euglena*, but this is the only systematic study. There are notable strain differences; strongly heterotropic strains utilize amino acids as nitrogen sources more readily than the "weak" heterotrophs.

Table V

AMINO ACIDS AS NITROGEN SOURCES FOR *Euglena*[a,b,c]

Neutral aliphatic amino acids
Glycine
Alanine
Valine
Leucine
Isoleucine
Serine
Threonine
Dicarboxylic amino acids and their amides
Aspartic acid
Asparagine
Glutamic acid
Glutamine

[a] Concentrations were $10^{-3}\ M$ (L-form) except for glycine, which was $2 \times 10^{-3}\ M$. Aspartic acid at $10^{-2}\ M$ gave a 10-fold increase in final cell density. Salt medium at pH 6.8 with $0.061\ M$ acetate as the major carbon source.
[b] *Euglena gracilis* var. *bacillaris* SM-L1.
[c] From Buetow (1966).

While there has apparently been no rigorous analysis of the amino acid composition of *Euglena* proteins, there is no reason to believe that it differs from other microorganisms. Kempner and Miller (1965c) identified 15 of the common amino acids in protein hydrolysates. In that study, glutamate served as the principal carbon source and trace amounts of radioactive amino acids were added to the media; the fate of these amino acids was then determined by chromatography of protein hydrolysates. The results were similar to those found for *Escherichia coli*, some (asparate, glycine, serine, threonine) being converted to a number of other amino acids, while others (alanine, histidine, proline, phenylalanine, tyrosine, valine) were incorporated into protein but not further metabolized. This work is summarized in Table VI.

Table VI

Fate of Exogenously Supplied Amino Acids[a]

^{14}C-Amino acid added to medium[b]	Major ^{14}C-amino acids found in protein hydrolysates[c]
Alanine	Alanine
Aspartic Acid	Aspartic
	Glycine
	Serine
	Alanine
	Isoleucine
	Threonine
Glycine	Glycine
	Serine
	Cysteine
	Lysine
Histidine	Histidine
Proline	Proline
Phenylalanine	Phenylalanine
Serine	Serine
	Glycine
	Cysteine
Threonine	Threonine
	Isoleucine
	Leucine
	Glycine
Tyrosine	Tyrosine
Valine	Valine

[a] From Kempner and Miller (1965c).
[b] Trace amounts of uniformly labeled ^{14}C-amino acids. Glutamic acid was the principal carbon source.
[c] *Euglena gracilis* Z. Heterotrophic growth.

6. *Phosphorous*

Albaum *et al.* (1950) identified about a dozen phosphates in *Euglena*, including several phosphorylated intermediates of glycolysis as well as inorganic metaphosphate and inorganic pyrophosphate. Polyphosphate (volutin) granules can be detected throughout the cytoplasm (but not the nucleus) following toluidine blue staining (Keck and Stich, 1957).

The energy of the P-O-P bond in the polyphosphate molecule is about the same as the "high-energy" adenylate phosphate bonds (Meyerhof *et al.*, 1953; Yoshida, 1955). It has been suggested that polyphosphates may serve to store both phosphate and energy from excess ATP (Hoffman-Ostenhof and Weigert, 1952). Some heterotrophic microorganisms possess enzymes that can reversibly transfer phosphate from ATP to polyphosphate (see e.g., Kornberg, 1957). In an extensive review of inorganic phosphate metabolism in algae, however, Kuhl (1962) concluded that there is no direct indication that algal cells are able to satisfy even a part of their energy requirement by using polyphosphate as a source of stored energy.

Stich (1953) found that volutin granules increased in size and number in *Acetabularia* in the light, and decreased in the dark. Respiratory poisons also suppressed granule formation. He concluded that polyphosphate synthesis in *Acetabularia* can proceed only if photosynthesis and oxidative phosphorylation are operating normally (Stich, 1955). In marked contrast to this, Smillie and Krotkov (1960) found polyphosphate levels in *E. gracilis* Z to be considerably greater when plastid synthesis was repressed. Previously unpublished work of the author confirms these findings (Table VII). While the distribution of polyphosphates may be widespread (Keck and Stich, 1957), it does not follow that they are always metabolized in the same way.

The total phosphorous content of *Euglena* also varies with growth conditions (Smillie and Krotkov, 1960), in general being negatively correlated with the growth rate when expressed on a dry weight basis. The dry weight also varied with the conditions of growth in these experiments, and the data when expressed on a per cell basis (referred to DNA content) show

Table VII

DISTRIBUTION OF RADIOPHOSPHOROUS IN *E. gracilis* Z[a]

Fraction	Total activity (%)	
	Dark-grown	Light-grown
RNA	35.0	32.0
Lipid	10.5	11.0
Acid-soluble	16.0	35.0
Polyphosphate	38.5	22.0

[a] *Euglena gracilis* Z was grown from inoculation in the salt medium of Cramer and Myers (1952) with orthophosphate-^{32}P (10 μ ci/ml) in the light or in the dark. Acetate was present as a carbon source in both cases. Cells were harvested in log growth for analysis.

less striking differences. Thus, Smillie and Krotkov (1960) found that autotrophic *Euglena* contained 5.1 pg phosphorous per cell, heterotrophic *Euglena* 7.1 pg phosphorous per cell, and cells grown in the light in the presence of organic carbon sources 6.0 pg phosphorous per cell. It is of interest to note that these authors found an approximate doubling of total phosphorous in streptomycin-bleached *Euglena* grown in the light with an organic carbon source (viz., 11.8 pg phosphorous per cell). Most of the difference in the latter case was the result of increased amounts of acid-soluble and polyphosphate phosphorous. Smillie and Krotkov (1960) suggest that these differences may arise from reduced levels of RNA synthesis in nongreen cells. RNA certainly contains a major fraction of the total phosphorous in *Euglena*. However, the reported values for RNA-P, which range between 2.3 and 2.8 pg per cell under the various culture conditions used by these authors, do not differ enough to explain the increase in total phosphorous found in nonphotosynthesizing cells, which is quite massive in streptomycin-bleached *Euglena*.

In light-synchronized populations of *Euglena*, the rate of total incorporation of added ^{32}P is decreased slightly by onset of the dark period (Fig. 6). However, lipid- and RNA-P, two of the principle phosphorous-containing groups in *Euglena*, increase in a linear manner during the entire light–dark

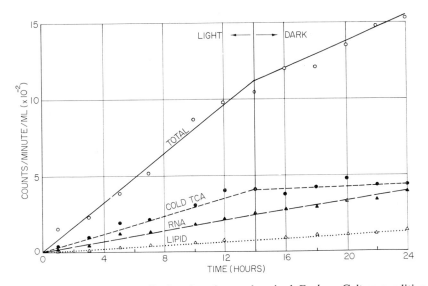

Fig. 6. Incorporation of radiophosphate by synchronized *Euglena*. Culture conditions the same as those described under Fig. 4. Orthophosphate-^{32}P (10 Ci/liter) was added at the beginning of the light period. The cold 8% trichloroacetic acid extract showed little increase during the dark period, which is reflected in the reduced total incorporation. Polyphosphate levels are not shown. All radioactive counts have been multiplied by 10^{-2}.

period. The acid-soluble pool, which contains most of the radiophosphorous, does not increase much in size during the dark period. This doubtless accounts for the slightly reduced rate of total incorporation found in the dark period, and suggests that active incorporation of phosphate into this pool may be associated with photosynthesis. Intracellular orthophosphate-^{32}P levels parallel those of the total acid-soluble, showing little increase in the dark, while incorporation of ^{32}P into polyphosphate (not shown in Fig. 6) is most vigorous in the latter half of the light period. Protein, DNA, and free nucleotides accounted for an insignificant part of the total activity at all stages of growth in synchronized *Euglena*.

Phosphate deprivation induces formation of an acid phosphatase in *Euglena* (Blum, 1965; Price, 1962). The induced enzyme is localized at the surface of the cell, in the notch region of each pellicle complex (Sommer and Blum, 1965). Data presented by Blum (1966) suggests that phosphate uptake is an active-transport process, with the further possibility of the inducible phosphatase being involved. Thus, although conditions could be found that inhibit phosphate uptake without interfering with phosphatase activity, the converse could not be demonstrated—inhibition of enzyme activity was always associated with inhibition of phosphate uptake.

7. *Carbon Dioxide**

It was shown by van Dreal and Padilla (1964) that heterotrophic CO_2 fixation is required for cell division in temperature-synchronized *A. longa* grown with acetate as sole carbon source. A similar requirement may exist in asynchronous succinate-grown *Euglena* (Tremmel and Levedahl, 1966).

Euglena gracilis Z grown with either glucose or acetate remains in log growth for a much longer period of time when flushed with air enriched to 5% with CO_2 than with air alone (Heinrich, 1966). A part of this effect is doubtless due to the buffering action of CO_2 (Cook and Heinrich, 1966; Section VI,A). Heterotrophic cultures of *E. gracilis* grown with glucose as carbon source fix CO_2 about twice as rapidly as cells grown on acetate, in spite of the fact that the activities of the malic enzyme are considerably greater in acetate-grown cells (Heinrich and Cook, 1966). Qualitative differences can also be seen in the first products of heterotrophic CO_2 fixation depending on whether the *Euglena* are grown with glucose or acetate as carbon source. In growth on acetate, CO_2 fixation is probably less important as a means of regenerating Krebs' intermediates drained off into amino acids, etc. (Heinrich and Cook, 1966; Section VIII,C), since the glyoxylate by-pass—induced by acetate—serves the same end (Heinrich and Cook, 1966; Section VIII,C).

* *Editor's Note:* This topic is further covered in Vol. II, Chapter 4.

8. *Oxygen*

Oxygen requirements for optimum growth of *Euglena* depend on the carbon source supporting growth. Thus, acetate- or ethanol-grown *Euglena* consume three to five times as much oxygen during log growth as glucose-grown *Euglena* in spite of the fact that glucose supports the same growth rate and yields cells of essentially the same biochemical profile (Cook and Heinrich, 1965; see Section VIII,C). Boehler and Danforth (1964) suggest that glucose reduces oxygen consumption by catabolite repression. However, many substrates other than glucose also do not stimulate respiration to levels significantly greater than the endogenous rate (Cook and Carver, 1966; Wilson *et al.*, 1959).

Wilson *et al.* (1959) have suggested that anaerobic sources of energy may be important during growth on substrates that do not stimulate respiration. Lindeman (1942) found that *E. deces* (*deses?*) survives anaerobiosis for as long as 30 days under nonphotosynthesizing conditions, when held at 0°C or 5°C, but not at 10°C. The respiration of acetate-grown *E. gracilis* Z decreases to 30% of its original level after 2 days of vigorous flushing with nitrogen at 30°C (Cook and Heinrich, 1966). The rate of cell division continues unchanged for a short while after initiation of anaerobiosis, but soon becomes negligible.

9. *Other Inorganic Requirements*

Although *Euglena* has requirements for trace amounts of other inorganic nutrients, there apparently has been no systematic study of minimal levels. They would be difficult to demonstrate; while some media call for addition of copper, molybdate, manganese, boron, etc., others do not, any requirement for these apparently being satisfied as contaminants of the other salts.

C. VITAMIN REQUIREMENTS

1. *Vitamin B_{12}*

Since the demonstration that B_{12} is a requirement for growth of *Euglena*, a voluminous amount of literature has accumulated on the use of this organism in bioassay. The earlier papers were discussed by Hendlin (1953). Bernhauer *et al.* (1964) have recently reviewed the various structural forms of vitamin B_{12} as well as biological aspects and metabolic functions. The latest edition (1965) of Smith's "Vitamin B_{12}" contains a wealth of information, often clinically oriented, on all aspects of B_{12} chemistry and physiology.

Although *Euglena* grows more slowly than the B_{12}-requiring bacteria, it has the advantage of greater sensitivity. The original assay (Hutner *et al.*, 1949) has been subsequently modified (see e.g., Hutner *et al.*, 1956; Cooper, 1959). *Euglena gracilis* Z provides a faster assay than *E. gracilis* var. *bacillaris*. Figure 7, redrawn from Robbins *et al.* (1953) is a typical dose-response curve.

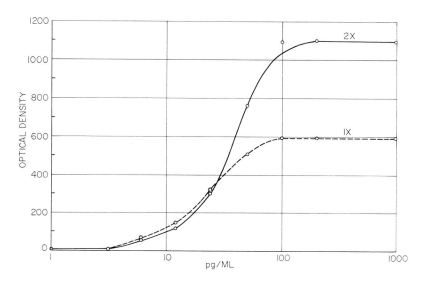

Fig. 7. Typical dose-response curve of *E. gracilis* to vitamin B_{12}. Tubes were grown for 7 days after inoculation in single-strength (1 ×) or double-strength (2 ×) media containing B_{12} concentrations shown on the abscissa. Optimal growth response is attained at 100–150 pg B_{12}/ml (redrawn from Robbins *et al.*, 1953).

Vitamin B_{12} (cyanocobalamin) can be autoclaved at 120°C with the loss of only 1–2% of the activity. Light splits off cyanide, yielding hydroxocobalamin, but this effect is reversed in the dark; prolonged exposure to sunlight, however, causes irreversible damage (Smith, 1965). The coenzyme form (5'-deoxyadenosyl cobalamin), which may be the functional form in many of the reactions mediated by vitamin B_{12}, is extremely sensitive to light however.

Many analogs of vitamin B_{12} occur naturally. Only those that contain a benzimidazole derivative as the "nucleotide" moiety are effective in antipernicious anemia (and in growth of *Ochromonas*), but several others can support growth of *Euglena* (Ford, 1959; Robbins *et al.*, 1952). This lack of specificity may possibly indicate that not all the enzymes required for B_{12} synthesis have been lost by *Euglena*. Indeed, Muto (1957) has shown that homogenates of aged *Euglena* can synthesize B_{12} from cobalt porphyrin and 5,6-dimethylbenzimidazole. However, Nathan and Funk (1962) found that factor B (i.e., B_{12} without the "nucleotide") plus benzimidazole or 5-methylbenzimidazole would not substitute for B_{12} in the growth of *Euglena*.

Arnstein and White (1962) have made a detailed study of B_{12} metabolism in *Ochromonas malhamensis*, but no comparable analysis with *Euglena* has been made. B_{12} is known to participate in a broad spectrum of reactions, including metabolism of some amino acids (Arnstein and Neuberger, 1953; Guest, 1959), fatty acids (Stern and Friedman, 1960; Swick and Wood, 1960), and nucleosides (Downing and Schweigert, 1956). Its suggested role in protein biosynthesis (Wagle *et al.*, 1958) could not be confirmed by later workers (Arnstein and Simkin, 1959; Fraser and Holdsworth, 1959).

Soldo (1955) found that suboptimal concentrations of B_{12} resulted in *Euglena* having reduced levels of RNA and DNA. The B_{12}-deficient cells showed no obvious accumulation of paramylum or carotenoids as compared with normal cells. The more recent work of Venkataraman *et al.* (1965) confirms these results. In this brief but important paper, it was also shown that B_{12} stimulated formate-^{14}C incorporation into RNA, DNA, protein, serine, methionine, and thymine of whole *Euglena*. Thymidylate formation by homogenates was also enhanced by B_{12} ; the enhancement was greater when uridine rather than deoxyuridine served as precursor, indicating that the vitamin may play a greater role in ribotide reduction than it does in the reduction of formate to thymine methyl.

Epstein *et al.* (1962) found that the multiplication rate of *Euglena* was directly proportional to B_{12} concentration, while the size of the cells was inversely proportional. Severe B_{12} depletion, however, resulted in gigantism.

The growth-promoting properties of B_{12} suggest that antagonists of B_{12}—"anti-B_{12}"—could stop growth. Epstein and Timmis (1963) screened over 300 compounds, and found examples of competitive B_{12} antagonism among varied structures, including purines, pteridines, and nicotinamides, none of which bear any structural analogy to cyanocobalamin. Methylamine-, ethylamine-, and aniline-substituted amides of B_{12} were reported to have anti-B_{12} activity (Cuthbertson *et al.*, 1956), but re-examination by Baker *et al.* (1959) showed that these compounds actually satisfy the B_{12} requirement of *E. gracilis* (but not *Ochromonas*).

While *Euglena* can use as a substitute for B_{12} a variety of analogs that differ with respect to the benzimidazole moiety, benzimidazoles alone are toxic (Epstein, 1960; Funk and Nathan, 1958).

2. *Vitamin B_1*

Physiological effects of thiamine deficiency in *Euglena* have received almost no attention, compared to the interest shown in B_{12} . Thiamine, which functions as thiamine pyrophosphate in the decarboxylation of α-keto acids, is not universally required among *Euglena* species; e.g., Leedale *et al.* (1965) could find no requirement for B_1 in the growth of *E. spirogyra*. On the other hand, Cramer and Myers (1952) found B_1 to be required by *E. gracilis* var. *bacillaris* and the Vischer strain, confirming the earlier results of Hutner *et al.* (1949). Attempts by Cramer and Myers to replace B_1 by a variety of organic and amino acids were not successful.

D. EXOGENOUS CARBON SOURCES

Even within the species *E. gracilis*, wide differences exist in the ability of various strains to utilize exogenous carbon sources for growth. Some strains are obligate phototrophs, while others will grow in the dark only on a very restricted number of carbon sources. Some strains will incorporate acetate in the light but not the dark. Many euglenoid flagellates can exhibit very strong heterotrophic growth on a wide variety of carbon sources.

Inability to utilize organic carbon sources for heterotrophic growth cannot be strictly the result of a lack of permeability. At least one obligately phototrophic strain of *E. gracilis* (strain L) will vigorously incorporate acetate-^{14}C into all major insoluble cell fractions during growth in the light, although incorporation ceases abruptly upon transfer to the dark (Table VIII; Cook,

Table VIII

DISTRIBUTION OF ^{14}C IN *E. gracilis* L AFTER LONG-TERM
GROWTH WITH ACETATE-^{14}C[a]

Fraction	Total incorporated (%)
Lipid-soluble	64
Paramylum	21
Protein	15

[a] The obligate phototroph *E. gracilis* L was grown from inoculation on the salt medium of Cramer and Myers (1952) containing sodium acetate-^{14}C. Culture conditions included a light–dark cycle (12:12 hours) at room temperature. Cells were harvested in late log growth (Cook, 1966f).

1966f). The failure of acetate to induce the formation of malate synthase in this strain suggests that acetate metabolism may be more closely integrated with the photosynthetic process than with oxidative metabolism. Furthermore, acetate does not stimulate oxygen consumption in this strain as it does in heterotrophic strains adapted to growth on acetate. These characteristics of this strain (which we refer to as strain L) would appear to make it peculiarly appropriate for a comparative study of acetate metabolism in *Euglena* spp. *Euglena gracilis* L is not a unique cell in this respect; Pringsheim and Wiessner (1960) have described a species of *Chlamydobotrys* that incorporates acetate in the light but not in the dark; R. A. Lewin (1954) has described a mutant *Chlamydomonas dysosmos* in which acetate replaces CO_2 in the light but does not support growth in the dark. Joyce Lewin (1953) reported comparable results with several diatoms.

Cramer and Myers (1952), in their comprehensive study of the growth characteristics of *E. gracilis*, found that the Vischer strain will grow heterotrophically on acetate or butyrate, but not on glucose or several Krebs' intermediates or amino acids. Failure to grow on glucose in some strains of *Euglena* can be explained by the absence of hexokinase (Ohmann, 1963) which, however, is found in the strongly heterotrophic strains (Belsky and Schultz, 1962). *Euglena gracilis* var. *bacillaris* grown on glucose contains the enzymes of the classic glycolytic pathway (Hurlbert and Rittenberg, 1962). It is of interest to note that of the possible substrates for heterotrophic growth of the Vischer strain tested by Cramer and Myers (1952), the two that sufficed (acetate and butyrate) are the only two that stimulate oxygen consumption in strains with stronger heterotrophic tendencies (Danforth, 1953); those not supporting heterotrophic growth of the Vischer strain do not stimulate respiration of *E. gracilis* Z. This correlation may be only coincidental.

Euglena gracilis Z, or *E. gracilis* var. *bacillaris*, strains of *Euglena* that can exhibit strong heterotrophic growth, utilize a wide variety of exogenous carbon sources (Table IX). Apparently there has been no exhaustive study of the suitability of various organics to support growth of *Euglena* in the dark, and studies of physiological and biochemical responses to different carbon sources are woefully behind those dealing with bacteria. Such studies could prove fruitful. To show that *Euglena* metabolizes acetate in the exact way that *E. coli* metabolizes acetate would not be very exciting, but to show that it does so in a different way might be very exciting indeed. Differences found in the utilization of acetate in heterotrophic and obligately phototrophic strains of *Euglena*, which were mentioned above, indicate that fundamental biochemical differences in energy metabolism exist even within the species *E. gracilis*. Taxonomically, *Euglena* is often assigned a key transitory position between the plants and animals. As we learn more about the energy metabolism of *Euglena*, it may well be that it will assume a similar role in concepts of the origin and evolution of metabolic types. Indeed, the whole gamut of biological energy transformation is now found in *Euglena*, save for the demonstration of anaerobic growth. Perhaps a careful study of suitable culture conditions (and the correct strain of *Euglena*) would permit even this possibility.

Table IX

CARBON SOURCES SUPPORTING HETEROTROPHIC GROWTH OF *E. gracilis*[a]

Substrate	E. gracilis var. *bacillaris*	E. gracilis Z	E. gracilis "Vischer"	E. gracilis L
Sucrose	+(c)	0	0	0
Glucose	+(b)	+(e)	−(b)	−(d)
Fructose	+(c)	+(d)	0	0
Galactose	+(c)	−(f)	0	0
Lactate	0	−(f)	0	0
Pyruvate	+(b)	+(g)	0	−(f)
Citrate	−(b)	−(f)	0	0
Succinate	+(b)	+(g)	−(b)	−(f)
Fumarate	+(b)	+(g)	0	−(f)
Malate	+(b)	+(g)	−(b)	−(f)
Glycerol	−(b)	0	0	0
Ethanol	+(h)	+(f)	0	−(d)
Propanol	0	+(f)	0	0
Butanol	0	+(f)	0	0
Acetate	+(b)	+(f)	+(b)	−(d)
Propionate	0	0	0	0
Butyrate	+(b)	+(f)	+(b)	0
Glycolate	−(b)	−(f)	0	0
Glyoxylate	0	−(f)	0	0
Glycine	0	−(f)	0	0
Alanine	+(b)	0	−(b)	0
Aspartate	+(b)	0	−(b)	0
Glutamate	+(b)	+(f)	−(b)	0

[a] Key: +, growth; −, no growth; 0, not tested.
[b] Cramer and Myers (1952).
[c] Belsky (1955).
[d] Cook (1966f).
[e] Cook and Heinrich (1965).
[f] Cook (1966h).
[g] Cook and Carver (1966).
[h] Buetow and Padilla (1963).

VI. Influence of *Euglena* on the Environment

A. pH CHANGES

Removal by *Euglena* of medium constituents required for growth can lead to significant changes in pH, to a degree that can become limiting for further growth. With an initial pH of 5.5, for example, heterotrophic culture with glucose as the only carbon source results in a final pH of less than 3.0 (Fig. 8; Cook and Heinrich, 1966). On the medium used for these studies, which was initially phosphate-buffered by $(NH_4)_2HPO_4$ and KH_2PO_4, $(NH_4)_2HPO_4$ also serves as the sole source of nitrogen. The decreasing pH during growth on glucose is doubtless due to the removal of ammonia. Conversely, growth on the same medium but with sodium acetate rather than glucose as the sole carbon source results in an increase in pH as the population grows (Fig. 8; Cook and Heinrich, 1965). In this case, the removal of ammonia is more than compensated for by acetate removal. Indeed, Pringsheim and Wiessner (1960) have used the pH increase as a measure of the degree of acetate uptake. *Euglena* grows poorly in alkaline media, and extended growth with acetate as the carbon source is possible only when precautions are taken to maintain a neutral or acid pH.

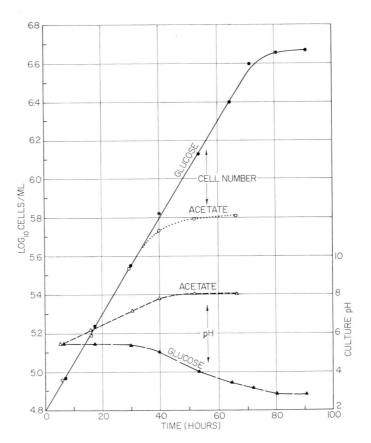

Fig. 8. Change in the medium pH during growth of *E. gracilis* Z in aerated cultures. Cells were grown heterotrophically at 30°C in the salt medium of Cramer and Myers (1952) with either glucose (solid symbols) or acetate (open symbols) as sole carbon source. Ammonia served as nitrogen source (Cook and Heinrich, 1966).

B. Depletion of Nutrients

In general, the rate at which nutrients are removed from the medium will be proportional to the rate of mass increase. It is not usually known which nutrient becomes limiting as the population enters stationary phase, assuming this is not the result of some other effect, e.g., a change in pH. Cramer and Myers (1952) showed that the maximum density attained by *E. gracilis* var. *bacillaris* increased with the initial concentration of glucose up to at least 1%. Ludwig *et al.* (1951) followed the rate at which calcium, magnesium, potassium, manganese, phosphate, and nitrogen disappeared from the medium in *Euglena* cultured on sewage. Nitrogen, phosphate, and magnesium all became limiting, but in these studies little control could be exerted over initial concentrations. Amino acids are completely removed from the medium only when the cells are in stationary phase (McCalla, 1963).

C. Excretion

We have found that the concentration of glucose in the medium of Cramer and Myers (1952) has little effect on maximum cell densities over the range 0.03–0.3 *M* (i.e., 0.5–5%) using *E. gracilis* Z (Cook and Heinrich, 1965). Used media—which had supported growth of *Euglena* to 4.5×10^6 cells per milliliter (late stationary phase)—failed to support growth even with the pH adjusted and with the addition of carbon and all the salts of the original medium (Heinrich, 1966). It was suspected that inhibitory products were being excreted to the medium. Mast and Pace (1942) have discussed the possibility of such products being formed by *Chilomonas paramecium*, and review the earlier literature of this problem. Heinrich (1966) grew *Euglena* in media containing acetate-^{14}C or glucose-^{14}C, and looked for carbon-containing excretory products after stationary phase was reached. Scans of one-dimensional radio chromatograms of the filtered media showed no significant activity in the glucose-containing medium, confirming an earlier report of Belsky (1955). Broad bands of activity were found in the acetate-containing medium, significantly above background, but no definite peaks were apparent. The report that a dozen or so amino acids are excreted by *Euglena* (McCalla, 1963) suggests the possibility that one or more products inhibiting growth or division may also be excreted.

VII. Changes in *Euglena* with Culture Age

Several authors have noted changes in the physiology, biochemistry, and morphology of the average *Euglena* as the population expands through the lag, exponential, and stationary phases of growth in noncontinuous cultures.

A few studies have been made in an effort to quantify the degree of this change, especially during steady log growth. In general, the "average" *Euglena* becomes smaller in almost every parameter measured during this period; i.e., the biochemical growth rate is less than the division rate, and biochemical growth is thus never balanced with division in log phase except in continuous culture.* This generalization of course does not hold for the DNA content.

Buetow and Levedahl (1962) showed that a streptomycin-bleached strain of *E. gracilis* var. *bacillaris* decreased some 76% in average dry weight during logarithmic growth on acetate. At the same time, the total protein decreased by 45% and RNA by 37%. These data imply an even greater decrease in lipid or paramylum. DNA levels in the average cell were constant in log growth, and actually increased slightly in early stationary phase, as did RNA levels; the latter results were interpreted to be the result of a sudden reduced rate of cytokinesis without a concomitant reduction in the rate of DNA and RNA synthesis. Similar results were reported by Wilson and Levedahl (1964) with *Euglena* grown on acetate, succinate, or ethanol.

Changes of this sort have been reported during exponential expansion of several cell types (*Tetrahymena*, Scherbaum and Rasch, 1957; *Aerobacter aerogenes*, Dean and Hinshelwood, 1959; *Azotobacter agile*, Belozersky, 1960). On the other hand, long-term growth in continuously diluted cultures yields exponentially expanding populations of *Euglena* with an invariant biochemical profile (Cook, 1966e). This suggests that if nondiluted cultures could be maintained in exponential expansion for relatively long periods of time, the average cell would have to come to some limiting value of biochemical constituents below which it could not go as long as the division rate did not change. As a matter of fact, the data of Buetow and Levedahl (1962), obtained with cells grown on acetate as sole carbon source, imply that such might be the case. This possibility was examined with *E. gracilis* Z grown with glucose as sole carbon source (Cook and Heinrich, 1966). In nonaerated cultures, glucose maintains *Euglena* in exponential expansion to cell densities 10 times or more as great as acetate (Cook and Heinrich, 1965). It was found, however, that RNA, protein, and the dry weight of the average cell decrease continuously during log growth on glucose (Fig. 9; Cook and Heinrich, 1966). Wacker (1962b) also reported a decrease in cellular RNA and protein with culture age in *Euglena*, when either cell mass or DNA was used as a baseline. Pogo *et al.* (1966) reported higher amounts of RNA in log-phase *Euglena*, with slight differences in absolute levels depending on whether the cells were grown in complex or synthetic media. In either case, particulate RNA comprised about 90% of the total. Smillie

---

* *Editor's Note:* This topic is further covered in Chapter 7 and in Vol. II, Chapter 20.

and Krotkov (1960), however, found only minor differences in RNA levels during autotrophic growth, *Euglena* actually showing a slight increase of 10–15% in RNA toward the end of log growth and a decrease in stationary phase. These authors reported that light-grown cells cultured in the presence of exogenous organic carbon showed a steady increase in RNA levels throughout log phase, to values about 1.5 times greater than those found in early log phase, with a comparable decrease occurring in stationary phase.

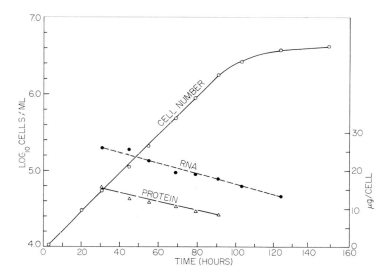

Fig. 9. Change in the amount of RNA and protein contained by the average *E. gracilis* Z during the logarithmic and early stationary phases of growth. Protein levels have been multiplied by 0.1 to facilitate graphing. The cells were grown heterotrophically with glucose as sole carbon source. The dry mass changed in the same manner as protein and RNA, but for purposes of clarity is not shown here (Cook and Heinrich, 1966).

The reason for the discrepancy between this report and those of other authors, who find a decrease in RNA during log growth, is not clear. The RNA content of *Euglena* can assume rather wide values as a function of culture conditions and growth rate, and light is particularly effective in altering patterns of RNA synthesis (Brawerman and Chargaff, 1959a; Cook, 1966e). The study by Smillie and Krotkov (1960) is apparently the only such attempt to follow RNA levels over the growth curve in light-grown *Euglena*.

It is of interest to note that the oxygen consumption of glucose-grown *Euglena* also decreases with culture age (Cook and Heinrich, 1966). As shown in Fig. 10, the decrease is continuous, and is correlated with the rate of growth as measured by mass or protein, but not with the division rate. This implies that division processes in *Euglena* are not very demanding on energy supplies, at least not directly. Figure 10 also shows that the specific activity of some of the respiratory enzymes (expressed as enzyme units per milligram of protein) are constant during the log and early stationary periods, indicating that while the concentration of these enzymes does not change during these periods, the total amount per cell decreases continuously at a rate strictly proportional to the total protein content. On the other hand,

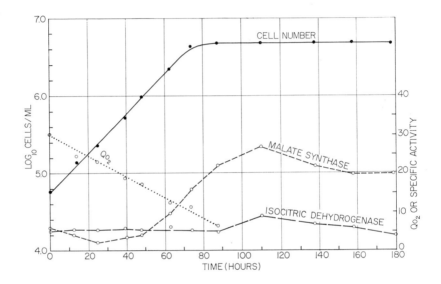

Fig. 10. Changes in Q_{O_2} (μliters 10^6 cells) and the activities of isocitrate dehydrogenase and malate synthase during growth of *E. gracilis* Z with glucose as sole carbon source. Malate synthase activity expressed as millimicromoles acetyl-CoA disappearing per minute per milligram of protein; isocitric dehydrogenase activity as the rate of NADP reduction. The massive increase in malate synthase activity may be associated with a shift to utilization of fats; the respiratory quotient is 1.0 during log growth but drops as low as 0.3 in stationary phase (Cook and Heinrich, 1966).

the activity of malate synthase (and the malic enzyme, not shown in Fig. 10) show significant increases in stationary phase. The latter enzymes are intimately involved in operation of the glyoxylate by-pass, important in the utilization of fat reserves (see Section VIII,C), and their increased activities in stationary phase of glucose-grown *Euglena* is indicative of a shift from carbohydrate to fat oxidation. The respiratory quotient of such cells decreases from a level of 1.0 in log growth to about 0.3 in stationary phase (Heinrich, 1966).

Albergoni and Pranzetti (1963) have shown that the endogenous respiration of *E. gracilis* var. *bacillaris* decreases during the log phase of growth. Respiration stimulated by acetate decreases at a proportionate rate. These cultures were maintained at $22° \pm 3°C$. Studies at $30°C$ with *E. gracilis* Z cultured with acetate show quite different results (Cook and Heinrich, 1966). The endogenous rate drops rather steadily during log and stationary growth, but the oxygen consumption stimulated by acetate is nearly constant in log

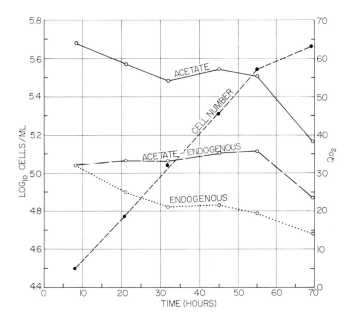

Fig. 11. Changes in Q_{O_2} (microliters oxygen per hour per 10^6 cells) of *E. gracilis* Z during heterotrophic growth with acetate as sole carbon source. Both the endogenous respiration and the respiration in the presence of acetate decrease during log growth, with perhaps a tendency toward constancy in late log growth. The oxygen consumption actually stimulated by acetate above the endogenous level is very nearly constant over the whole of the logarithmic portion of growth, but drops precipitously early in stationary phase (Cook and Heinrich, 1966).

growth, with a precipitous drop occurring just as the population enters the stationary phase of growth (Fig. 11). These results do not necessarily contradict those of Albergoni and Pranzetti (1963), who worked at a lower temperature and, more importantly, grew the cells in a medium that did not contain acetate or ethanol. Both these parameters tend to reduce the oxygen consumption of *Euglena* in the presence of substrate to a value more nearly equal that of the endogenous level; elevated temperatures and adaptation to growth on acetate increase the difference in endogenous oxygen consumption and that due to acetate, at least in *Euglena* (Cook, 1966b; Cook and Heinrich, 1965).

An important study by Gross and Villaire (1960) showed that the average number of chloroplasts in *E. gracilis* var. *bacillaris* changes with time after inoculation, decreasing slightly in early log growth (from 5.4 to 4.5 in 3 days) and then increasing to an average number of 8.6 after 15 days. Under the culture conditions employed, 15 days would correspond to stationary phase of growth, and the elevated number of chloroplasts may have been the result of continued chloroplast division without cell division. This study, in which *Euglena* were grown in dim light on a complex organic medium, showed a wide range in chloroplast numbers among individual cells. This inconstancy has been confirmed more recently in another strain of *Euglena* grown with strict phototrophic nutrition (Cook, 1966c). Since *Euglena* apparently cannot divide at rates that permanently outstrip the rate of chloroplast multiplication, these studies tend to indicate that the number of chloroplasts in the progeny need not equal that of the parent cells; i.e., the filial cell may contain more or fewer chloroplasts than the parent cell contained. Gross and Villaire (1960) found that in early log growth (up to day 4), in which the number of plastids per cell ranged from 0 to 20, the largest class was the one that had no chloroplasts at all. This class disappeared as the culture aged, and after 15 days the smallest number of chloroplasts found was 2 per cell. This study, which is very interesting from the standpoint of plastid inheritance, does not imply that plastids arise *de novo*, in spite of the fact that the frequency of cells without plastids decreases in number as the culture ages. Statistically, it shows that the rate of chloroplast multiplication is not directly proportional to the rate of cell division. Other workers have shown that the number of chloroplasts in *Euglena* is proportional to the light intensity of culture (Brawerman and Chargaff, 1959b). These basic facts of chloroplast replication are not all accounted for by current hypotheses of cytoplasmic inheritance.

Corbett (1957a) found that the average volume of *Euglena* also decreases during log growth, finally becoming stable in stationary phase. Transfer to fresh media gave rise to volume expansion, the magnitude of the expansion being a function of temperature, pH, and osmotic pressure (Corbett, 1957b).

Dilution of stationary phase *Euglena* with fresh media results in an almost immediate stimulation of RNA synthesis (Fig. 12; Cook and Heinrich, 1966). The burst of RNA synthesis occurs before any detectable increase in cell number. The amount of protein per cell remains constant until after the burst of RNA synthesis is complete, and then increases as the population enters the log phase of growth (Fig. 12). The lag period is thus important as a time when the individual *Euglena* can actively synthesize material, in the absence of cell division, to levels compatible with exponential expansion of cell numbers. Blum and Buetow (1963) have made a detailed analysis of biochemical changes in *Euglena* as a function of the length of time spent in stationary phase (i.e., following acetate deprivation) and in the log phase.

Changes in the average *Euglena* that occur as the culture grows clearly make it risky business to reproduce experiments that could be influenced by such changes. Harvest of different cultures at the same population density

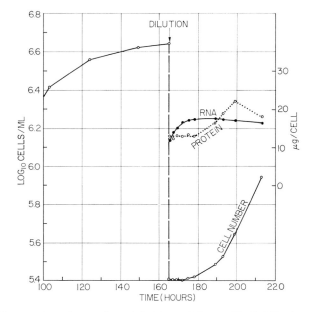

Fig. 12. Resumption of growth and division of *E. gracilis* Z following dilution of a culture in early stationary phase. No "inoculation synchrony" was observed in division. The synthesis of RNA was initiated immediately upon dilution, but that of protein was delayed until RNA levels stabilized. Cellular protein levels (10 times greater than those plotted here) were proportional to cell number during the lag period, and began declining immediately after entry into log phase. Salt medium of Cramer and Myers (1952) with glucose as carbon source (Cook and Heinrich, 1966).

is not the only precaution to be observed; the entire regimen of culture should be reproduced in successive runs, including inoculations with like numbers of cells from populations in the same stage of growth. Implicit in this level of attention given to cultures is an accurate knowledge of the division rate. The propensity of many who work with *Euglena* to express the division rate as "days of growth" seems hardly compatible with some of the sophisticated aims of their experiments, particularly when the cells may have a generation time of no more than 10–12 hours.

VIII. Effect of the Environment on *Euglena*

In the strictest sense, no studies are made with *Euglena* that exclude the effect of the environment, whether this be the bleaching action of ultraviolet light or changes in biochemical profiles as the culture ages. In this section, however, attention will be focused on some of the more easily controlled environmental variables that directly affect the growth and physiology of *Euglena*.

A. INITIAL pH

Within limits, *Euglena* grows better in acid media. The range of pH values permitting optimal growth is heavily influenced by the composition of the media, and particularly the source of carbon. Thus, Wilson *et al.* (1959) found heterotrophic growth rates of a streptomycin-bleached *Euglena* to have different pH optima depending on whether the carbon source was acetate, malate, fumarate, or succinate (Table VIII). Furthermore, the pH permitting maximum oxygen consumption was not always the pH yielding optimum growth rates (Table X); in some cases, respiration proceeded vigorously at acidities that did not permit growth (e.g., malate at pH 3.5); in other cases, respiration above the endogenous was negligible at pH values permitting optimal growth rates (e.g., succinate at pH 7.0).

Table X

INFLUENCE OF pH ON GROWTH AND RESPIRATION OF *Euglena*[a]

	Initial pH							
	7.0		6.0		5.0		3.5	
	G.T.[b]	Qo_2[c]	G.T.[b]	Qo_2[c]	G.T.[b]	Qo_2[c]	G.T.[b]	Qo_2[c]
Acetate	100	400	90	430	55	490	0	0
Malate	0	120	60	—	45	160	0	160
Fumarate	0	110	—	—	60	140	75	150
Succinate	100	110	95	—	80	200	90	190

[a] A streptomycin-bleached strain of *E. gracilis* var. *bacillaris* SM-L1 was grown on the medium of Cramer and Myers (1952) with the substrates listed as the sole carbon source and with the pH variously adjusted. Cells used for respiration studies were grown on succinate at pH 7, fumarate at pH 5.0, malate at pH 5.0, or acetate at pH 7.0 (from Wilson *et al.*, 1959).

[b] All generation times are expressed relative to that of cells grown on acetate at pH 7 (i.e., 100 times generation time of acetate-grown *Euglena*, pH 7, divided by the generation time under the other conditions).

[c] All respiration data are expressed as percent of endogenous consumption.

Cook and Heinrich (1965) compared the effect of initial pH on growth of *E. gracilis* Z cultured with acetate or glucose as sole carbon source. The optimal pH for growth on glucose was pH 3–4, although pH 7 did not greatly extend the generation time. The generation time increased significantly above pH 7 (Table XI). An earlier report (Cramer and Myers, 1952) indicated that negligible growth of *E. gracilis* var. *bacillaris* occurs on glucose at pH 6.8, while growth is vigorous at pH 4.5.

Table XI

EFFECT OF INITIAL pH ON GROWTH OF *Euglena*[a]

pH	Glucose		Acetate	
	G.T.[b] (hours)	Density[c]	G.T.[b] (hours)	Density[c]
3	11.5	4.2	—	—
4	11.5	3.7	—	—
5	12	3.8	15	0.32
6	12	3.3	11	0.24
7	13	3.6	12	0.20
8	16	3.2	11	0.14
9	26	3.4	12	0.13

[a] *Euglena gracilis* Z was grown in the dark on the salt medium of Cramer and Myers (1952), with 0.03 M glucose or acetate as the sole carbon source (Cook and Heinrich, 1965).

[b] Generation time.

[c] Maximum density of cells per milliliter $\times$ 10^6. No growth occurred on acetate below pH 5.

Little change in the generation time was found with acetate as the carbon source with initial pH values of 6–9. However, the maximum density attained was much less when the initial pH was above 7 (Cook and Heinrich, 1965). The growth rate is reduced at pH 5, and no growth is found below pH 5 when acetate is the sole carbon source, confirming the earlier results of Wilson *et al.* (1959).

Oxygen consumption of glucose-grown *Euglena*, cultured from an initial pH of 6.8, was essentially constant over the pH range 3–9 (Table XII). A comparable study with acetate-grown cells showed no pH effect on respiration in the range 3–8, but this dropped precipitously at pH 9. It is of interest to note that pH 9 will permit near optimal growth rates (but reduced final densities, see Table XI). Wilson *et al.* (1959) have commented on the separation of growth and respiration on succinate by pH; thus, growth is optimal at pH 7, but the respiration is not significantly greater than the endogenous level at this pH; lower pH values (3.5–5) cause a doubling of oxygen consumption without a concomitant increase in growth rate. Glucose causes no stimulation of respiration above endogenous levels at any pH investigated (Cook and Heinrich, 1965).

Table XII

Effect of pH on Respiration of *Euglena*[a]

pH	Glucose	Acetate
3	18.4	89.6
4	21.0	84.6
5	21.4	85.8
6	19.7	93.2
7	21.4	92.0
8	21.4	93.2
9	21.4	31.4

[a] *Euglena gracilis* Z was grown heterotrophically at 30°C with either glucose or acetate as sole carbon source, with an initial pH of 6.8. Respiration of log-phase cells was followed with the oxygen electrode in the presence of substrate. Values listed are microliters oxygen per hour per 10^6 cells (Cook and Heinrich, 1965).

The maximum population density decreases with increasing initial pH in cultures of *Euglena* grown with acetate as sole carbon source (Cook and Heinrich, 1965). This is doubtless because of the fact that the pH of such cultures increases still further during population growth, as a result of acetate utilization (Section VI,A). Thus, growth on ethanol, which yields cells having essentially the same biochemical and physiological profiles, is considerably more dense than it is on acetate (Buetow and Padilla, 1963). Acetate cultures can be made relatively dense by starting with the lowest possible pH (5.1–5.3) coupled with aeration, particularly when the air is enriched to 5% with CO_2; the buffering effect of CO_2 also serves to decrease the rate of pH change. Such conditions routinely yield populations in excess of 10^6 cells per milliliter, considerably greater than those obtained in non-aerated standing cultures started at pH 6–7 (Buetow and Padilla, 1963; Cook and Heinrich, 1965).

B. Nutritional Effects on the Generation Time

The influence of temperature and light intensity on average generation times are considered below. Other conditions being equal, the growth rate can still be varied by culture on different carbon sources. Wilson *et al.* (1959) found that malate and fumarate support growth rates equal to about half those found with acetate and succinate, using a bleached mutant of *E. gracilis* var. *bacillaris* (at optimal pH in all cases).

The fastest growth rates observed on defined media are comparable to those found on complex organic media. However, the spread of individual generation times in complex media is considerably less than is found with defined media (Fig. 13; Cook and Cook, 1962). Presumably, the ready availability of exogenous intermediates renders unnecessary many reaction steps, reducing the possibility of cells getting out of phase with one another in those processes leading to division.

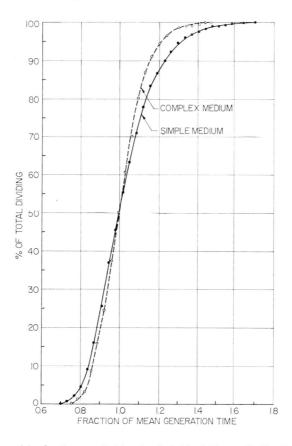

Fig. 13. Spread in the time to division by individual *E. gracilis* Z cells. The ordinate shows the fraction of the total number that have divided by the time indicated on the abscissa. The cells were grown heterotrophically in a salt medium with acetate as sole carbon source (solid circles) or in 2% proteose peptone containing acetate (open circles); 790 cells were followed in the former case and 548 cells in the latter. The spread in individual generation times is significantly less in the complex media (Cook and Cook, 1962).

Hoogenhout and Amesz (1965) have compiled a large number of references in comparing growth rates of various photosynthetic cells. While the survey is useful from a comparative standpoint, it is recognized that the growth rate is subject to considerable modification by culture conditions.

C. The Carbon Source

Exogenous carbon sources known to be utilized for heterotrophic growth of *Euglena* were reviewed in Section V,D. The final population density is proportional to the initial concentration of succinate, ethanol, and acetate between the levels of 1 and 5 mM, with no significant change in growth rate (Levedahl and Wilson, 1965; see Chapter 7). Concentrations of glucose or acetate greater than 0.15 M inhibit the growth rate, markedly so at 0.3 M (Cook and Heinrich, 1965).

The carbon source used has a marked effect on the respiratory physiology of *Euglena*. Two-carbon compounds like ethanol or acetate result in very high levels of oxygen consumption, while the TCA intermediates (Danforth, 1953) as well as glucose (Cook and Heinrich, 1965), cause negligible stimulation of respiration, although they can support approximately equivalent growth rates. The respiratory response to acetate is adaptive, and in cells previously grown with autotrophic nutrition, about five generation times are required for complete adaptation (Cook, 1965). The oxygen consumption of acetate-grown *Euglena* can be partially reduced by concomitant culture with glucose (Cook and Heinrich, 1965) or, in green cells, by light (Cook, 1965). Propionate, butyrate, lactate, and pyruvate stimulate oxygen consumption to intermediate levels (Danforth, 1953).

The glyoxylate by-pass (Kornberg, 1959) is intimately involved in utilization of acetate and ethanol in *Euglena* (Reeves *et al.* 1962; Cook and Carver, 1966; Heinrich and Cook, 1966). This sequence, which cleaves isocitrate to succinate and glyoxylate, followed by the condensation of glyoxylate with acetyl-CoA to yield malate, is anaerobic. Since growth on acetate causes no increase in the activity of isocitric dehydrogenase (and therefore presumably none in the TCA cycle), the anomolous situation arises in which the induction of an anaerobic series of reactions results in increased oxygen uptake by *Euglena*, without a concomitant increase in activity of the TCA cycle. It may be noted that activities of malate dehydrogenase, an oxygen-requiring isozyme of *Euglena* (Chancellor-Maddison and Noll, 1963) common to both the TCA cycle and the glyoxylate cycle, are greater in acetate-grown *Euglena* than in glucose-grown or autotrophically grown cells, but the differences do not seem adequate to account for the total oxygen consumption of acetate-grown *Euglena* (Cook, 1966h). Furthermore, malate dehydrogenase activities in acetate-grown *Euglena* can be separated from oxygen consumption by making the cells anaerobic; oxygen consumption decreases in parallel with activity of the glyoxylate by-pass, but malate dehydrogenase activities remain unchanged (Cook and Heinrich, 1966). It has been suggested (Cook and Carver, 1966; Heinrich and Cook, 1966) that oxygen may be involved in some (unknown) reactions directly associated with the glyoxylate by-pass in acetate-grown *Euglena*, in addition to serving as electron acceptor in the cytochrome chain. In this context it is of interest to note that propionate and butyrate, which will condense with glyoxylate in much the same way as acetate (Rabin *et al.*, 1965), also stimulate oxygen consumption in *Euglena* (Danforth, 1953). Goldfine and Bloch (1963) have reviewed studies of nonrespiratory pathways that require oxygen.

D. EFFECTS OF LIGHT

1. *Phototrophic Growth*

The growth rate of *E. gracilis* Z cultured with strict phototrophic nutrition is optimal over the range of 400–1200 ft-c (Cook, 1963a). Light intensities lower than 400 ft-c support a reduced growth rate, while 3000 ft-c is slightly inhibitory. The chlorophyll content of the average cell is inversely proportional to the light intensity. Curiously, the cell volume increases at very low intensities (65 ft-c), although the mass of the cell is reduced. These adaptations to low light intensities—increased chlorophyll concentration and dispersion of the chloroplasts—serve to increase the amount of light that can be absorbed, permitting optimum growth rates at reduced light intensities.

While these adaptations for increased photosynthetic capacity maintain optimal growth rates, they are not adequate for the support of optimal biosynthetic rates. Cell mass decreases below 1200 ft-c, proportionate to the observed rate of photosynthesis (Cook, 1963a). Most of the variation in mass with the light intensity of culture is due to variable degrees of paramylum accumulation; interestingly enough, protein and RNA levels remain constant down to very low light intensities (120 ft-c). Lipid levels are actually increased slightly at lower light levels. Presumably most of the lipids are associated with the photochemical apparatus. Some of these relationships are summarized in Fig. 14.

Certain species of *Euglena* (e.g., *E. pisciformis*) are obligate phototrophs (Lwoff, 1951) while some strains of *E. gracilis* (e.g., the Vischer strain) will grow in the dark on a very restricted number of carbon sources, and this even with low growth rates (Cramer and Myers, 1952). One obligate phototrophic strain of *E. gracilis* will vigorously incorporate acetate in the light but not in the dark (Cook, 1966f). In this strain, acetate does not induce formation of the glyoxylate by-pass enzymes isocitrate lyase and malate synthase, nor does acetate stimulate respiration.

In those strains of *Euglena* that can exhibit strong heterotrophic growth, the presence of acetate during culture in the light results in markedly altered patterns of CO_2 fixation. Over 90% of bicarbonate-^{14}C fixed by *E. gracilis* Z grown under such conditions was found in protein, while strict autotrophic culture leads to significant incorporation of bicarbonate-^{14}C into lipid and

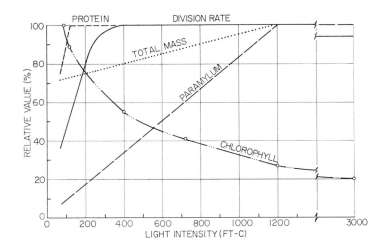

Fig. 14. Some characteristics of *E. gracilis* Z grown at 25°C with phototrophic nutrition, during log growth in light of different intensities. The 100% values were as follows: mass, 820; protein, 240; paramylum, 380; and chlorophyll, 33, all values in picograms per cell. The maximum division rate (ln 2/generation time) was 0.065 (Cook, 1963a).

paramylum as well (Cook, 1965). Indeed, more label is found in paramylum than in protein under the latter conditions. Acetate labeled with radiocarbon is largely routed into lipid and paramylum during growth in the light, only 6% of the label appearing in protein, while almost 70% of the radioactive acetate is incorporated into protein by strict heterotrophic cultures. These results are in line with those of Russell and Gibbs (1966) who found that acetate repressed many photosynthetic enzymes of the obligately phototrophic alga *Chlamydomonas mundana*. Control mechanisms affecting concomitant utilization of photosynthates and exogenous carbon sources for growth, a subject that could have important overtones relating to the origin of metabolic types, have not been much studied.

2. *Inhibition of Division*

Visible light inhibits cell division of *Euglena* (Cook, 1960). In autotrophic cultures, this inhibition is hardly detectable, but the inhibition observed when nongreen cells are exposed to light can be severe. Figure 15, for example, shows the effect of incandescent light on a heat-bleached strain of *E. gracilis* Z cultured in the chemostat. Under these special conditions

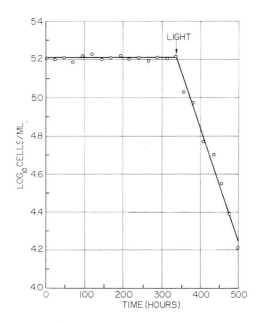

Fig. 15. Effect of incandescent light (1000 ft-c) on cell division of a permanently heat-bleached strain of *E. gracilis* Z. Cells were grown in the chemostat (cf. Fig. 3) with acetate as the limiting nutrient; the dilution rate was such that the generation time during balanced growth was 35 hours. Imposition of light caused immediate inhibition of cell division, with the cells then being washed out of the vessel by continued dilution. After 500 hours, all cells were still motile.

of culture, in which acetate limits growth, it is anticipated that many of the biosynthetic pathways in *Euglena* operate at liminal levels, so that cellular responses to environmental perturbations are maximized. In etiolated *Euglena* able to green, the inhibition of division is overcome after a few generations; the recovery is associated with development of the photosynthetic apparatus (Cook, 1966d,e). In growth not limited by any nutrilite, etiolated cells also recover more rapidly than permanently bleached strains (Cook, 1960).

The observed degree of inhibition is temperature-dependent, being less obvious at elevated temperatures. Figure 16 shows this relationship for the colorless flagellate *A. longa;* steady-state growth rates in the light approach those in the dark as the temperature becomes optimal for division.

The photoinhibition of division in *Euglena* is largely relieved if the cells are grown on a complex organic medium, e.g., proteose peptone, rather than a salt medium (Cook, 1960). In this context it is also of interest to note that light inhibits division of the ciliate *Tetrahymena pyriformis*, an inhibition

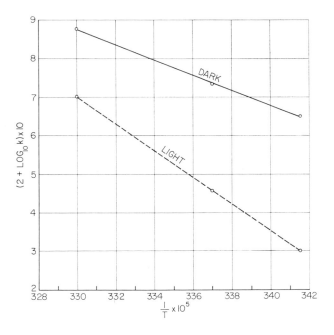

Fig. 16. Influence of temperature on the degree of photoinhibition of division in *A. longa*. The growth rate $k = \ln 2/\text{generation time}$. The cells were grown in nonaerated Erlenmeyer flasks on the salt medium of Cramer and Myers (1952) with acetate as carbon source, and cultures illuminated by 400 ft-c of fluorescent light. The inhibitory nature of light is less severe at the higher temperatures (Cook, 1960).

that persists even if the cells are transferred to the dark; the inhibition can be lifted by the addition of excess lipoic acid, however, a requirement of *Tetrahymena* known to be photolabile (Fig. 17). Taken together, these data suggest that light destroys or prevents the formation of some intermediate necessary for division of *Euglena*. The effect is less severe in green cells either because of shielding by the pigments or because the affected pathways become less important during autotrophic growth.

Casual observations indicate that fluorescent light may be more inhibitory than incandescent light. Powell (1955) has briefly mentioned that growth of *Streptococcus faecalis* is stopped by light (red light being less inhibitory than the blue end of the spectrum,) and that addition of sheep serum minimizes the effect. Epel and Krauss (1966) have recently demonstrated a similar effect for other cells (*Prototheca zopfii*, *Saccharomyces cerevisiae*, and *T. pyriformis*), with the action spectrum implicating one of the cytochromes as the chromophore.

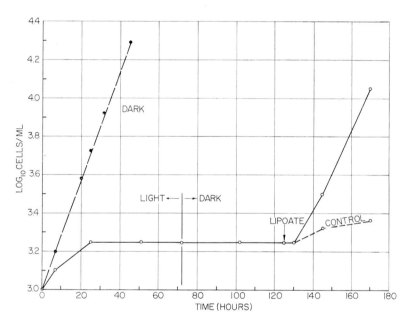

Fig. 17. Photoinhibition of division in *T. pyriformis* grown on Elliott's defined medium. The inhibition persisted even after transfer to dark, but was lifted by the addition of lipoate; a culture handled in the same way but not receiving lipoate ("control") showed little recovery in the dark. Solid circles show the normal growth pattern under these conditions when kept in the dark (Cook, 1960).

Photoinhibition may have widespread effects. Thus, Manson and Defendi (1961) reported that light reduced the rate of DNA synthesis in a mammalian cell grown in tissue culture, perhaps by interfering with the ribonucleotide reductase system, which requires the light-sensitive cobamide coenzyme. Even some green cells are not immune to the inhibitory effects of light. Pirson (1957) has shown that continuous visible light causes irreparable damage to the photosynthetic apparatus of the coenocytic alga *Hydrodictyon*. Grown on light–dark cycles, *Hydrodictyon* has a characteristic pattern of photosynthetic rates. These go through a maximum and are faithfully repeated in successive light periods. When placed in continuous light, however, photosynthetic rates go into a gradual decline. The former level of photosynthesis is not recovered even when *Hydrodictyon* is returned to the light–dark cycle.

3. *Chloroplast Formation*

The most dramatic effect of light on dark-grown *Euglena* is the induction of chlorophyll synthesis and chloroplast development. Associated with this induction is the synthesis of large amounts of RNA and protein, largely confined to the plastids (Brawerman *et al.*, 1962). Light may also stimulate the synthesis of chloroplast DNA in *Euglena* (Cook, 1966c). Light is known to stimulate the "turnover" of cytoplasmic DNA in *Chlorella* (Iwamura, 1960). These topics are discussed in detail in other chapters of this treatise.

4. *Light-Synchronized Euglena*

In continuous light, population expansion of adapted *Euglena* is exponential, but culture on light–dark cycles results in periodic or synchronous increases in cell number with strict phototrophic nutrition. This phenomenon, with associated physiological aspects of growth and division, may be viewed as an effect of light.

a. Gross Biochemical Changes. Two strains of *E. gracilis* have been synchronized by light–dark cycles, namely the Z strain (Cook and Hess, 1964; Edmunds, 1964) and a variety erroneously called *bacillaris* (Cook and James, 1960), now referred to as strain L. The latter is an obligate phototroph. Growth in the major biochemical fractions has been followed in both strains.

Strain L was synchronized with light having an incident intensity of 300 ft-c. This intensity is optimal for multiplication, but permits little accumulation of storage products (cf. Fig. 14). Consequently, protein and RNA synthesis are confined to the light periods (Cook, 1961b). Similar patterns were described for strain Z synchronized with light of comparable intensity (Edmunds, 1965). When light of ca. 1000 ft-c, optimal for biosynthesis as well as division, is used to synchronize strain Z, large amounts of paramylum accumulate in the light period and are utilized in the dark period (Cook, 1966a). RNA and protein are synthesized in both the light and dark periods, paramylum presumably serving both as carbon and energy source for these processes in the dark (Fig. 18). Entirely comparable results with synchronized *Chlorella* were obtained by Pirson *et al.* (1963). The continued incorporation of $^{35}PO_4{}^{3-}$ into RNA and $^{35}SO_4{}^{2-}$ into protein, in parallel with the total incorporation of these isotopes by the cell, indicates that not all the components of these macromolecules are supplied endogenously in the dark.

b. Respiratory Activity. The respiratory quotient of synchronized *E. gracilis* Z is about 1.1, comparable to asynchronous cells grown photo-trophically, and shows no significant changes over the light–dark cycle (Cook, 1966g). This value suggests that some of the oxygen required by

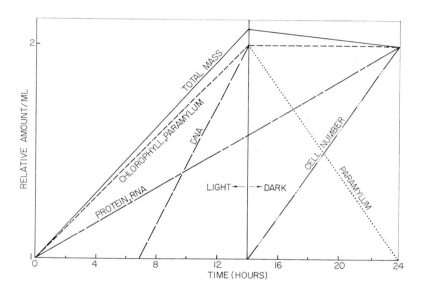

Fig. 18. Schematic summary of growth and division patterns in *E. gracilis* Z syn-chronized on a light–dark cycle at 21.5°C; the light intensity was ca. 1000 ft-c. Paramylum shows a 7-fold linear increase during the 14-hour light period, but is largely consumed in the 10-hour dark period. The other constituents exhibit a doubling over the division cycle (Cook, 1966a).

phototrophic *Euglena* may be supplied endogenously, as e.g., in the con-version of carbohydrates to fat. The total oxygen consumption increases rather rapidly to a doubling in the first half of the light period, and then shows no further significant change throughout the remainder of the light–dark cycle. Specific activities of at least two TCA-cycle enzymes, although changing in a regular manner over the synchronous cycle, show no correla-tion with total oxygen uptake (Cook, 1966g). In view of the fact that oxygen consumption is the terminal event in a multienzyme complex, this lack of correlation does not seem surprising.

c. Photosynthetic Activity. The photosynthetic capacity of synchronized *E. gracilis* Z, measured as oxygen production, increases in a linear manner over the entire 24-hour light–dark cycle (Cook, 1966a). The capacity for photosynthesis is thus better correlated with protein content than with chlorophyll content; synthesis of the latter is confined to the light period, while protein levels increase throughout the light–dark cycle, in parallel with the growth in photosynthetic capacity. Under these conditions of culture, it seems likely that the enzymic complement of the photosynthetic apparatus, rather than chlorophyll concentrations, limits the rate of oxygen evolution.

Lovlie and Farfaglio (1965) have examined the increase in photosynthetic rate in single *Euglena* by means of a Cartesian diver. The increase in rate over the cell cycle was found to occur in several discrete steps, with the average increase following a sigmoidlike curve. These results do not differ significantly from those found with mass populations of synchronized *Euglena*, and such differences as there are can be ascribed to culture procedures, particularly the light intensity used. Certainly the increase in photosynthetic capacity in *Euglena* shows no dramatic maxima such as those found in single *Acetabularia* (Sweeney and Haxo, 1961).

Euglena is synchronized on a light–dark regime that offers just enough light for optimum growth and division (Cook, 1966a; Cook and James, 1960). Such a system offers the possibility of determining the amount of energy required to produce a single cell. Estimates of this amount are in the range of 5×10^{-6} cal (Cook, 1966a).

d. DNA Synthesis. In contrast to most of the biochemical constituents of synchronized *Euglena*, DNA synthesis is singularly discontinuous. The S period, if the term may be applied to mass populations, occurs in the latter half of the light period. The time course does not appear to be altered whether the cells are synchronized with light of 350 ft-c (Edmunds, 1964) or 1000 ft-c (Cook and Hess, 1964). Edmunds (1964) showed that if the normal 14-hour light period is terminated early (by 6 hours), a doubling of DNA still occurs, but is not followed by the usual burst of cell division. It was concluded that DNA replication is a necessary but insufficient condition for cell division. These very interesting studies were conducted with cultures grown on relatively low light intensities (350 ft.-c.), conditions that resulted in termination of protein synthesis at the onset of the "early" dark period. The implication is that DNA synthesis, once initiated, is an "all-or-none" process in synchronized *Euglena*. It would be of considerable interest to know if all the immediate prerequisites for DNA replication are assembled prior to initiation of synthesis. Preliminary studies of the kinases for monophosphate deoxynucleotides (dGMP, TMP) show significant but transient activities, occurring synchronously early in the light period (Fig. 19). Repeated studies of this transient phenomenon have shown that the exact time at which the activity appears is quite sensitive to prior history of culture; furthermore, it is not known at this time whether nucleotide reduction in *Euglena* occurs at the monophosphate level or (as is more common) at the diphosphate level. The results shown in Fig. 19, therefore, cannot be taken for more than a rough indication that the initiation of DNA synthesis in synchronized Euglena may have synchronous roots early in the division cycle.

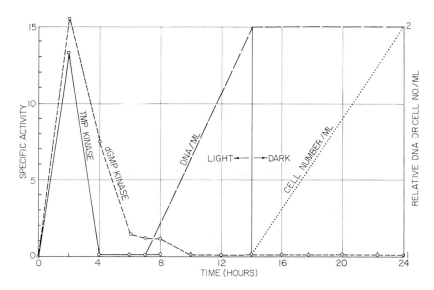

Fig. 19. Transient activities of dGMP and TMP kinases in light-synchronized *E. gracilis* Z. The actual time at which the activity appears is quite sensitive to previous history of culture.

Nuclear DNA comprises about 96–97% of the total DNA in *Euglena*, while chloroplast DNA makes up about 3% (Edelman *et al.*, 1964). In synchronized *Euglena*, there is some indication that the replication of chloroplast DNA occurs very early in the light period—so early indeed that it appears to be stimulated by light—with a second wave of cytoplasmic DNA synthesis coming at the end of the light period, synchronous with the synthesis of nuclear DNA (Cook, 1966c; Cook and Hunt, 1965). Two species of DNA have been described in *Euglena* cytoplasm (Brawerman and Eisenstadt, 1964; Edelman *et al.*, 1965; Ray and Hanawalt, 1964). It has been suggested that the two peaks of cytoplasmic DNA synthesis in synchronized *Euglena* represent separate S periods for these two species of DNA (Cook, 1966c,d). Significant incorporation by the mitochondria is of course not excluded.

f. Frequency Distribution of Cell Volumes. Ideally, the average cell in synchronized populations will mimic the single cell. This hope is probably never fully realized; certainly never in "division-oriented" synchrony, and rarely perhaps in "cycle-oriented" synchrony, as e.g., in cells selected for a common age by their size (terminology from James, 1966). The volume of a cell is proportional to its age, notwithstanding some variation resulting principally from the spread in individual generation times. The distribution of cell volumes in asynchronous populations thus ranges between $\bar{v}$ and $2\bar{v}$. In synchronized *Euglena*, however, the frequency distribution of cell volumes always shows at least a twofold range between the small and large cells. The average volume doubles over the light–dark cycle, but repeated cycles of synchronous division do not reduce the spread in cell volumes (Cook, 1961a). These data suggest that the synchronizing procedure synchronizes cell division (DNA synthesis?) but not cell growth. More importantly, perhaps, these data suggest that knowledge gained with synchronized populations, which may be quite valid in context, should not be projected directly into interpretation of events that occur in single cells.

g. Light As a Synchronizer. Edmunds (1966) has described a persistent rhythm of cell division in previously synchronized populations of *Euglena* following transfer to dim light (ca. 80 ft-c). The bursts of division are of low amplitude, and the rhythm is abolished by brighter light or addition of exogenous carbon sources. Edmunds concludes that an endogenous biological clock operates in *Euglena*, with cell division occurring only at certain allowed times when the "gate" is open.

At light intensities ordinarily used in studies of synchronous division (i.e., 300–1200 ft-c), rhythmic cell divisions can be made to occur at any desired time by appropriate shifts of the light–dark periods with reference to solar time. The "reason" for synchronous division under these conditions is not yet clear, but is probably the result of a complex interplay between photoinhibition of division coupled with periodic starvation (in the dark periods). Phasing of DNA synthesis by the periodically imposed light periods very likely also synchronizes associated events related to mitosis and cytokinesis. The important question is really whether DNA synthesis is discontinuous because the cells are synchronized, or whether the cells are synchronized because DNA synthesis is discontinuous. The role of light in controlling the initiation of DNA synthesis in synchronized *Euglena* is quite unknown. The development of chloroplasts, a massive job for synchronized *Euglena*, begins immediately with the onset of the light period, and involves the synthesis of both DNA and RNA. The growth of chloroplasts may be more than casually related to the discontinuous synthesis of nuclear DNA in synchronized *Euglena* (Cook, 1966d).

E. EFFECTS OF TEMPERATURE

1. *Division Rate*

Those strains of *Euglena* that can show strong heterotrophic growth also appear to tolerate higher temperatures (Baker *et al.*, 1955). Correlated with "high-temperature" characteristics are the ability to utilize a wider range of substrates and a somewhat greater osmotic tolerance than is found in "low-temperature" strains.

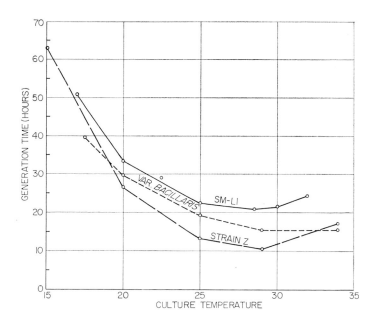

Fig. 20. Effect of temperature on the generation times of three strains of *E. gracilis* (Z, var. *bacillaris*, and var. *bacillaris* SM-L1). The cells were grown heterotrophically with acetate as the carbon source (from Buetow, 1962, and Cook, 1966b).

Even very closely related strains of *Euglena* differ quantitatively in their response to temperature (Cook, 1966b). Figure 20 is a comparison of the generation times of *E. gracilis* strain Z, var. *bacillaris*, and *bacillaris* SM-L1, grown heterotrophically at various temperatures. Although all three have about the same optimum for division rate (29–30°C), var. *bacillaris* is more sluggish in response to incubation temperature than is strain Z; the strepto-mycin-bleached SM-L1 grows even more slowly at a given temperature. An Arrhenius plot of the growth rate of the streptomycin-bleached *E. gracilis* var. *bacillaris* shows that the activation energy for division changes con-tinuously over the temperature range 13.3–28.5°C (Table XIII; Buetow, 1962).

Table XIII

ACTIVATION ENERGY FOR THE MULTIPLICATION RATE OF *Euglena*[a]

Temperature interval (°C)	Activation energy (calories)
25 –28.5	3,500
20 –25	14,000
17 –20	24,000
13.3–17	49,000

[a] A streptomycin-bleached strain (*E. gracilis* var. *bacillaris* SM-L1) was grown in the dark on the defined medium of Cramer and Myers (1952) with acetate as the carbon source. Multiplication rates were calculated from the exponential portions of the growth curves (Buetow, 1962).

2. *Biosynthetic Rates*

While high temperatures favor cell division, low temperatures favor protoplasmic growth of *Euglena* (Buetow, 1962). This condition results because the Q_{10} for protoplasmic growth is nearly constant over physiological temperature ranges, while the Q_{10} for division decreases continuously with increasing temperature (Cook, 1966b). Figure 21 is a semilog plot of the rate of division and the rates of synthesis of protein, RNA, and total mass in *E. gracilis* Z, showing this relationship.

3. *Respiration*

In contrast to most other cell types, including the cryptomonad flagellate *Chilomonas paramecium* (Johnson, 1962), the Q_{O_2} of acetate-grown *Euglena* increases in a linear rather than an exponential manner as the incubation temperature is increased (Buetow, 1963). In marked contrast, the endogenous respiratory level is almost unchanging in acetate-grown *Euglena* (Fig. 22). The temperature of incubation can thus separate endogenous oxygen con-sumption and that stimulated by acetate. It would be of interest to know how temperature affects the respiration of *Euglena* cultured heterotrophically on substrates that do not stimulate oxygen consumption. The effect of temperature on respiration of acetate-grown _Euglena_ has been interpreted to mean that under such culture conditions _Euglena_ may consume oxygen via two principal routes: one as terminal electron acceptor in the cytochrome chain and the other in some nonrespiratory reaction(s) associated with growth on acetate (Cook and Heinrich, 1966).

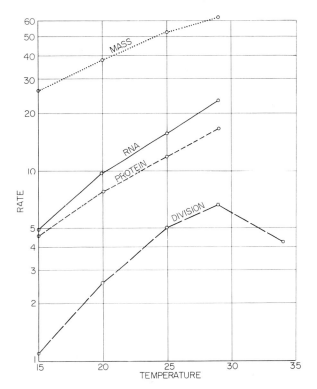

Fig. 21.   Effect of temperature on the rates of division and biosynthesis in _E. gracilis_ Z. Values for division rate ($k = \ln 2/\text{generation time in hours}$) have been multiplied by 10^2, and those for RNA synthesis (picograms per cell per hour) by 10. Rates of mass and protein accumulation are also expressed as picograms per cell per hour. The cells were grown heterotrophically on the salt medium of Cramer and Myers (1952) with acetate as the carbon source, and harvested at 10^5 cells per milliliter (Cook, 1966b).

4. _Supraoptimal Temperatures_

As usually defined, supraoptimal temperatures are those higher than the temperature permitting optimal division rates. In _E. gracilis_, temperatures above about 30°C are generally supraoptimal (Buetow, 1962; Cook, 1966b), although some low-temperature strains multiply optimally at about 26°C (Baker _et al._, 1955).

While supraoptimal temperatures retard division processes, they do not necessarily reduce the capacity for protoplasmic growth, at least up to 36°–37°C (in *E. gracilis* Z). The ability to accumulate mass at temperatures prohibitive to division indicates that the differing Q_{10} values for these two processes observed at lower temperatures (Fig. 21) also extend to supraoptimal temperatures. A similar relationship has been described in *Tetrahymena* by Thormar (1962). This phenomenon has made it possible to synchronize cell division in *Euglena* by repeated shifts to elevated temperatures, which retard cell division but permit protoplasmic growth (Pogo and Arce, 1964).

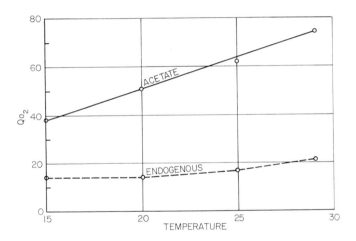

Fig. 22. Effect of incubation temperature on the oxygen consumption of *E. gracilis* Z. Cells were grown in a salt medium with acetate as carbon source, and the respiration measured at the temperature of incubation. The endogenous Q_{O_2} (μ liters oxygen per hour per 10^6 cells) does not change much with temperature, but that due to acetate increases in a linear manner with incubation temperature (Cook, 1966b; see also Buetow, 1963).

Gross and Jahn (1962) and Gross (1962) have examined some of the physiological and morphological effects of elevated temperatures on *Euglena*, effects compounded by the simultaneous imposition of light. The *Euglena* in these experiments were grown in a complex organic medium, which probably would minimize adverse effects, but 35°C produced giant multi-nucleated monster euglenas. At this temperature, the growth rate was inversely proportional to light intensity (between 15 and 150 ft-c). Anamolous effects of light were found in apochlorotic substrains at 35°C, being inhibitory in some cases (as in the parent strain) but stimulatory in other cases (Gross, 1962). It was postulated that heat denatures a protein essential for cell division, with light having some indirect synergistic role in the phenomenon.

The most striking effect of supraoptimal temperatures is permanent bleaching of *Euglena*, first studied in detail by Pringsheim and Pringsheim (1952). Strain differences exist in the bleaching response, with the low-temperature strains being more resistant, in some cases completely so (e.g., the "Mainx" strain). Even among the high-temperature strains, however, can be found some that resist bleaching by elevated temperatures. It may be of some evolutionary significance to note that the low-temperature strains are in general less permeable to exogenous carbon sources.

Brawerman and Chargaff (1960) have studied in some detail the bleaching response to temperature in *E. gracilis* Z, a high-temperature strain (Baker *et al.*, 1955). These authors showed that the loss of chlorophyll-synthesizing ability in heat-shocked cells is intimately related to the rate of cell multiplication.

Digitonin extraction of *Euglena* yields "chloroplastin," a pigment–macromolecule complex containing chlorophyll, proteins, carotenoids, lipids, and lipoproteins. Chloroplastin is bleached by heat or light or both, with an activation energy of 48.3 kcal/mole (Wolken and Mellon, 1957). Bleaching of chlorophyll in intact *Euglena* has a somewhat higher activation energy, 67 kcal/mole. Both values are within the range found for denaturation of proteins. An earlier study by Wolken *et al.* (1955) indicated that elevated temperatures caused the removal of magnesium from chlorophyll and its transformation to pheophytin.

IX. Summary

In writing this chapter the author was privileged to examine many papers on *Euglena* which had hitherto escaped his notice. Doubtless many others of interest have been missed. As a final exercise, a plot was made of the number of those papers published yearly since 1950; it is shown in Fig. 23. While it does not represent an exhaustive literature search, and indeed includes only papers of interest to the author, it does indicate that research on *Euglena* is continuing apace. Much has been learned about euglenoid physiology in the last 15 years, and clearly much more remains to be discovered. A cross section of papers appearing now indicates that *Euglena* is already enjoying tremendous popularity as a model system in studies of cytoplasmic inheritance. Nevertheless, it appears that too little attention is being paid to the fundamental aspects of euglenoid physiology. The ease with which *Euglena* is cultured under defined conditions, and its ready adaptability to growth in synchrony or in continuous culture, makes the cell ideally suited for studies of the interaction between *Euglena* and its environment—which, broadly interpreted, includes the whole of euglenoid physiology. Particular stress should be placed on energy metabolism. Few species exhibit such a wide range of nutritional types, or the ability to exploit

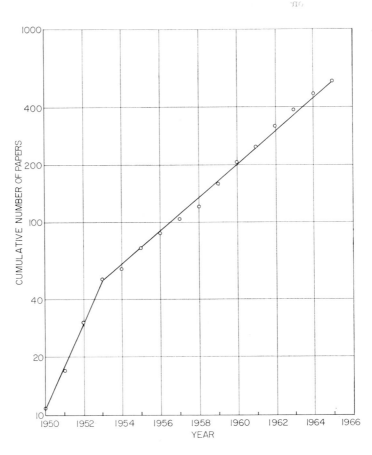

Fig. 23. A cumulative logarithmic plot of the papers published on *Euglena* since 1950. This does not represent an exhaustive search for all pertinent papers. The doubling time is a little more than 3 years.

such wide sources of free energy, as *E. gracilis*. It may be anticipated that the next 15 years will see *Euglena* contributing not only to our knowledge of subcellular morphogenesis, but also to our knowledge of the origin and evolution of metabolic types. To the euglenaphiles, at least, the next decade should prove even more exciting than the last.

ACKNOWLEDGMENTS

Previously unpublished work was supported in part by N.I.H. Grant GM-12179 to the author and in part by a N.A.S.A. grant (NsG-338) to the University of Maine. The author was ably assisted in this work by Mrs. Mary Carver, Mrs. Judith Whitney, and Mr. Bernd Heinrich. Illustratic s are by Prof. C. Z. Westfall. Stenographic assistance was offered by Miss Bonnie Marshall and Mrs. Kay Littlefield. The author is particularly indebted to Mrs. Jennie Boynton, whose conscientious attention to detail improved the manuscript.

References

Abraham, A., and Bachhawat, B. K. (1963). *Biochim. Biophys. Acta* **70**, 104.
Abraham, A., and Bachhawat, B. K. (1965). *Indian J. Biochem.* **1**, 192.
Albaum, H. G., Schatz, A., Hutner, S. H., and Hirschfeld, A. (1950). *Arch. Biochem.* **29**, 210
Albergoni, V., and Pranzetti, P. (1963). *Studi Sassaresi Sezione III* [3] **11**, 453.
Anonymous (1958). *J. Protozool.* **5**, 1.
Arnstein, H. R. V., and Neuberger, A. (1953). *Biochem. J.* **55**, 259.
Arnstein, H. R. V., and Simkin, J. L. (1959). *Nature* **183**, 523.
Arnstein, H. R. V., and White, A. M. (1962). *Biochem. J.* **83**, 264.
Bach, M. K. (1960). *J. Protozool.* **7**, 50.
Baker, H. S., Hutner, S. H., and Sobotka, H. (1955). *Ann. N. Y. Acad. Sci.* **62**, 351.
Baker, H. S., Frank, O., Pasher, I., Hutner, S. H., Herbert, V., and Sobotka, H. (1959). *Proc. Soc. Exptl. Biol. Med.* **100**, 825.
Belozersky, A. N. (1960). *In* "Nucleoproteins" (R. Stoops, ed.), p. 199. Wiley (Interscience), New York.
Belsky, M. M. (1955). *Dissertation Abstr.* **15**. 678.
Belsky, M. M., and Schultz, J. (1962). *J. Protozool.* **9**, 195.
Benson, A. A., Daniel, H., and Wiser, R. (1959). *Proc. Natl. Acad. Sci. U. S.* **45**, 1582.
Benson, A. A., Cook, J. R., and Yagi, T. (1962). *Plant Physiol.* **37**, Suppl., 44.
Bernhauer, K., Müller, O., and Wagner, F. (1964). *Advan. Enzymol.* **26**, 233.
Bernstein, E. (1960). *Science* **131**, 1528.
Blum, J. J. (1965). *J. Cell Biol.* **24**, 223.
Blum, J. J. (1966). *J. Gen. Physiol.* **6**, 1125.
Blum, J. J., and Buetow, D. E. (1963). *Exptl. Cell Res.* **29**, 407.
Blum, J. J., and Padilla, G. M. (1962). *Exptl. Cell Res.* **28**, 512.
Boehler, R. A., and Danforth, W. F. (1964). *J. Cell Biol.* **23**, 11A.
Bowne, S. W., Jr. (1964). *Nature* **204**, 801.
Brawerman, G., and Chargaff, E. (1959a). *Biochim. Biophys. Acta* **31**, 164.
Brawerman, G., and Chargaff, E. (1959b). *Biochim. Biophys. Acta* **31**, 178.
Brawerman, G., and Chargaff, E. (1960). *Biochim. Biophys. Acta* **37**, 221.
Brawerman, G., and Eisenstadt, J. M. (1964). *Biochim. Biophys. Acta* **91**, 477.
Brawerman, G., Pogo, A. O., and Chargaff, E. (1962). *Biochim. Biophys. Acta* **55**, 326.
Buetow, D. E. (1962). *Exptl. Cell Res.* **27**, 137.
Buetow, D. E. (1963). *Nature* **199**, 196.
Buetow, D. E. (1965). *J. Cellular Comp. Physiol.* **66**, 235.
Buetow, D. E. (1966). *J. Protozool.* **13**, 585.
Buetow, D. E., and Levedahl, B. H. (1962). *J. Gen. Microbiol.* **28**, 579.
Buetow, D. E., and Padilla, G. M. (1963). *J. Protozool.* **10**, 121.

Carell, E. F., and Price, C. A. (1965). *Plant Physiol.* **40**, 1.

Chancellor-Maddison, J., and Noll, C. F. (1963). *Science* **142**, 60.

Cook, J. R. (1960). Ph.D. Thesis, Univ. of California, Los Angeles, California.

Cook, J. R. (1961a). *Plant Cell Physiol. (Tokyo)* **2**, 199.

Cook, J. R. (1961b). *Biol. Bull.* **121**, 277.

Cook, J. R. (1963a). *J. Protozool.* **10**, 436.

Cook, J. R. (1963b). *Photochem. Photobiol.* **2**, 407.

Cook, J. R. (1965). *Plant Cell Physiol. (Tokyo)* **6**, 301.

Cook, J. R. (1966a). *Plant Physiol.* **41**, 821.

Cook, J. R. (1966b). *Biol. Bull.* **131**, 83.

Cook, J. R. (1966c). *J. Cell Biol.* **29**, 369.

Cook, J. R. (1966d). *In* "Cell Synchrony—Studies in Biosynthetic Regulation" (I. L. Cameron and G. M. Padilla, eds.), pp. 153–168. Academic Press, New York.

Cook, J. R. (1966e). *Nature* (in press).

Cook, J. R. (1966f). *J. Protozool.* **14**, 382.

Cook, J. R. (1966g). *Exptl. Cell Res.* (in press).

Cook, J. R. (1966h). Unpublished data.

Cook, J. R., and Carver, M. (1966). *Plant Cell Physiol. (Tokyo)* **7**, 377.

Cook, J. R., and Cook, B. (1962). *Exptl. Cell Res.* **28**, 524.

Cook, J. R., and Heinrich, B. (1965). *J. Protozool.* **12**, 581.

Cook, J. R., and Heinrich, B. (1966). *J. Gen. Microbiol.* (in press).

Cook, J. R., and Hess, M. (1964). *Biochim. Biophys. Acta* **80**, 148.

Cook, J. R., and Hunt, W. (1965). *Photochem. Photobiol.* **4**, 877.

Cook, J. R., and James T. W. (1960). *Exptl. Cell Res.* **21**, 583.

Cooper, B. A. (1959). *J. Clin. Pathol.* **12**, 153.

Corbett, J. J. (1957a). *J. Cellular Comp. Physiol.* **50**, 309.

Corbett, J. J. (1957b). *J. Protozool.* **4**, 71.

Cramer, M., and Myers, J. (1952). *Arch. Mikrobiol.* **17**, 384.

Cuthbertson, W., Gregory, J., O'Sullivan, P., and Pegler, H. F. (1956). *Biochem. J.* **62**, 15p.

Danforth, W. (1953). *Arch. Biochem. Biophys.* **46**, 164.

Daniel, H., Miyano, M., Mumma, R. O., Yagi, T., Lepage, M., Shibuya, I., and Benson, A. A. (1961). *J. Am. Chem. Soc.* **83**, 1765.

Davies, W. H., Mercer, E. I., and Goodwin, T. W. (1965). *Phytochemistry* **4**, 741.

Davies, W. H., Mercer, E. I., and Goodwin, T. W. (1966). *Biochem. J.* **98**, 369.

Dean, A. C. R., and Hinshelwood, Sir C. (1959). *In* "Regulation of Cell Metabolism" (G. E. W. Wolstenholme and C. M. O'Connor, eds.), p. 311. Little, Brown, Boston, Massachusetts.

Downing, M., and Schweigert, B. S. (1956). *J. Biol. Chem.* **220**, 521.

Dusi, H. (1933). *Ann. Inst. Pasteur* **50**, 550, 840.

Edelman, M., Cowan, C. A., Epstein, H. T., and Schiff, J. A. (1964). *Proc. Nat. Acad. Sci. U. S.* **52**, 1214.

Edelman, M., Schiff, J. A., and Epstein, H. T. (1965). *J. Mol. Biol.* **11**, 769.

Edmunds, L. N., Jr. (1964). *Science* **145**, 266.

Edmunds, L. N., Jr. (1965). *J. Cellular Comp. Physiol.* **66**, 159.

Edmunds, L. N., Jr. (1966). *J. Cell. Physiol.* **67**, 35.

Elliott, A. M. (1949). *Trans. Am. Microscop. Soc.* **68**, 228.

Epel, B., and Krauss, R. W. (1966). *Biochim. Biophys. Acta* **120**, 73.

Epstein, S. S. (1960). *Nature* **188**, 143.

Epstein, S. S., and Timmis, G. M. (1963). *J. Protozool.* **10**, 63.

Epstein, S. S., Weiss, J. B., Causey, D., and Bush, P. (1962). *J. Protozool.* **9**, 336.

Evans, H. J., and Nason, A. (1953). *Plant Physiol.* **28**, 233.

Fogg, G. E. (1965). "Algal Cultures and Phytoplankton Ecology," pp. 1–126. Univ. of Wisconsin Press, Madison, Wisconsin.

Ford, J. E. (1959). *J. Gen. Microbiol.* **21**, 693.

Fraser, M. J., and Holdsworth, E. S. (1959). *Nature* **183**, 519.

Funk, H. B., and Nathan, H. A. (1958). *Proc. Soc. Exptl. Biol. Med.* **99**, 394.

Goldfine, H., and Bloch, K. (1963). *In* "Control Mechanisms in Respiration and Fermentation" (B. Wright, ed.), pp. 81–103. Ronald Press, New York.

Goodman, N. S., and Schiff, J. A. (1964). *J. Protozool.* **11**, 120.

Goodwin, C. M. (1951). *Proc. Iowa Acad. Sci.* **58**, 451.

Goodwin, T. W., and Gross, J. A. (1958). *J. Protozool.* **5**, 292.

Gorham, E. (1957). *Tellus* **9**, 174.

Greenblatt, C. L., and Schiff, J. A. (1959). *J. Protozool.* **6**, 23.

Gross, J. A. (1962). *J. Protozool.* **9**, 415.

Gross, J. A., and Jahn, T. L. (1962). *J. Protozool.* **9**, 340.

Gross, J. A., and Villaire, M. (1960). *Trans. Am. Microscop. Soc.* **79**, 144.

Gross, J. A., and Wolken, J. J. (1960). *Science* **132**, 357.

Guest, J. R. (1959). *Biochem. J.* **72**, 5 pp.

Hall, R. P., Johnson, D. F., and Loefer, J. B. (1935). *Trans. Am. Microscop. Soc.* **54**, 298.

Hase, E., Mihara, S., and Tamiya, H. (1960). *Plant Cell Physiol. (Tokyo)* **1**, 131.

Hase, E., Mihara, S., and Tamiya, H. (1961). *Plant Cell Physiol. (Tokyo)* **2**, 9.

Heinrich, B. (1966). M. S. Thesis, Univ. of Maine, Orono, Maine.

Heinrich, B., and Cook, J. R. (1966). *J. Protozool.* (in press).

Hendlin, D. (1953). *Ann. N. Y. Acad. Sci.* **56**, 870.

Hoffman-Ostenhof, O., and Weigert, W. (1952). *Naturwissenschaften* **39**, 303.

Hoogenhout, H. (1963). *Phycologist* **2**, 136.

Hoogenhout, H., and Amesz, J. (1965). *Arch. Mikrobiol.* **50**, 10.

Huling, R. T. (1960). *Trans. Am. Microscop. Soc.* **79**, 384.

Hurlbert, R. E., and Rittenberg, S. C. (1962). *J. Protozool.* **9**, 170.

Hutner, S. H., Provasoli, L., and Stokstad, E. L. R. (1949). *Proc. Soc. Exptl. Biol. Med.* **70**, 118.

Hutner, S. H., Provasoli, L., Schatz, A., and Haskins, C. P. (1950). *Proc. Am. Phil. Soc.* **94**, 152.

Hutner, S. H., Bach, M. K., and Ross, G. I. M. (1956). *J. Protozool.* **3**, 101.

Hutner, S. H., Zahalsky, A. C., Aaronson, S., Baker, H. S., and Frank, O. (1966). *In* "Methods in Cell Physiology" (D. M. Prescott, ed.), Vol. 2, pp. 217–228. Academic Press, New York.

Huzisige, H., and Satoh, K. (1960). *Biol. J. Okayama Univ.* **6**, 71.

Iwamura, T. (1960). *Biochim. Biophys. Acta* **42**, 161.

James, T. W. (1960). *Pathol. Biol. Semaine Hop.* [N. S.] **9**, 510.

James, T. W. (1961). *Ann. Rev. Microbiol.* **15**, 27.

James, T. W. (1966). *In* "Cell Synchrony—Studies in Biosynthetic Regulation" (I. L. Cameron and G. M. Padilla, eds.), pp. 1–13. Academic Press, New York.

James, T. W., and Anderson, N. G. (1963). *Science* **142**, 1183.

Johnson, B. F. (1962). *Exptl. Cell Res.* **28**, 419.

Karali, E. F., and Price, C. A. (1963). *Nature* **198**, 708.

Keck, K., and Stich, H. (1957). *Ann. Botany, N. S.* **21**, 611.

Kempner, E. S., and Miller, J. H. (1965a). *Biochim. Biophys. Acta* **104**, 11.

Kempner, E. S., and Miller, J. H. (1965b). *Biochim. Biophys. Acta* **104**, 18.

Kempner, E. S., and Miller, J. H. (1965c). *Biochemistry* **4**, 2735.

Korn, E. D. (1964). *J. Lipid Res.* **5**, 352.

Kornberg, H. L. (1959). *Ann. Rev. Microbiol.* **13**, 49.

Kornberg, S. R. (1957). *Biochim. Biophys. Acta* **26**, 294.

Kott, Y., and Wachs, A. M. (1964). *Appl. Microbiol.* **12**, 292.

Krinsky, N. I., and Goldsmith, T. H. (1960). *Arch. Biochem. Biophys.* **91**, 271.

Kuhl, A. (1962). *In* "Physiology and Biochemistry of Algae" (R. Lewin, ed.), pp. 211–229. Academic Press, New York.

Kylin, A. (1964a). *Physiol. Plantarum* **17**, 384.

Kylin, A. (1964b). *Physiol. Plantarum* **17**, 422.

Leedale, G. F. (1959). *Biol. Bull.* **116**, 162.

Leedale, G. F., Meeuse, B. J. D., and Pringsheim, E. G. (1965). *Arch. Mikrobiol.* **50**, 133.

Levedahl, B. H., and Wilson, B. W. (1965). *Exptl. Cell Res.* **39**, 242.

Lewin, J. (1953). *J. Gen. Microbiol.* **9**, 305.

Lewin, R. A. (1954). *J. Gen. Microbiol.* **11**, 459.

Lewis, M. H. R., and Spencer, B. (1962). *Biochem. J.* **85**, 18p.

Lindeman, R. L. (1942). *Ecology* **23**, 1.

Loefer, J. B., and Mefferd, R. B., Jr. (1955). *Proc. Intern. Congr. Microbiol., 6th, Rome, 1953* **5**, 357.

Lorenzen, H. (1964). *In* "Synchrony in Cell Division and Growth" (E. Zeuthen, ed.), pp. 571–578. Wiley (Interscience), New York.

Lovlie, A., and Farfaglio, G. (1965). *Exptl. Cell Res.* **39**, 418.

Ludwig, H. F., Oswald, W. J., Gotaas, H. B., and Lynch, V. (1951). *Sewage Ind. Wastes* **23**, 1337.

Lwoff, A., ed. (1951). *In* "Biochemistry and Physiology of Protozoa," Vol. 1, pp. 1–26. Academic Press, New York.

Lyman, H., Epstein, H. T., and Schiff, J. A. (1961). *Biochim. Biophys. Acta* **50**, 301.

McCalla, D. R. (1963). *J. Protozool.* **10**, 491.

Mainx, F. (1928). *Arch. Protistenk.* **60**, 355.

Manson, L. A., and Defendi, V. (1961). *Abstr. Am. Soc. Cell Biol., 1st Meeting*, p. 134.

Mast, S. O. and Pace, D. M. (1942). *J. Cellular Comp. Physiol.* **20**, 1.

Meyerhof, O., Shafas, R., and Kaplan, A. (1953). *Biochim. Biophys. Acta* **12**, 121.

Millar, E., and Price, C. A. (1960). *Plant Physiol.* **35**, Suppl., xxiii.

Muto, T. (1957). *J. Vitaminol. (Kyoto)* **3**, 50.

Myers, J., and Clark, L. B. (1945). *J. Gen. Physiol.* **28**, 103.

Nason, A. (1962). *Bacteriol. Rev.* **26**, 16.

Nathan, H. A., and Funk, H. B. (1962). *Proc. Soc. Exptl. Biol. Med.* **109**, 213.

Nishimura, M. (1959). *J. Biochem.* **46**, 219.

Novick, A., and Szilard, L. (1950). *Proc. Natl. Acad. Sci. U.S.* **36**, 708.

O'Brien, A. A., and Shibuya, I. (1964). *J. Lipid Res.* **5**, 432.

Ohmann, E. (1963). *Naturwissenschaften* **50**, 552.

Padilla, G. M. (1960). Ph.D. Thesis, Univ. of California, Los Angeles, California.

Padilla, G. M., and Cook, J. R. (1964). *In* "Synchrony in Cell Division and Growth" (E. Zeuthen, ed.), pp. 521–536. Wiley (Interscience), New York.

Padilla, G. M., and James, T. W. (1960). *Exptl. Cell Res.* **20**, 401.

Padilla, G. M., and James, T. W. (1964). *In* "Methods in Cell Physiology" (D. M. Prescott, ed.), pp. 141–157. Academic Press, New York.

Pappas, G. D., and Hoffman, H. (1952). *Ohio J. Sci.* **52**, 102.

Perini, F. (1963). *Natl. Acad. Sci.—Natl. Res. Council, Misc. Publ.* **145**, 291.

Perini, F., Kamen, M. D., and Schiff, J. A. (1964a). *Biochim. Biophys. Acta* **88**, 74.

Perini, F., Schiff, J. A., and Kamen, M. D. (1964b). *Biochim. Biophys. Acta* **88**, 91.

Petropolous, S. F. (1964). *Science* **145**, 268.

Pirson, A. (1957). *In* "Research in Photosynthesis" (H. Gaffron, ed.), p. 127. Wiley (Interscience), Inc., New York.

Pirson, A., Lorenzen, H., and Ruppel, H. G. (1963). *In* "Microalgae and Photosynthetic Bacteria" (E. Hase, ed.), pp. 127–139. Japan. Soc. Plant Physiol. Tokyo.

Pogo, A. O., and Arce, A. (1964). *Exptl. Cell Res.* **36**, 390.

Pogo, B. G. T., Ubero, I. R., and Pogo, A. O. (1966). *Exptl. Cell Res.* **42**, 58.

Powell, E. D. (1955). *Biometrika* **42**, 16.

Prescott, D. M. (1959). *Exptl. Cell Res.* **16**, 279.

Price, C. A. (1961). *Biochem. J.* **82**, 61.

Price, C. A. (1962). *Science* **135**, 46.

Price, C. A., and Carell, E. F. (1964). *Plant Physiol.* **39**, 862.

Price, C. A., and Millar, E. (1962). *Plant Physiol.* **37**, 423.

Price, C. A., and Vallee, B. L. (1962). *Plant Physiol.* **37**, 428.

Pringsheim, E. G. (1914). *Beitr. Biol. Pflanz.* **12**, 1.

Pringsheim, E. G., and Pringsheim, O. (1952). *New Phytologist* **51**, 65.

Pringsheim, E. G., and Wiessner, W. (1960). *Nature* **188**, 919.

Provasoli, L., Hutner, S. H., and Schatz, A. (1948). *Proc. Soc. Exptl. Biol. Med.* **69**, 279.

Rabin, R., Reeves, H. C., Wegener, W. S., Megraw, R. E., and Ajl, S. J. (1965). *Science* **150**, 1548.

Ray, D. S., and Hanawalt, P. C. (1964). *J. Mol. Biol.* **9**, 812.

Reeves, H. C., Kadis, S., and Ajl, S. (1962) *Biochim. Biophys. Acta* **57**, 403.

Richards, O. W., and Jahn, T. L. (1933). *J. Bacteriol.* **26**, 385.

Robbins, P. W., and Lipmann, F. (1956). *J. Am. Chem. Soc.* **78**, 2652.

Robbins, W. J., Hervey, A. H., and Stebbins, M. E. (1952). *Nature* **170**, 845.

Robbins, W. J., Hervey, A. H., and Stebbins, M. E. (1953). *Ann. N. Y. Acad. Sci.* **56**, 818.

Rodhe, W. (1948). *Symbolae Botan Upsalienses* **101**, 1.

Russell, G. K., and Gibbs, M. (1966). *Plant Physiol.* **41**, 885.

Rutner, A. C., and Price, C. A. (1964). *Proc. Intern. Congr. Biochem.* **6**, 331.

Sakai, H. (1962). *J. Gen. Physiol.* **45**, 411.

Scherbaum, O., and Rasch, G. (1957). *Acta Pathol. Microbiol. Scand.* **41**, 161.

Schiff, J. A. (1959). *Plant Physiol.* **34**, 73.

Schildkraut, C. L., Mandel, M., Levisohn, S., Smith-Sonneborn, J. E., and Marmur, J. (1962). *Nature* **196**, 795.

Schmidt, R. R. (1966). *In* "Cell Synchrony—Studies in Biosynthetic Regulation" (I. L. Cameron and G. M. Padilla, eds.), pp. 189–235. Academic Press, New York.

Serenkov, G. P. (1962). *Izv. Akad. Nauk SSSR, Ser. Biol.* **27**, 857.

Shibuya, I., Yagi, T., and Benson, A. A. (1963). *In* "Microalgae and Photosynthetic Bacteria" (E. Hase, ed.), pp. 627–636. Japan Soc of Plant Physiol., Tokyo.

Smillie, R. M. (1963). *Can. J. Botany* **41**, 123.

Smillie, R. M., and Krotkov, G. (1960). *Arch. Biochem. Biophys.* **89**, 83.

Smith, E. L. (1965). "Vitamin B_{12}," 3rd Ed., pp. 1–180. Wiley, New York.

Soldo, A. T. (1955). *Arch. Biochem. Biophys.* **55**, 71.

Sommer, J. R., and Blum, J. J. (1965). *J. Cell Biol.* **24**, 235.

Stern, J. R., and Friedman, D. L. (1960). *Biochem. Biophys. Res. Commun.* **2**, 82.

Stich, H. (1953). *Z. Naturforsch.* **8b**, 36.

Stich, H. (1955). *Z. Naturforsch.* **10b**, 282.

Sweeney, B. M., and Haxo, F. T. (1961). *Science* **134**, 1361.

Swick, R. W., and Wood, H. G. (1960). *Proc. Natl. Acad. Sci. U. S.* **46**, 28.

Ternetz, C. (1912). *Jahrb. Wiss. Botan.* **51**, 435.

Thormar, H. (1962). *Exptl. Cell Res.* **28**, 269.

Tremmel, R. D., and Levedahl, B. H. (1966). *J. Cell. Physiol.* **67**, 361.

van Dreal, P. A., and Padilla, G. M. (1964). *Biochim. Biophys. Acta* **93**, 668.

Venkataraman, S., Netrawali, M. S., and Sreenivasan, A. (1965). *Biochem. J.* **96**, 552.

Wacker, W. E. C. (1962a). *Federation Proc.* **21**, 379.

Wacker, W. E. C. (1962b). *Biochemistry* **1**, 859.

Wagle, S. R., Mehta, R., and Johnson, B. C. (1958). *J. Biol. Chem.* **230**, 137.

Wedding, R. T., and Black, M. K. (1960). *Plant Physiol.* **35**, 72.

Whitney, J. S. (1966). M. S. Thesis, Univ. of Maine, Orono, Maine.

Wilson, B. W., and Levedahl, B. H. (1964). *Exptl. Cell Res.* **35**, 69.

Wilson, B. W., Buetow, D. E., Jahn, T. L., and Levedahl, B. H. (1959). *Exptl. Cell Res.* **18**, 454.

Wilson, L. G., and Bandurski, R. S. (1958). *J. Biol. Chem.* **233**, 975.

Wolken, J. J., and Gross, J. A. (1963). *J. Protozool.* **10**, 189.

Wolken, J. J., and Mellon, A. D. (1957). *Biochim. Biophys. Acta* **25**, 267.

Wolken, J. J., Mellon, A. D., and Greenblatt, C. L. (1955). *J. Protozool.* **2**, 89.

Yoshida, A. (1955). *J. Biochem.* **42**, 165.

CHAPTER 7

SYNTHETIC AND DIVISION RATES OF *EUGLENA*: A COMPARISON WITH METAZOAN CELLS

Barry W. Wilson and Blaine H. Levedahl

I. Introduction

Elucidation of the major pathways of cell metabolism, the mechanisms of protein synthesis, and the molecular bases of inheritance have laid the foundations for renewed interest in the integrated functioning of the cell. One area of research stimulated by these advances involves the regulation of cell composition. With respect to bacterial cells, Neidhardt (1963) points out that "for years it has been appreciated... that a cell grown in one medium can differ in important respects from the genetically identical cell grown in a different medium.... Until very recently, quantitative descriptions and analyses of these differences were sparse, and the few published papers attracted little attention.... Until microbiologists became interested in the orderliness of metabolism and growth, and in the biochemical bases of this orderliness, phenotypic variations in cell size and composition were phenomena that could be of only peripheral interest to them." Similarly, there have been few studies of the factors controlling the composition of nucleated cells, and many of the experiments performed have been qualitative in nature (Harris, 1964).

The purpose of this article is to examine briefly some studies on the gross composition of exponentially growing cells of a colorless strain of *E. gracilis* var. *bacillaris* and compare the results and conclusions of these experiments with those obtained on the growth of avian and mammalian cells *in vitro*.

II. Methods

The information on cell composition to be discussed was gained from observation of organisms grown in batch culture. Such techniques are probably the most common methods of growing cells, although continuous culture systems have been gaining in popularity (James, 1961). Batch-culture methods involve the inoculation of cells into a specific volume of growth medium in which they grow and divide. With the exception of experiments involving growth of monolayers of metazoan cells, the medium is neither replenished nor replaced during the course of the experiments. After a lag period, the length of which depends upon the cell type and the experimental conditions, the cells usually divide exponentially until some component in the medium required by the cells becomes exhausted and growth ceases. Many who work with protozoa study the peak populations reached by the cultures and do not determine the rate at which the cells multiply during their exponential growth phase. Regardless of whether growth rate or peak population is studied, the growth of cells is rarely described by more than a single parameter. Studies involving the determination of more than one cellular component, e.g., simultaneous measurement of cell number, dry weight, and protein, have usually involved sampling of the cultures at one or two times during the growth cycle of the cells, such as at the midpoint of the exponential growth phase and at the stationary growth phase (Pogo *et al.*, 1966; Neff, 1960). These kinds of experiments are useful in studying the effects of nutrition and environment on cells, or in comparing the responses of different cells to similar conditions, but they reveal little concerning the mechanisms regulating cell composition. Most studies are based on the implicit assumption that logarithmically dividing cells double their components with each division.

The lag phase preceding exponential growth of bacteria in batch culture has long been recognized as a period of cell syntheses without cell division, and it is known that cells may decrease in size during the early part of the logarithmic growth phase (Toennies *et al.*, 1961). However, there have been few studies of the constancy of the gross composition of nucleated cells during exponential growth in batch culture. Those that have been performed suggest the existence of an interesting and widespread problem in the regulation of cell growth.

III. Cell Composition

A. EUGLENA

Table I lists data similar to that first noted by Buetow and Levedahl (1962); these results were obtained in a later study reported by Wilson and

Table I

GROWTH CHARACTERISTICS OF *Euglena* CULTURED ON ACETATE[a]

Time[b] (hours)	No. of cells (cells/ml)	Dry weight (μg/10⁶ cells)[c]	Protein (μg/10⁶ cells)[c]	DNA (μg/10⁶ cells)[c]	RNA (μg/10⁶ cells)[c]
28	29,400	2510 ± 30	563 ± 68	—	—
45	47,300	2440 ± 30	539 ± 25	—	—
56	60,400	2130 ± 30	510 ± 12	—	30.0 ± 0.2
61	73,600	1230 ± 50	426 ± 51	4.44 ± 0.24	—
76	117,000	1310 ± 30	425 ± 56	4.62 ± 0.22	—
78	122,000	—	—	4.56 ± 0.41	22.5 ± 0.2
85	176,400	1170 ± 30	411 ± 41	5.14 ± 0.57	21.6 ± 1.0

[a] From Wilson and Levedahl (1964).
[b] Measured from time of inoculation.
[c] Standard deviation, ±1.

Levedahl (1964). The data indicate that *Euglena* cells grown with acetate as the sole carbon source progressively declined in dry weight, protein content, and RNA content during the exponential growth phase. Further experiments showed that the content of the cells also changed during growth on succinate and ethanol as sole carbon sources as shown in Tables II and III, respectively. Other cell types also showed similar changes during multiplication in batch culture. The examples presented here will be limited to metazoan cells in culture.

B. METAZOAN CELL COMPOSITION

Swaffield and Foley (1960) examined the RNA, DNA, and protein composition of three established mouse tumor lines. The results of some of their experiments are shown in Table IV. The data indicate that the

Table II

GROWTH CHARACTERISTICS OF *Euglena* CULTURED ON SUCCINATE[a]

Time[b] (hours)	No. of cells (cells/ml)	Dry weight (μg/10^6 cells)[c]	Protein (μg/10^6 cells)[c]	Carbohydrate (μg/10^6 cells)[c]	DNA (μg/10^6 cells)[c]	RNA (μg/10^6 cells)[c]
36	31,200	—	—	836 ± 56	4.32 ± .16	—
47	45,300	2270 ± 10	538 ± 15	805 ± 22	—	—
54	55,300	—	—	856 ± 42	4.37 ± .77	39.4 ± 6.7
56	60,000	2150 ± 80	515 ± 34	—	—	—
60	68,000	—	—	750 ± 52	—	—
63	74,600	1350 ± 50	504 ± 10	—	—	—
71	98,000	1140 ± 30	463 ± 34	718 ± 52	4.42 ± .09	29.0 ± 1.4
76	114,000	960 ± 90	457 ± 56	639 ± 40	4.22 ± .32	27.6 ± 1.7
81	135,000	—	—	629 ± 37	4.88 ± .48	—
83	142,000	1060 ± 20	460 ± 37	—	—	—
84	151,000	—	—	557 ± 15	4.11 ± .51	25.6 ± 1.5
89	175,000	—	—	536 ± 14	4.42 ± .10	24.1 ± 1.1
94	202,000	—	—	451 ± 24	—	—
96	217,000	850 ± 20	444 ± 34	—	—	—

[a] From Wilson and Levedahl (1964).

[b] Measured from time of inoculation.

[c] Standard deviation, ±1.

Table III

GROWTH CHARACTERISTICS OF *Euglena* CULTURED ON ETHANOL[a]

Time[b] (hours)	No. of cells (cells/ml)	Dry weight (μg/10⁶ cells)[c]	Protein (μg/10⁶ cells)[c]	Carbohydrate (μg/10⁶ cells)[c]	DNA (μg/10⁶ cells)[c]	RNA (μg/10⁶ cells)[c]
16	35,300	2360 ± 10	552 ± 7	974 ± 26	3.17 ± .69	27.4 ± 1.3
24	44,500	2590 ± 12	516 ± 39	825 ± 63	3.54 ± .50	26.2 ± 0.3
43	77,900	1660 ± 00	523 ± 18	—	4.11 ± 0.28	—
67	156,000	1640 ± 30	464 ± 16	664 ± 39	3.58 ± 0.32	22.0 ± 0.7
77	207,000	—	445 ± 16	732 ± 45	4.64 ± 0.61	23.6 ± 1.4
82	237,000	—	413 ± 10	831 ± 30	3.37 ± 0.32	21.0 ± 0.5
95	348,000	1400 ± 40	—	617 ± 29	3.57 ± 0.30	20.6 ± 1.4
115	605,000	1480 ± 10	—	—	—	—

[a] From Wilson and Levedahl (1964).
[b] Measured from time of inoculation.
[c] Standard deviation, ± 1.

Table IV

PROTEIN, RNA, AND DNA CONTENT OF MAMMALIAN CELLS[a]

Time (hours)	No. of cells ($\times 10^5$/flask)	Protein (mg $\times 10^{-9}$/cell)	RNA (mg $\times 10^{-9}$/cell)	DNA (mg $\times 10^{-9}$/cell)
L Cell strain MF–929				
0	3.00	455	30.9	14.9
24	2.70	624	51.0	26.1
48	6.39	508	40.2	24.1
72	13.20	492	39.2	23.2
96	19.80	391	31.5	21.0
144	32.93	446	29.4	16.9
S–180 Mouse tumor cell				
0	3.00	502	41.7	21.2
24	2.53	706	69.2	34.9
48	5.50	482	60.0	29.5
72	8.67	445	52.3	23.4
96	12.50	477	51.4	21.1
144	19.20	492	51.6	23.4
Ascites tumor cell				
0	4.50	388	29.6	18.5
24	1.70	625	58.4	30.8
48	3.10	443	38.8	25.0
72	5.54	311	30.0	16.1
96	12.53	417	28.8	19.9
144	27.80	329	25.6	16.6

[a] Adapted from Swaffield and Foley (1960).

composition of the three cell lines was not constant during logarithmic growth *in vitro*. Swaffield and Foley state that the "... total percentage increments of RNA, DNA and protein during the ... growth period are similar to the percentage increases in cell number. At intervening stages of growth, however, marked fluctuations occur in the mean cellular content of RNA, DNA and protein which are not consistent with the changes in cell count." The results of an experiment by Paul (1959) on another strain of the L cell are shown in Table V. The data reveal what he interpreted as "systematic fluctuations" in cell composition during logarithmic growth. The above-mentioned experiments on mammalian cells were performed with established cell lines, that is, cells that had adapted to long-term growth in culture.

Table V

PROTEIN, RNA, AND DNA CONTENT OF A STRAIN OF L CELL[a]

Time (hours)	No. of cells ($\times 10^6$/flask)	Protein (nitrogen $\times$ 6.25 mg $\times 10^{-9}$/cell	RNA ($mg \times 10^{-9}$/cell)	DNA (mg $\times 10^{-9}$/cell)
0	5.0	256	29	17
26	6.3	218	36	16
45	10.8	206	31	12
71	17.5	218	25	11
91	18.6	268	31	14
116	22.3	268	27	13
138	21.3	232	29	16

[a] Adapted from Paul (1959).

The data in Table VI show the increase in cell number and the protein content per cell in primary cultures of chick embryo breast muscle cells (Wilson *et al.*, 1966). These cells were obtained by enzymically dispersing 14-day-old embryo muscles from two genetic strains of chickens and growing the cells liberated by the treatment *in vitro*. The results indicate that the protein content of these cells also progressively decreased during exponential growth.

Table VI

PROTEIN CONTENT OF CHICK EMBRYO BREAST CELLS[a]

Time (hours)	No. of cells ($\times 10^6$/flask)	Protein (mg $\times 10^{-9}$/cell)
	Line 200	
21	1.13	544
34	2.19	356
44	3.05	306
54	4.59	246
93	4.43	296
	Line 304	
21	0.94	653
34	1.91	425
44	2.58	366
54	4.52	244
93	5.18	280

[a] Adapted from Wilson *et al.* (1966).

C. RELATIVE ALTERATIONS IN CELL COMPOSITION

Thus growth experiments on *Euglena*, established mammalian cell lines, and primary cultures of avian embryo cells all indicate that large variations, or progressive decreases in the gross content of nucleated cells, can occur during their exponential growth in batch culture.

It is of interest to attempt to determine whether or not the decreases and fluctuations in the amounts of cell material noted above are accompanied by relative alterations in the composition of the cells during growth. To do this, the dry weight, RNA, and DNA content of the cells tabulated in Tables I–V were recalculated as percentages of their protein content. The results of this operation are listed in Tables VII, VIII, and IX.

Table VII

RELATIVE COMPOSITION OF *Euglena*[a]

Time (hours)	Dry weight/protein (%)	RNA/protein (%)	DNA/protein (%)
		Acetate	
28	449	—	—
45	454	—	—
56	419	5.88	—
61	288	—	1.04
76	310	—	1.09
78	—	5.29	1.07
85	284	5.25	1.25
		Succinate	
47	423	—	—
56	417	—	—
63	268	—	—
71	247	—	0.956
76	209	6.35	0.923
83	230	—	—
96	192	—	—
		Ethanol	
16	426	4.96	0.574
24	504	5.07	0.692
43	317	—	0.787
67	354	4.74	0.772
77	—	5.30	1.04
82	—	5.07	0.865

[a] Calculated from Wilson and Levedahl (1964).

Table VIII

RELATIVE COMPOSITION OF MAMMALIAN CELLS *in Vitro*[a]

Time (hours)	RNA/protein (%)	DNA/protein (%)
L Cell strain MF–929		
0	6.80	3.28
24	8.20	4.16
48	7.90	4.73
72	7.96	4.71
96	8.05	5.37
144	6.57	3.78
S–180 Mouse tumor cell		
0	8.33	4.22
24	9.80	4.94
48	12.50	6.13
72	11.70	5.25
96	10.80	4.42
144	10.50	4.75
Ascites tumor cell		
0	7.63	4.76
24	9.30	4.93
48	8.77	5.65
72	9.60	5.15
96	6.90	4.76
144	7.80	5.05

[a] Calculated from Swaffield and Foley (1960).

Table IX

RELATIVE COMPOSITION OF A STRAIN OF L CELL *in Vitro*[a]

Time (hours)	RNA/protein (%)	DNA/protein (%)
0	11.3	6.64
26	16.5	7.35
45	15.0	5.80
71	11.5	5.05
91	11.6	5.21
116	10.1	4.84
138	12.5	6.90

[a] Calculated from Paul (1959).

The protein content of the *Euglena* cells (Table VII) increased relative to dry weight during logarithmic growth on all three carbon sources. The RNA: protein ratio showed little change during growth; however, more values are needed to draw meaningful conclusions. The DNA: protein ratio was relatively constant during growth of the cells n acetate, but the ratio may have increased with time when the cells were grown with ethanol as the sole carbon source.

The calculated relative composition of the mammalian cells examined by Swaffield and Foley (1960) is shown in Table VIII; that of the L cells studied by Paul (1959) is shown in Table IX. The results of the calculations indicate that the RNA : protein ratio fluctuated during growth of the four cell types and that there was less change in the DNA : protein ratio than there was in the RNA : protein ratio. Thus, studies on *Euglena* and on mammalian cells show that these widely different cell types varied in total amounts and relative proportions of their constituents during multiplication in batch culture.

The primary lesson gained from these experiments was pointed out by several investigators: Physiologists and biochemists should exercise caution in assuming that cells harvested from the exponential growth phase possess temporally independent properties. Cells in a steady state of multiplication

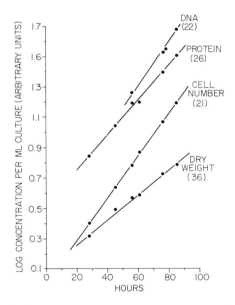

Fig. 1. Growth and division of *Euglena* on acetate. Cramer–Meyers medium, pH 7.0 with acetate as the sole carbon source for growth. Numbers in parentheses indicate the doubling time in hours of the parameters measured. (From Wilson and Levedahl, 1964.)

need not be constant in their characteristics. It is less clear what information can be obtained from these studies concerning the causes of decreases and fluctuations in cell composition. The *Euglena* cells were grown in suspension cultures, unstirred and unaerated. The metazoan cells were grown upon glass or treated plastic surfaces; in each case it was possible that some constituent of the medium became limiting to cell syntheses as the cell number increased in the growth vessels, or that the cells themselves produced some toxic product or products that inhibited their growth but not their division, and that the fluctuations in cell composition found in some of the studies represented adaptations to the changed environmental conditions.

IV. Synthetic and Division Rates

The data themselves permit a test of this class of explanations since such hypotheses predict that the synthetic rates of cells change during the exponential multiplication phase of the cultures. Thus, if the amounts of

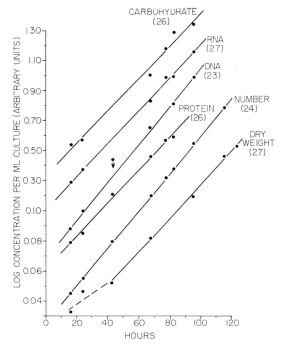

Fig. 2. Growth and division of *Euglena* on ethanol. Cramer–Meyers medium, pH 7.0 with ethanol as the sole carbon source for growth. Numbers in parentheses indicate the doubling time in hours of the parameters measured. (From Wilson and Levedahl, 1964.)

protein, RNA, DNA, etc., are plotted as the logarithm of the amounts per culture rather than as the amounts per cell, the curves should not be linear with time if toxic substances are excreted or if some component in the medium has become limiting. The data for *Euglena* are shown plotted in this way in Figs. 1, 2, and 3, and the results of the experiments of Swaffield and Foley (1960) and Paul (1959) for mouse tumor cells are shown in Figs. 4, 5, and 6. The cell number and protein content per culture for the avian embryo breast cells (Wilson *et al.*, 1966) are shown in Fig. 7.

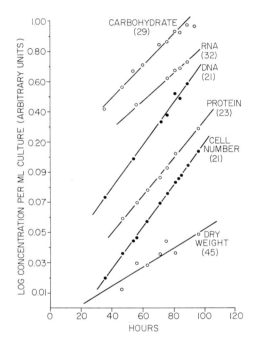

Fig. 3. Growth and division of *Euglena* on succinate. Cramer–Meyers medium, pH 7.0 with succinate as the sole carbon source for growth. Numbers in parentheses indicate the doubling time in hours of the parameters measured. (From Wilson and Levedahl, 1964.)

In the case of the *Euglena* cells grown on acetate, ethanol, or succinate as sole carbon sources Wilson and Levedahl (1964) pointed out that "... the syntheses of cell materials proceeded at constant rates during the exponential increase in cell number of *Euglena* grown on all three substances. The curves are loglinear; there was no evidence for drastic changes in the synthetic rates...." The result was that the rates of syntheses of the cell constituents tended to be slower rather than equal to the rates of increase in cell number, particularly for cells grown on acetate and succinate. DNA was the only cell material manufactured at rates comparable to the rates of multiplication of the cells.

Results similar to those described for *Euglena* are obtained with mouse tumor and avian cells, although there are differences between the patterns of growth of the various cell types. The L cells in Fig. 4 exhibited relatively "balanced growth" during the exponential growth phase; the doubling

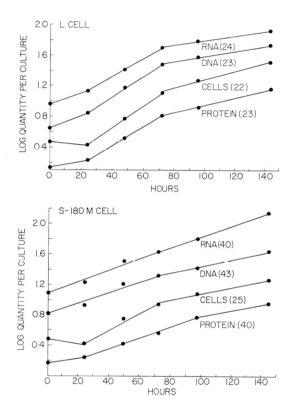

Fig. 4. Growth and division of two established mammalian cell lines in batch culture. Numbers in parentheses indicate the doubling times of the parameters measured during the period of most rapid cell division. (Adapted from data of Swaffield and Foley, 1960.)

times in the culture ranged from 22 to 24 hours. The S-180M cells shown in Fig. 4, however, tended to divide more rapidly than they synthesized RNA, DNA, and protein during their exponential growth phase, and the rates of syntheses of these cell constituents continued unchanged after the cells ceased to divide rapidly. Separation of division and synthesis is even more marked in the growth of the ascites cells shown in Fig. 5, in which

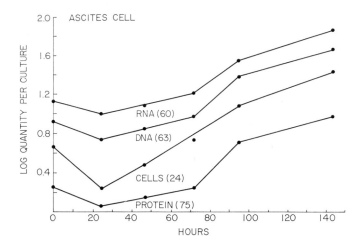

Fig. 5. Growth and division of an ascites tumor cell line grown in batch culture. Numbers in parentheses indicate the doubling times of the parameters measured during the period of most rapid cell division. (Adapted from data of Swaffield and Foley, 1960.)

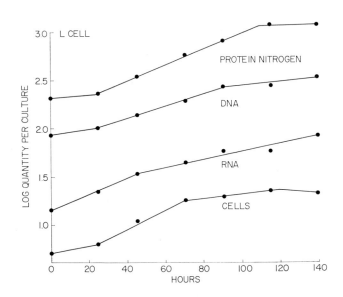

Fig. 6. Growth of a strain of L cell in batch culture. (Adapted from data of Paul, 1959.)

rapid increases in RNA, DNA, and protein occurred at the end of the logarithmic multiplication of the cells.

When used to calculate the RNA, DNA, and protein production per culture shown in Fig. 6, the data of Paul (1959) on a strain of L cells (Table V) yield results that differ from those obtained by Swaffield and Foley (1960) shown in Fig. 4. The data indicate that the rates of synthesis occur in a

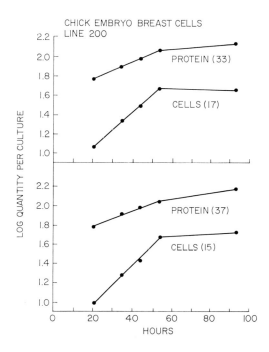

Fig. 7. Growth of chick embryo muscle fibroblasts in batch culture. Line 200 and line 304 represent different genetic strains of chickens. Numbers in parentheses indicate doubling times of the parameters measured. (Adapted from data of Wilson *et al.*, 1966.)

regular manner in the culture and that pronounced fluctuations in the composition of the cells were the result of relative changes in the cell division and synthetic rates.

The rates of protein synthesis and cell multiplication of chick embryo breast cells, illustrated in Fig. 7, were markedly different; the doubling time of the cells ranged from 15 to 17 hours while the doubling time for protein synthesis ranged from 33 to 37 hours, one-half the rate of increase of the cells in the cultures.

V. Discussion

The data indicate that decreases and fluctuations in the contents of the several cell types investigated during their growth in batch culture cannot readily be attributed either to the exhaustion of growth-limiting nutrients or to the production of toxic metabolites. The calculations carried out suggest that changes in cell composition were the result of mismatchings of steady-state rates of cell division and cell syntheses. Each cell studied showed a different pattern of synthesis and division. In the case of *Euglena*, the relative rates of syntheses and division of cells could be altered by changing the carbon source added to the medium. These results indicate that the attainment of steady-state kinetics during cell multipication in batch culture does not guarantee the maintenance of constant cell composition.

However, the data should not be interpreted as implying that balanced growth of *Euglena* cells in batch culture is difficult to achieve. For example, Kempner and Miller (1965) found that *E. gracilis* Klebs cells maintained rates of syntheses of protein, paramylum, RNA, DNA, and chlorophyll equal to their rate of cell division during logarithmic growth. It is difficult to compare directly the data obtained with these chlorophyll-containing cells with the results presented in this chapter for streptomycin-bleached *E. gracilis* var. *bacillaris* cultures, since the *E. gracilis* Klebs cells were grown on a light–dark cycle. It would be interesting to find out whether or not *E. gracilis* Klebs cells would show synthetic and division rates similar to the streptomycin-bleached cells if they were grown in the dark.

The analysis presented here sheds some light upon the nature of the fluctuations in cell composition observed during exponential growth. It does not suggest, however, what mechanisms are operative in producing these changes. The elucidation of these regulatory mechanisms must wait for the results of future investigations. Most of the experiments reviewed here were not specifically designed to examine this question; they were performed to describe the growth of cells in general. There are several difficulties in interpreting data available in the literature on synthetic and division rates of cells. One of the problems involves the nature of the measurements themselves; the data obtained represent average values of the sum of the compositions of many single cells, supposedly randomly distributed over their life cycles. The behavior of a single cell in the growing culture is masked by the way the experiments are performed. For example, although large decreases in the average content of *Euglena* cells during their growth suggest that the cells did not fully replicate themselves before dividing, the data cannot be used to distinguish a situation in which there was a decrease in the content of all of the cells from one in which there was a large decrease in the content of some of the cells.

Other difficulties in interpretation arise if the cells were not randomly distributed with respect to their cell cycles during growth. For example, the growth curves for the ascites cells examined by Swaffield and Foley (1960) are those one might expect if the cells were partially synchronized with respect to their divisions. If this were the case the "fluctuations" noted by these workers would have been the result of differences in the synthetic rates of the cells at different stages in their life cycles.

Complexities in the nature of cell populations themselves must also be considered. Established lines of mammalian cells are heteroploid, and the populations are nonhomogeneous with respect to the number of chromosomes per cell. Little is known concerning the variability of these cells with respect to their individual division times and composition during growth. This nonhomogeniety becomes important when the DNA content of the cells is considered. The data on mammalian cells presented in Tables IV–VI shows that the amounts of DNA per cell changed during the growth of these metazoan cells *in vitro*. Although the DNA content of the *Euglena* cells was constant during growth, the data in Tables I–III indicate that the amount of DNA per cell varied from experiment to experiment. It would not be surprising to find that *Euglena* cells,* lacking the genetically stabilizing influence of sex, and heteroploid mammalian cells possess DNA molecules above and beyond their minimum needs. Indeed, evidence is accumulating that, at least in some cases, the amount of DNA per cell need not be constant (Maale and Kjeldgaard, 1966). Plausibility is weak evidence for events at the cellular level, however, and more work is needed to clarify the situation.

In conclusion, the studies discussed in this chapter illustrate the importance of measuring more than one parameter when describing the growth of a cell culture, and the results indicate the ways in which a multifactoral analysis of cell growth can be exploited. In addition, as was pointed out in a report on the *Euglena* experiments (Wilson and Levedahl, 1964) the results of the studies lead one "...to consider that the... cell plus its medium, rather than the cell itself, is the real unit of physiological activity of the growing culture."

* *Editor's Note*: For a discussion of polyploidy in *Euglena*, see Chapter 5.

References

Buetow, D. E., and Levedahl, B. H. (1962). *J. Gen. Microbiol.* **28**, 579.
Harris, M. (1964). "Cell Culture and Somatic Variation." Holt, New York.
James, T. W. (1961). *Ann. Rev. Microbiol.* **15**, 27.
Kempner, E. S., and Miller, J. H. (1965). *Biochim. Biophys. Acta* **104**, 11.

Maale, O., and Kjeldgaard, N. O. (1966). "Control of Macromolecular Synthesis." Benjamin, New York.

Neff, R. H. (1960). *J. Protozool.* **7**, 69.

Neidhardt, F. C. (1963). *Ann. Rev. Microbiol.* **17**, 61.

Paul, J. (1959). *J. Exptl. Zool.* **142**, 475.

Pogo, B. G. T., Ruiz Ubero, I., and Pogo, A. O. (1966). *Exptl. Cell Res.* **42**, 58.

Swaffield, M. N., and Foley, G. E. (1960). *Arch. Biochem. Biophys.* **86**, 219.

Toennies, G., Iszard, L., Rogers, N. B. and Shockman, G. D. (1961). *J. Bacteriol.* **82**, 857.

Wilson, B. W., and Levedahl, B. H. (1964). *Exptl. Cell. Res.* **35**, 69.

Wilson, B. W., Peterson, D. W., Stinnett, H. O., Nelson, T. K., and Hamilton, W. H. (1966). *Proc. Soc. Exptl. Biol. Med.* **121**, 954.

AUTHOR INDEX

Numbers in italics indicate the pages on which the complete references are listed.

333

SUBJECT INDEX

A

Acetabularia,
  photosynthetic capacity of, 300
  volutin in, 271
Acetate,
  carbon dioxide requirement and, 273
  chlorophyll synthesis and, 179
  generation time and, 288, 291
  growth on, 278, 279
    cell composition and, 317, 322, 324
    initial pH and, 288–290
    light and, 293–294
    respiration and, 292, 304–306
  incorporation, light and, 277–278, 294
  motility and, 53, 54, 73
  oxygen requirement and, 274
  pH changes and, 280
  respiration and, 285–286, 288
Acetic orcein, nucleus and, 189
Acetocarmine, nucleus and, 189, 191
Acid(s),
  euglenoid ecology and, 36–37
  organic, penetration into cell, 53–54
Acid phosphatase,
  constitutive, Golgi apparatus and, 168
  induction of, 273
  localization of, 96, 97, 136–137
  mitochondria and, 159
  paramylon and, 145
Actin,
  cilia and, 76, 78
  flagellar motion and, 81
Activation energy, division rate and, 304
Actomyosin, water and, 51
Adenosine triphosphatase,

flagella and, 81, 90
  gliding movements and, 98
Adenosine triphosphate,
  flagella and, 81–82
  polyphosphate and, 271
  sulfate reduction and, 263
Adenosylmethione, sulfur utilization and,
  262
Aerobacter aerogenes,
  composition, growth phase and, 282
Alanine, utilization of, 269–270, 279
Algae,
  euglenoids and, 8
  swarmers, pattern swimming by, 61
Altitude, euglenoids and, 40
Amino acid(s),
  depletion of, 281
  excretion of, 281
  as nitrogen sources, 268–270
p-Aminobenzoic acid, motility and, 56
Amitosis, occurrence of, 233–235
Ammonia, motility and, 61
Ammonium magnesium phosphate,
  euglenoids and, 36
Ammonium salts,
  pH changes and, 280
  as nitrogen source, 268
Amoeba proteus, hydrostatic pressure and,
  59
Anaerobiosis, motility and, 61
Anaphase, duration of, 223, 225
Animals,
  egesta and excretions, euglenoids and,
  35–36
Anisonema, classification of, 9
Anisonemidae, 8, 9

343

Printed in Belgium